The Man in the Ice

THE MAN IN THE ICE

The Discovery of a 5,000-Year-Old Body Reveals the Secrets of the Stone Age

KONRAD SPINDLER
and a Team of International Scientists

Doubleday Canada Limited

Originally published in Germany by C. Bertelsman, Verlag GmbH, Munchen and in Great Britain by Weidenfeld and Nicholson in 1994.

Canadian Cataloging-in-Publication Data

Spindler, Konrad, 1939–
The man in the ice

Translation of: Der Mann im Eis.
ISBN 0-385-25462-8

1. Neolithic period–Austria–Tyrol. 2. Man, Prehistoric–Austria–Tyrol. 3. Tyrol (Austria)–Antiquities. I. Title
GN776.22.A95S6513 1994 930.1'4 C94-9300080-2

Printed and bound in the U.S.A.

Published in Canada by
Doubleday Canada Limited
105 Bond Street
Toronto, Ontario
M5B 1Y3

Contents

Contents

Illustrations

List of Illustrations

Line drawings

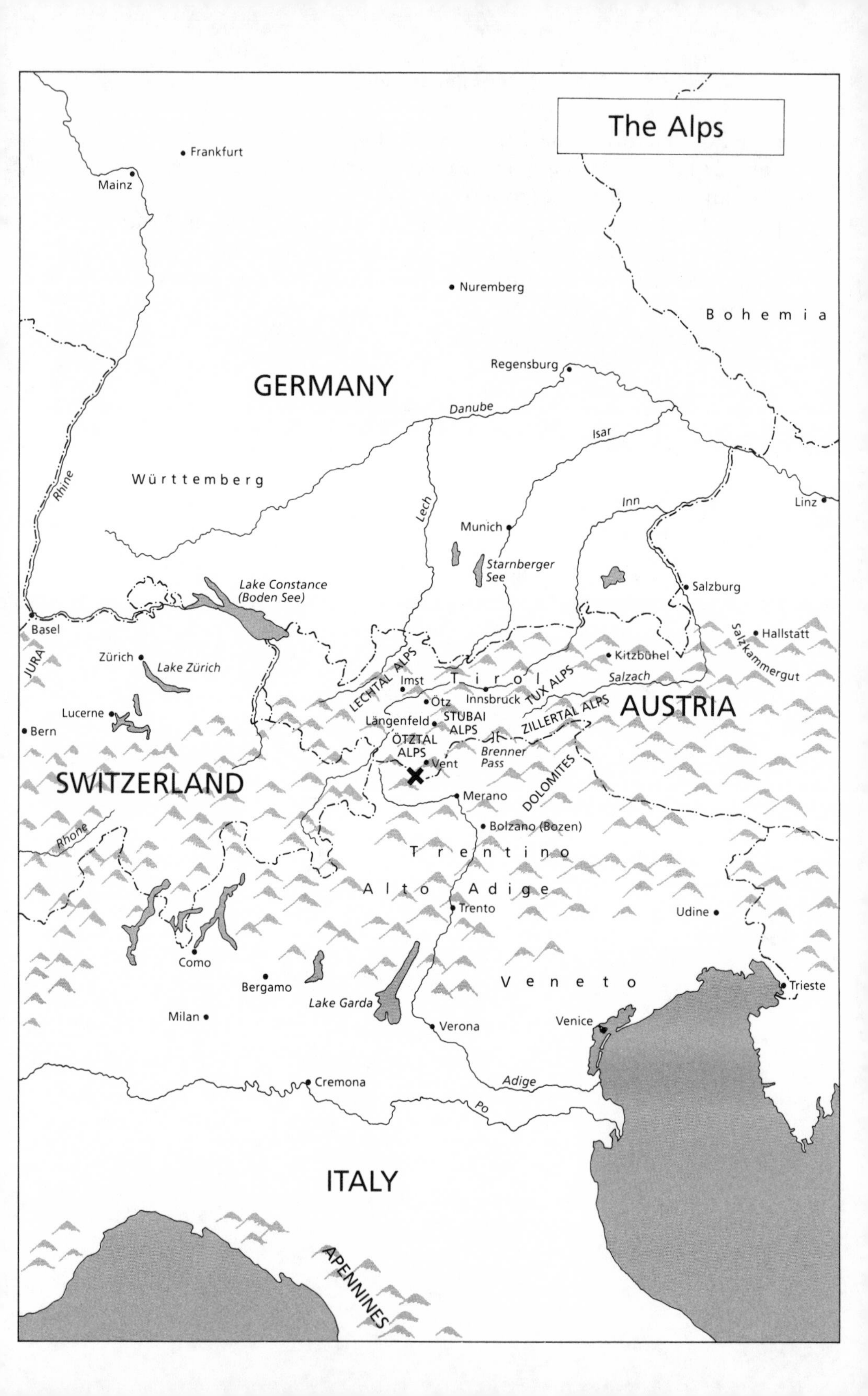
The Alps
Frankfurt
Mainz
Nuremberg
Bohemia
Regensburg
GERMANY
Danube
Isar
Württemberg
Rhine
Lech
Inn
Linz
Munich
Starnberger See
Lake Constance (Boden See)
Salzburg
Basel
Hallstatt
Salzkammergut
JURA
Zürich
Lake Zürich
Kitzbühel
LECHTAL ALPS
Imst
Tirol
Innsbruck
TUX ALPS
Salzach
Ötz
AUSTRIA
Lucerne
Längenfeld
STUBAI ALPS
ZILLERTAL ALPS
Bern
ÖTZTAL ALPS
Brenner Pass
Vent
DOLOMITES
SWITZERLAND
Merano
Rhone
Bolzano (Bozen)
Trentino
Alto Adige
Trento
Udine
Como
Bergamo
Veneto
Trieste
Lake Garda
Milan
Venice
Verona
Cremona
Adige
Po
ITALY
APENNINES

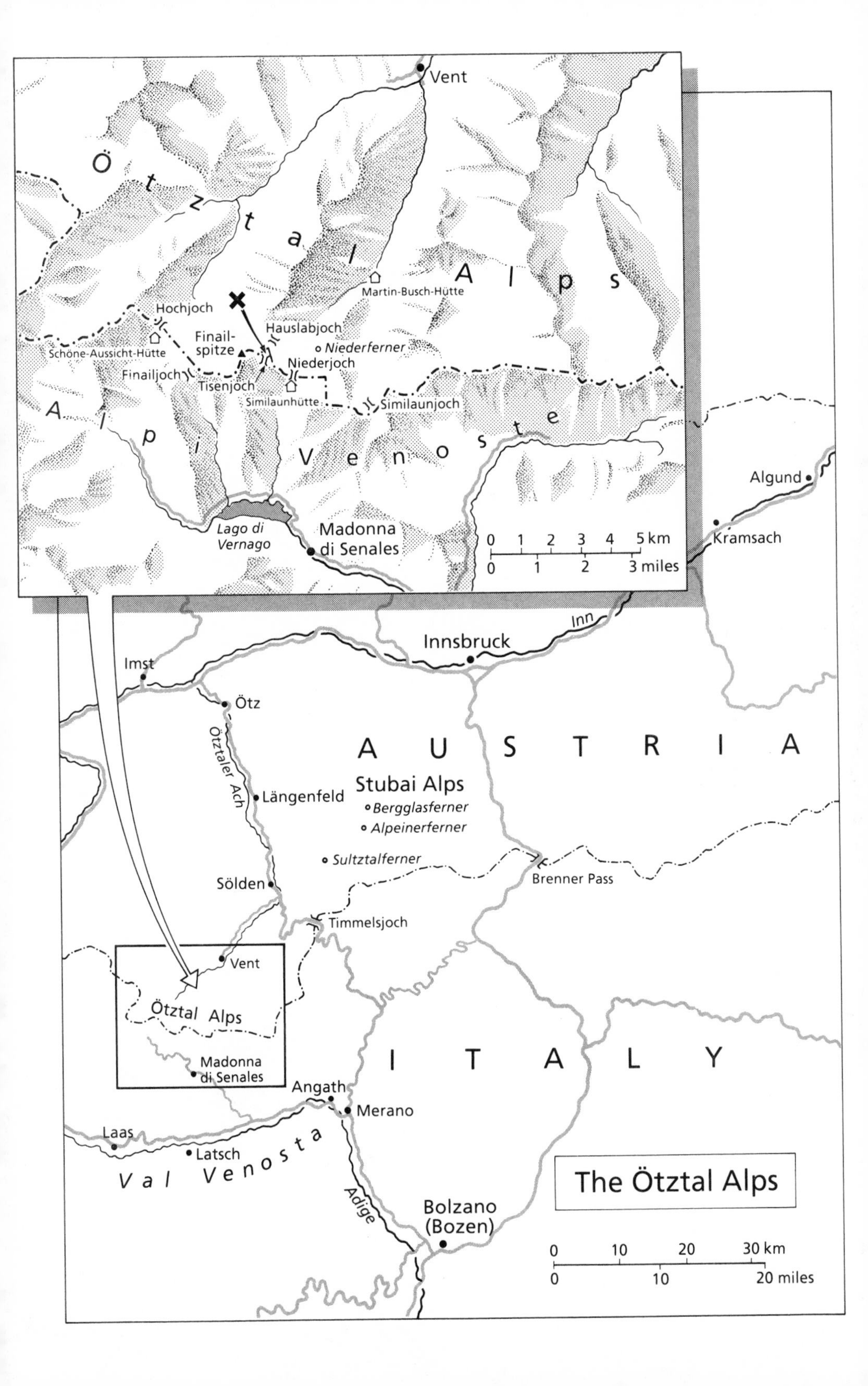
Vent
Ötztal Alps
Martin-Busch-Hütte
Hochjoch
Hauslabjoch
Finail-
spitze
Schöne-Aussicht-Hütte
Niederferner
Niederjoch
Finailjoch
Tisenjoch
Similaunhütte
Similaunjoch
Alpi Venoste
Lago di
Vernago
Madonna
di Senales
0 1 2 3 4 5 km
0 1 2 3 miles
Algund
Kramsach
Inn
Innsbruck
Imst
Ötz
AUSTRIA
Ötztaler Ach
Längenfeld
Stubai Alps
Bergglasferner
Alpeinerferner
Sultztalferner
Brenner Pass
Sölden
Timmelsjoch
Vent
Ötztal Alps
Madonna
di Senales
ITALY
Angath
Merano
Laas
Latsch
Val Venosta
Adige
Bolzano
(Bozen)
The Ötztal Alps
0 10 20 30 km
0 10 20 miles

Prologue

Today is Tuesday, 24 September 1991. My red Mercedes Kombi is moving down the tarmac strip of the Inn valley motorway, as it does every morning. Only this morning began differently. As I drive to my job in Innsbruck, I have the feeling that today will change a good many things – though what, exactly, I don't know. I steer the car onto the slip road. It's 7 a.m. and the carriageway in the direction of the Brenner Pass and Italy is still almost empty, with only a few trucks on the move. It's a good time for reflection, this early hour.

Three days ago, on Saturday, a brief report appeared in the *Tiroler Tageszeitung*:

During their descent from the Finailspitze on Thursday tourists discovered a body partially emerging from the glacier below the Hauslabjoch. Only its head and shoulders were clear of the ice. The warden of the refuge reported the find to the police at Sölden. Judging by the dead man's equipment, he was a mountaineer. It seems that the accident occurred some decades ago. The body has not yet been identified.

As a rule archaeologists do not concern themselves with bodies in glaciers. They are 'too recent'. But time and again I am moved by such a lonely, cold death. An ice-bound sleep for dozens of years, until the glacier again releases the mortal remains – almost as though nature were scoffing at a puny human, depriving him of fitting burial.

Now the sun has risen over the ridge of the Kitzbühel Alps, flooding the valley. To my right lies the small village of Kramsach with its popular open-air museum of Tyrolean farm life. A few years ago we worked on a dig for that museum. An ancient farmhouse had been discovered, dating back to the Middle Ages – to the fourteenth or possibly even the thirteenth century. The museum staff had wanted to excavate it and reassemble it at Kramsach. Everything was carefully planned, and archaeologists were brought in. We dug and discovered that in fact the house probably dated from the seventeenth century. Dendrological dating confirmed our verdict – 1621. The attractive old house ceased to be of interest, and it fell victim to the builder's axe. But that was not on my mind this Tuesday.

1991 had been a year of glacier bodies, victims of the mountains. One discovery followed another. Surely this one would be the last. Six bodies found in the glacier ice in a single year, the same number as for the entire thirty-nine years between 1952 and 1990. This year a normally rare event had become almost commonplace.

On 7 August 1991 the mountain-lodge keeper Horst Frankhauser found two bodies on the Alpeinerferner glacier in the Stubai Alps. Both were identified. They had been apprentice mountain guides, 28-year-old Odo Strolz and 31-year-old Otto Linher. A detailed statement on the incident was made at the time. On 2 May 1953, during a training course, two rope parties chose a route across the flank of the glacier. They crossed a crevasse, its lethal dangers hidden from their eyes by a treacherous snow-bridge. Two climbers crashed 10 metres to their death. Masses of snow and ice fell on top of them, burying the two men; a rescue attempt was unsuccessful. The accident took place at an altitude of approximately 3,000 metres. Thirty-eight years later, at an altitude a thousand metres less, and about 300 metres lower down the glacier, the slowly descending ice flow released the bodies.

On 24 August 1991, in the ice of the Bergglasferner, also in the Stubai Alps, a climber at the Franz-Senn-Hütte found the body of Dr Kurt Jeschke, a mountain guide who had lost his life on 5 March 1981, in the course of a guided ascent, at an altitude of 2,910 metres. The safety rope mechanism had failed when Jeschke fell through a snow-bridge into a 30-metre-deep crevasse. Trapped in the 25-centimetre gap between the ice walls, he was buried under the masses of snow that crashed down after him. At the time, unfavourable weather conditions prevented the recovery of the body. For ten years it remained encased in the ice, until it, too, was released by the glacier, 350 metres away from the site of the accident.

Only five days later, on 29 August 1991, the bodies of two other mountaineers appeared out of the Sulztalferner glacier, also in the Stubai Alps. They were still carrying their identity cards and were found to be Henna Schlager and Josef Schneider from Vienna. No one knows how that accident happened. They were reported missing on 8 August 1934 – evidently the day they died. Their bodies were held in the glacier ice for fifty-seven years.

On my left the wide mouth of the Ziller valley moves past. The light, as so often at this hour, is still muted by a delicate veil of morning mist. But already the peaks of the Zillertal Alps gleam in the distance. Another 30 kilometres to Innsbruck. Probably I would have completely forgotten about Saturday's brief newspaper item, had it not been for a strange article on Monday – yesterday.

I can't forget that piece. It was in the *Tiroler Tageszeitung* (23 September 1991). It seems that the media have not yet done with the latest glacier corpse:

> The world-famous South Tyrolean mountaineers Reinhold Messner and Hans Kammerlander, now on a circular tour of the South Tyrol peaks, took a closer look on Saturday at the body found at the Similaun glacier. At the Similaun refuge, the stopover for their twelfth stage, Messner remarked that the body might be a warrior from the days of Frederick IV, popularly nicknamed 'Empty-Purse'. On the man's back burn marks can be identified, his legs are wrapped in leather straps, and his footwear reminded Messner of that of the Eskimos. In his hand he was holding a metal axe. The well-preserved corpse, which is at least a hundred years old but might date back to the fourteenth century, reveals a deep wound on the back of the head. The body was yesterday recovered by helicopter and is now being examined at the Forensic Institute. Messner and Kammerlander had met the historian Hans Haid at the Similaun refuge for a discussion on the Ötztal and its significance as a border region between North and South Tyrol. However, the main topic of conversation was the body found at the Similaun glacier. Messner and Kammerlander, whose attention had been drawn to the find by a German mountaineer, believe that this is an exceptional archaeological discovery.

That made me sit up, albeit in what was to prove a totally wrong expectation.

On my right appears the ancient fortress of Burg Tratzberg, high on the slopes of the Karwendel range. Shaded by black conifers, the massive walls with their battlements stand sombre guard above the Inn valley. In the old days ships, rafts, carriages and carts had to pay a toll here. Now cars and trains streak past unhindered.

It occurs to me that Messner's conjectures have sometimes to be taken with a pinch of salt. After all, not so long ago he claimed to have tracked down the yeti in the Himalayas. Still, maybe there really is something to the Hauslabjoch find. Recently I had set up a Department for Medieval and Modern Archaeology at the Archaeological Institute of the University of Innsbruck. Suppose the body was really rather older than the usual glacier corpses? This would make it a case for Modern Archaeology. At any rate, yesterday I decided to telephone the Forensic Department, with which I had had no dealings up till now. At first the director, Professor Rainer Henn, was cautious. The body would only be brought in later in the day. He drew my attention to the fact that according to the police report it was probably a case of accidental death. Henn himself would rather not speculate about the age of the body.

Unfortunately, I couldn't join the recovery party because there wasn't enough room in the helicopter. But Professor Henn promised to phone me at the institute if there really was anything out of the ordinary about the body.

I waited for his call until the end of the working day; then, disappointed, I drove home. I didn't know what to think. Not until much later did I learn that Henn had indeed tried to reach me at the institute that evening, but I had already left.

Today things have begun to move towards a climax. The newspaper articles are getting longer, and the first pictures are appearing. This from the *Tiroler Tageszeitung* of 24 September 1991:

Mystery still surrounds the age and identity of the presumably historical glacier corpse recovered below the Hauslabjoch near the Finailspitze in the Upper Ötztal. The body was released by the glacier ice just on the state frontier. The remains were handed over yesterday to the Innsbruck Forensic Institute. The examination is to be performed today. The first impression of Prof. Rainer Henn, the head of department, was 'Old rather than recent!' According to a report by Markus Pirpamer, the warden of the Similaun refuge who took part in the recovery, further external evidence was discovered yesterday, pointing to the great age of the find. It seems that near the body pieces of birch bark stuffed with hay were found, which served as gloves. Also stuffed with hay was the footwear, of which only tatters of leather survive. An axe was also found, consisting of a wooden handle and an iron or copper blade attached to it with leather thongs, as well as a small stone knife. On the dead man's back there is said to be a kind of tattoo in the shape of ten lines arranged in three rows one above the other. Chamois hairs, also discovered according to eyewitnesses, suggest that the dead man may have been a hunter.

What excited me even more than the description was the pictures. A kind of tassel, described as 'plaited'. Not much of a clue. Likewise the wooden implement, which looks like the chewed stub of a carpenter's pencil and is described in the article as a 'spindle'. And then, scarcely recognizable in the coarse screen of newsprint, the 'stone knife'. Was that minute edge really made of stone? And what about the axe? Was the blade of 'iron or copper'? The fuzzy photograph didn't help. But the handle, only partially visible in the picture, is beyond doubt an elbow-shafting with leather-thong binding. Such shaftings existed in our area in the Neolithic period, and even into the Bronze Age and the Early Iron Age. This means that the axe must be at least 2,000 years old, if not a lot older. The other photograph shows the 'exceptionally well-preserved body in the ice', as the caption puts it. Again, not a lot to be made out clearly. On the left there's a man with white hair and glasses, giving an expert opinion on the corpse. This was Henn, the forensic expert, whom I had yet to meet.

At 8 a.m. the floors and corridors of the faculty building are still deserted. My room is on the sixth floor, high above the roofs of the city. Its windows face south and west. My eyes roam over the Tux and Stubai Alps. So much could have happened beyond those mountains. Before me the Berg-Isel ski-jump sits on the hillside where, not long ago, the Pope preached to a huge crowd of believers. I could see him in his white vestments as I sat at my desk.

The telephone rings. Professor Henn informs me that the recovered body is indeed an unusual case, a non-typical glacier corpse. Would I like to see it? What a question!

I get him to tell me the way; it's no more than two or three minutes' walk. The Institute of Forensic Medicine is boxed in by ugly hospital buildings. Its large brownish panels between slim concrete pillars are a cool contrast to the venerable, Schönbrunn-yellow building of the Anatomical Institute across the road. I can't see inside: the windows are made of reflecting glass.

Henn is at the entrance to welcome me. He offers me a worker's muscular forearm with a firm handshake. We introduce ourselves and take an instant liking to one another. A head shorter than me, he has a broad, open face with shrewd blue eyes that smile up at me. He is wearing a knee-length white medical overall with rolled-up sleeves. Sadly, our friendship is to be all too brief.

Polite formalities are dispensed with. Henn leads the way to the small dissecting room. At once I am surrounded by that indefinable hospital smell, probably phenol or some other disinfectant. The door can only be opened from inside. The room is large and light, cool but not cold, the walls covered with pale-green tiles. In front of me is a row of slabs with stainless steel tops on massive low columns; two of the slabs are covered with sheets. Over to the right, in the corner, is the frame for the refuse bag. Better not look into it. To the left, at the back, is a trolley, also covered with a sheet, which shows the outlines of a shapeless bundle underneath. Bare pale feet stick out. Some formalities remain to be completed. The head surgeon, Dr Hans Unterdorfer, is introduced to me. What a contrast to Professor Henn! Tall and heavily built, he shakes hands. A relaxed face smiles at me. So these are the forensic medics about whose work one tends to talk in whispers, if at all.

Henn gives a sign and everything falls silent. Only a chronometer can be heard ticking softly in the background. Unterdorfer and a porter remove the sheets. The time is exactly 8.05 a.m. A routine forensic case is about to become a major archaeological event.

On the slab lies the shrivelled corpse of a man, naked except for a strange, grass-filled shoe. Alongside him lies a long wooden stave with a fibrous end. Laid out on the same slab, on a piece of green cloth, are the other finds – the axe with its elbow-shafting and metal blade, a stone bead with a strange tassel of twisted hide thongs, the small dagger with its wooden handle, the 'pencil stub', a kind of strap-bag, a wooden stick with holes in it, a scrap of leather and a nut-sized stone. On the other slab are all kinds of round and flat pieces of wood, torn cords, plaited material, bunches of hair and grass, two objects suggesting tree fungi, numerous remains of leather and fur, and various other items.

I only need a few seconds. This is something any first-year archaeology student could identify. The items that are more or less readily datable are the metal blade of the axe, the flint blade of the dagger and another flint implement which peeps through a tear in the belt-bag. Because of its delicate water patina it is impossible to say whether the axe-head is of reddish copper or golden bronze. The flanged edges lead me to opt for the Early Bronze Age, since 99.9 per cent of all flanged axe-heads belong to that culture. At this early stage, it would be too bold to opt for its being one of the minute number of Neolithic – invariably copper – flanged axe-heads that survive.

I straighten up and give my verdict: 'Roughly four thousand years old.' And because the whole collection conveys a rather ancient impression even to the naked eye, I add a little more softly, 'And if the dating is revised it will be even earlier.' Henn and Unterdorfer look at me with total disbelief but quickly the tension vanishes and the fiercest discussion ever to be conducted in the Institute of Forensic Medicine erupts. This is only the beginning. What we would be faced with over the next few hours, days and nights remains indescribable. Though I will try to describe events as they unfolded, at best it will be only a pale imitation of what really happened. You just had to be there.

A fully-equipped prehistoric man – nothing like it had ever been seen by an archaeologist, or indeed by anyone else in modern times. Whenever he described my reaction later, Henn would claim with a grin: 'His jaw simply dropped.' But those who know me will also know that then, in the dissecting room, I had 'put on my inscrutable expression', as the German weekly *Die Zeit* was to report on 7 August 1992.

I
Discovery, Recovery, Investigation

1 Thursday, 19 September 1991

Erika and Helmut Simon live in Nuremberg, the provincial capital of Franconia. All around that ancient imperial city, which suffered such destruction in the war, extends the flat Nuremberg countryside, resembling nothing so much as a sandpit. Helmut Simon is a caretaker, his wife works on a newspaper. They are a modest couple who radiate contentment. Never in their wildest dreams did they imagine that one day they would be seen on television screens around the world. Although the Simons are no longer young, they go to the Alps every year.

From 15 to 23 September 1991 the couple were on holiday in the Italian South Tyrol, in the popular little resort village of Unserfrau im Schnalstal. From there it is not far to the peaks of the main Alpine ridge. As always they planned their daily excursions with care. Both of them are experienced mountaineers, unafraid of even the most difficult routes, but they like to be back in their village by nightfall.

Their destination for Wednesday, 18 September, was the Similaun summit, at 3,607 metres above sea level a normal day's climb of medium difficulty for a trained mountaineer. The Simons are a capable pair. They chose a little-used route across an icefield. Unexpectedly, they found themselves faced with wide crevasses, disappearing down into darkness. Laborious detours became necessary. Already the midday sun was beating down on them. Their schedule was falling apart. But like all true mountaineers they wanted to reach their objective. At last they stood on the summit – but by then it was 3.30 p.m.

It became clear that a full descent would be impossible before dusk. They consulted their tourist map to locate the nearest refuge. An unplanned night in the mountains was inevitable, and the descent to the Similaunhütte, the Similaun refuge, took about two hours. On the way they were joined by two Austrian mountaineers, who had made reservations there. The sun was just dipping below the crest as they reached the hut.

The early morning heralded a radiant day. This is Thursday, 19 September 1991. The Simons had really intended to return directly to the car park, but such brilliant mountain weather is irresistible. The peaks beckon. Their programme has been abandoned anyway. On the spur of the moment the Simons decide to climb the 3,516-metre Finailspitze. At the peak they rest for about an hour, beginning their descent around 12.30 p.m.

Once more they leave the marked path and choose a route to the Similaun refuge, where they have left their rucksacks. They cross a wide, slightly inclined snowfield that ends at a small rocky ridge, behind which lies a narrow gully. Its base is filled partly with glacier ice and partly with meltwater. To avoid it they make an arc towards the left. The time is 1.30 p.m.

Later, Helmut Simon would describe their discovery to us like this: 'From a distance of 8 or 10 metres we suddenly saw something brown sticking out of the ice. Our first thought was that it was rubbish, perhaps a doll, because by now there is plenty of litter even in the high mountains. As we came closer, Erika said: "But it's a man!"' Helmut Simon immediately ran back to get the Austrian couple, from whom they had parted shortly before. But they were no longer in sight.

Sticking out of the ice is a leather-brown round bald skull with a medallion-sized injury. Also visible are the shoulders and back, draped against a rock. The face is immersed in water, with dirt around the chin. The arms cannot be seen, and seem to be missing. Because of its delicate proportions, Erika suggests that it is the body of a woman.

'We thought it was a mountaineer who died here. We were shocked and didn't touch the body. There was a blue ski-clip lying nearby, the rubber strap used for tying skis for transport. We thought the accident probably happened ten or twenty years ago. Not far from the body we saw a piece of birch bark, which used to be a tube but had been squashed flat, wound round with string or leather and open at both ends. Helmut picked it up, looked at it carefully and put it back. We memorized the exact position of the body before we left. Helmut took a photograph as a record, in case the place couldn't be found again from our description.' It was the last frame of his film – the photograph of the year.

The Simons still have a descent of about an hour ahead of them. After ten minutes they meet a walker who intends to climb the Finailspitze. They exchange greetings – but they do not tell him of their discovery. For a long while they watch him move on. He sticks closely to the path and so is unlikely to see the body. Some twenty minutes later they meet another man who has lost his way on the ridge. They do not reveal anything to him either.

The Simons reach the Similaun refuge towards 2.30 p.m. The guests are sitting outside, enjoying the sunshine. Only the innkeeper, Markus Pirpamer, is busy inside. Tormented by thirst, but even more so by the need to recover from their shock, they go into the kitchen and order a beer. Then Helmut Simon asks if anyone from the hut is missing. Pirpamer says no, and the Nuremberg couple tell him of their discovery. Immediately there is great excitement. Pirpamer questions them meticulously about the spot. He even steps outside with Helmut Simon and gets him to describe the route. A few tourists prick up their ears and ask what has happened, and the Simons tell of their find below the Hauslabjoch. Markus Pirpamer undertakes to notify the police.

The Nuremberg couple wait for the rescue helicopter. But when it fails to arrive they set off again at about 4.30 p.m. They spend a few more enjoyable days in the mountains before returning home on Monday, 23 September. The reporters are already lying in wait.

Markus Pirpamer knows his mountains. He is still quite young and only recently took over the running of the hut from his father. In the summer he is very busy, but in winter, when the refuge is buried under deep snow, it stays closed. The exact spot of the find was said to be close to the main ridge. Every Tyrolean knows that the international frontier runs along the watershed. But in the high mountains, who cares about the exact line of the frontier? The discovery of a dead body, however, does give rise to formalities. Better get in touch with the Italians as well as the Austrians. Markus Pirpamer informs the police in Schnals (Senales) in South Tyrol and their counterparts in Sölden in North Tyrol. Since the

Italians decline any interest, the incident becomes the responsibility of the Austrians.

To be on the safe side, Markus Pirpamer also phones his father, Alois, who is in charge of mountain rescue in the Upper Ötztal as well as manager of the Hotel Post in Vent. At one time Vent was just a small mountain village where life was tough for the inhabitants, but now tourism provides a reliable income. At an altitude of 1,893 metres Vent is the highest permanent settlement anywhere in the Tyrol, but with only four snow-free months in the year summer there is very short. Vent is also the last spot in the Ötztal accessible by car. Any further progress into rock and ice has to be on foot – or by helicopter.

Alois Pirpamer begins to research the likely identity of the dead man. There is Carlo Capsoni, a music professor from Verona, who has been missing roughly since the outbreak of the Second World War. He had been on a walk from the Schöne Aussicht to the Similaun refuge by way of the Hauslabjoch. The identity of the victim seemed, for the time being, established.

Shortly after the Simons' departure Markus Pirpamer gets ready to set out. He asks his Yugoslav helper, Blaz Kulis, to go with him. After a short search he finds the corpse, its upper body sticking out of the ice. Under the chin he spots the remains of fur or hair. It looks like a beard. The skin reminds him of leather and is otherwise totally smooth. The bones protrude clearly. Pirpamer notices, too, the birch-bark container lying some 1.5 metres from the head. It contains a little wet grass. About 5 metres away the two men discover more items, this time on a raised rocky ledge. There is an implement with a long wooden handle bound with leather thongs and a metal blade rather like an ice-pick or an axe. Naturally, a mountaineer will think of an ice-pick first. How could one cross a glacier without one?

All kinds of small round and flat pieces of wood are lying nearby. Pirpamer thinks they are broken snowshoes, while Kulis suggests they might be bits of a wrecked sledge. Scattered about are torn cords and squeezed into a crevice in the rock are clumps of hair and scraps of fur, presumably a chamois-skin. Leaning at an angle against the rocky ledge is a long pointed stave, its lower end held firmly in the ice. They pick up the loose fragments and examine them carefully, then they put them back. Everything is exceedingly odd, exceedingly mysterious. They take another look at the corpse.

On his back they can make out black streaks. Could they be burn marks? After about half an hour they leave, and meet no one on their return journey.

That same afternoon officer Florian Bauernfeind is sitting in his gloomy office at the Sölden police station catching up on a backlog of reports and statements. At 3.10 p.m. the telephone rings – it is Markus Pirpamer. The two men know each other well. Bauernfeind receives the report of the discovery of a body below the Hauslabjoch. At that precise moment no Alpine officer is available in Sölden to follow up the matter. Pirpamer's information is checked against an Alpine Club map. The spot in question is indeed close to the Austro-Italian frontier. Helmut Hager, the station chief, phones back for more details. But he can only reach Adolfine Pirpamer, the wife of Alois, the mountain rescue chief.

Information so far is vague. There is talk of a certain Capsoni. As a precaution Inspector Hager contacts his colleague in Imst and requests the dispatch of Alpine police by helicopter to recover the body. Police officer Anton Koler takes the call. It is late in the afternoon, too late to do anything today, he says. But tomorrow the Ministry of the Interior helicopter will be sent. Koler asks for a copy of the official report to be sent over.

At Sölden the police officer on duty is relieved by his colleague on the evening shift. Officer Helmut Holzknecht writes up the report and at 7.18 p.m. faxes it to Imst where it arrives four minutes later. This is the report which the forensic examiner Dr Henn will find on his desk four days later, on Monday, 23 September 1991:

> Alpine incident: body discovered at Hauslabjoch (Niederjochferner) – preliminary report. On 19 September 1991 towards 12.00 mountaineers descending from the Finailspitze, in the area of the Hauslabjoch (a little below), found a body half emerged from the glacier, almost upright in the ice. Only head and shoulders projecting from the ice. Markus Pirpamer, the keeper of the Similaunhütte, thereupon notified the Sölden police. He was at the location himself in the afternoon. Judging by the equipment, this was an alpine accident going back many years. As was established at Vent, a music professor from Verona by name of Capsoni, on a walk from the Schöne Aussicht to the Similaunhütte via the Hauslabjoch about the year 1938, is still missing. Recovery of the body is expected to take place on 20.09.1991. A further report will be submitted.

Koler also makes contact with the air operations centre in

Innsbruck. The recovery flight is scheduled for the following day.

And so the day moves towards its close. Markus Pirpamer serves the guests at his mountain lodge. In Sölden the police carry on with their duties. Alois Pirpamer has the dining room of his hotel prepared for breakfast. The Simons are on their way to Unserfrau. No one talks about the incident any more. Later in the evening, however, towards 9 p.m., the phone rings again. It is the Italian police in Senales asking whether anything is known about the body found that day. They are told that everything is under control. So things take their normal course, and the case of the body found in the glacier follows the official procedure.

2 Friday, 20 September 1991

Subsequently I was to be asked time and again by reporters what exactly I felt when I first saw the Iceman lying in front of me on the dissection table at the Institute of Forensic Medicine. In my desperate search for an original answer, I said frivolously to one of them: 'It must have been more or less what the British archaeologist Howard Carter felt when, in Tutankhamun's tomb, he opened the lid of the sarcophagus and gazed into the golden features of the pharaoh.' The media gleefully picked up this remark – only I have to admit that it wasn't the truth. I didn't think of anything. But one thing was immediately clear to me: a lot of work would come our way.

In September 1991 Tyrolean archaeology was virtually at a standstill: all the university teachers were either on holiday or on some far-flung dig. Myself apart, only the research assistant Dr Gerhard Tomedi, our secretary Julia Tschugg and a bare handful of students were left at the Innsbruck Institute. We had completed a seven-week dig in the Pustertal a few days earlier and had only just recovered. We had also been busy with the tricky restoration of the glazing works of a pottery; the last glazing works in the Alps had escaped the threat of demolition. We had restored it and preserved it for posterity as a rare testimony to an ancient Tyrolean craft tradition. The formal opening was imminent.

Elisabeth Zissernig, along with our excavation technician Gerhard Sommer and the draftsman Michael Schick, was working at an archaeological dig on the Rennweg in Innsbruck's old town centre. They had penetrated the city's history to a depth of 6 metres. The dig had to be completed before the onset of winter, as the construction of an underground car park for the Congress Centre depended on it. These archaeologists were, therefore, more than fully occupied.

The winter term was due to start in a week's time. We had a lot of plans. Next year we wanted to celebrate the institute's fiftieth anniversary and were preparing a Festschrift and planning a symposium.

And then, on 24 September 1991, the Iceman came into our lives. No one was prepared, neither in terms of staffing, nor organization, nor finances. The infrastructure of one of the biggest interdisciplinary projects ever embarked upon in Austria, and maybe beyond, had to be built up almost from scratch.

We soon learned that the man had been visible in the ice of the Hauslabjoch for four days before we saw him, and that many people had been at the spot during the time between discovery and recovery. There had been digging with ice-picks and other activities, photos had been taken, and items of equipment were said to have disappeared. While reliable information was unobtainable, there was no lack of rumours. Among the many tasks facing us, one was essential – the events at the discovery site had to be established down to the last detail.

And so Elisabeth Zissernig was recalled from Rennweg. It was she who undertook to track down everyone who had visited the gully from the moment the corpse was discovered to the opening of the recovery operation at 12.37 p.m. on Monday, 23 September 1991. It was a laborious task and one which eventually took her six months. One fragment was added to another to build up the complete mosaic. Painstakingly she found out the names of anyone who might have been involved, examined the guest books of mountain refuges, and made appointments. Some of the testimonies did not tally and contradictions had to be resolved. Some of the people were by then abroad, one as far afield as Argentina. Others were not very forthcoming. But overall the sequence of events began to fall more sharply into place. Eventually her investigations filled two box-files and from their contents a picture emerged, often exciting, sometimes difficult to follow, but on the whole reliable, a picture that would provide the crucial basis on which the archaeological findings could be reconstructed.

During the four days prior to the official recovery of the body a total of twenty-two people visited the spot, some of them repeatedly. It cannot be ruled out that someone else – maybe more than one person – was also there, but there is no concrete evidence. Considering the enormous publicity which the discovery of the body subsequently attracted, we assume that such people would either have come forward on their own or that they would have been seen by others who would have informed Elisabeth Zissernig. After all, the mountain paths in the area of the find are not so very remote. It may, therefore, be assumed that the list is almost certainly complete. (Not included are those who arrived at the spot along with Professor Henn or after him.)

Nineteen people were personally interviewed by Elisabeth Zissernig. Two men, Osvald Lubomir and Anton Matthis, provided written accounts of what they had seen. Only Nicolas de la Cruz, a sound recordist, was unavailable for interview because he had left for South America shortly afterwards. However, this was not too vital a gap since his companion, the cameraman Fulvio Mariani, was available for extensive questioning. Besides, the two had not actually seen the corpse.

Reasons for visiting the spot varied. They ranged from mere chance, to curiosity, to semi-official and official recovery duties. Apart from the material shot by the television crew from the Austrian Broadcasting Service (ÖRF) during the recovery, numerous black-and-white and colour photographs, starting with the Simons' sensational first snap, were taken during those days. All of them are readily available and have been valuable aids in reconstructing the archaeological data. To my knowledge, no such thorough investigations have ever been undertaken before for the clarification of a prehistoric event.

But what did happen after the discovery of the body by the Simons and the first local inspection by Markus Pirpamer and Blaz Kulis? Only slowly did the 'case of the corpse', as the lawyers referred to it, develop into the archaeological show of the century.

During the night of 19/20 September 1991 there is a sudden drop in atmospheric pressure. On Friday morning, 20 September 1991, the sky is heavily overcast. Banks of mist drift through the mountains. A wind springs up and fierce gusts whip the peaks. Not exactly helicopter weather, but the pilots are used to worse. During the morning the wind abates a little. Towards 11 a.m. pilot Hermann Steiner is ordered to get the helicopter ready. Josef Leitgeb, police officer and air rescue expert, is to accompany him. At 12.35 p.m. they take off from Innsbruck airport in the direction of the Ötztal.

Koler, in his role as chief of the Alpine operations group in Imst, also sets out for Sölden, where he arrives at midday. Along with his colleagues Sieghart Schöpf and Bernhard Gruber, he continues his journey to Vent. Schöpf has already asked the local undertaker Anton Klocker from Längenfeld to meet them at the landing place with a coffin. Dr Edgar Wutscher from Sölden, the local medical officer, has also been notified so that he can issue a death certificate. And at 1.04 p.m. the helicopter touches down. Because of the weight problem at these altitudes the stretcher and other items are off-loaded. Schöpf and Gruber are also left behind. In any case, only one man at a time can operate the pneumatic chisel usesd to free bodies from the ice. Koler boards the helicopter and a minute later it lifts off again. Klocker drives up with the coffin and Alois Pirpamer arrives from his nearby hotel.

In Austria responsibility for the recovery of bodies found in glaciers rests, in the first instance, with the police and the air rescue service (sometimes the same people). In the absence of any suspected crime, and provided the body can be identified, the death certificate is made out by the district medical officer, who indicates the cause of death. Only then can the body be released for burial to the next-of-kin, who also have to meet the costs of recovery. However, if the suspicion of foul play arises, or if the body cannot be readily identified, the public prosecutor must be called in, and it is he who takes charge. As a rule he would order a forensic examination. In addition, the criminal investigation branch of the provincial police begins its work.

Even though the significance of the prehistoric find at the Hauslabjoch was only gradually being appreciated, at this point it may be as well to take a closer look at Austria's Ancient Monuments Act. This makes it obligatory to report any previously hidden 'monument', and lists penalties for non-compliance. The law specifically stipulates that 'the federal police ... must immediately notify the District Administration of the find, and the District Administration in turn the Federal Ancient Monuments Office'.

That Friday morning, however, one day after the discovery of the body, the possibility that there might be an 'ancient-monument' aspect to the case has not yet occurred to anyone. The evidence available leads only to talk of a mountain accident in the twentieth century. Capsoni's name is mentioned. As matters stand, the authorities acted quite properly when they ordered the body to be released from the ice and brought down from the mountain. As the dead man was lying in inaccessible terrain at an altitude of 3,210 metres above sea level, the recovery party had to approach by helicopter. The first stage of the transportation of the body, from the Hauslabjoch to Vent, had to be performed by the same means. In Austria only officially authorized undertakers' firms are allowed to transport dead bodies – a privilege which the undertakers' firms make sure is meticulously observed. Klocker had prepared a fine pinewood coffin – not the most expensive model, but one with a dignified carving of palm branches.

The pilot has some trouble finding the Hauslabjoch. Although over the next few days it will become the most frequented helicopter pad in the Alps, today he has to radio the mountain-lodge keeper for directions. Once again Markus Pirpamer ascends to the location of the find. A tall, lanky figure, he swiftly climbs the rocks and arrives at the spot at 1.16 p.m., at the same time as the helicopter. Koler gets out. Because of the uncertain weather – fog is closing in from the south and the wind has stiffened again – Steiner does not leave the helicopter but takes off at once in the direction of the Martin Busch refuge, where he will await the end of the operation.

Carrying the pneumatic chisel and a body bag Markus Pirpamer and Koler walk over to the corpse. Operated by compressed air, the tool used by the mountain rescue service to free bodies looks like a pistol with a chisel mounted in its barrel. First the police officer takes a location photograph. This reveals that during the twenty-four hours since his discovery, the dead man has emerged from the ice by a further 10 centimetres. (I shall come back to the reason for this rapid glacier melt during those days of late September.) From this we can deduce that the ice had actually begun to release the corpse some three days before its discovery. The same photograph also shows the as yet uncrushed birch-bark container lying near the dead man's head.

Koler, though used to seeing a good deal in his work as a policeman, is even so profoundly moved: 'Sticking out of the ice was the sunburnt, clean-shaven head and neck and shoulders of a man. The skin was just like leather. Unlike other glacier corpses, this one didn't give off a smell. The dead man was lying face-down on a rock. The meltwater formed a little lake around him.' These are the words Koler will use to Elisabeth Zissernig six weeks later.

Some of the work has to be done under water. Time and again the pneumatic chisel slips and cuts into the flesh of the corpse, especially its left hip. Although they remove stones to enable the meltwater to drain away, it only fills up again. They take turns with the chisel. After just over half an hour the tool runs out of air. The body is only half exposed, its arms extended towards the right. Despite vigorous tugging, one arm cannot be freed from the ice. Time is pressing – helicopters cost money. Besides, any further attempt is pointless without suitable tools, so by mutual agreement the recovery is suspended. Meanwhile the weather is rapidly deteriorating.

Koler climbs out of the gully in order to radio to the pilot that it is impossible to free the body that day and that he would like to be picked up. While he waits he has a chance to look round and take a few more photographs. Thoughtfully he contemplates the axe on the rocky ledge – it must be at least a hundred years old. In which case the body cannot be Capsoni. He decides to take the strange 'ice-pick' with him as evidence that the Hauslabjoch corpse is not, after all, the missing music professor from Verona. Unfortunately, his decision has disastrous consequences, because for the next few days the only datable item is missing from the investigations.

The helicopter arrives and Koler climbs aboard. Markus Pirpamer returns to the refuge on foot. At 2.50 p.m. the helicopter lands in Vent. Naturally enough, the axe causes quite a stir among the curious crowd. The general consensus is that it is an unusual ice-pick. Alois Pirpamer would like to acquire it for the mountain guides' corner in his Hotel Post. Although Koler feels that the ethnographic museum in Längenfeld is the right place for it, he nevertheless dutifully hands it over to his colleague Schöpf, who takes it to the police station in Sölden; it is for the moment, state property. Then the helicopter takes off again for Innsbruck. The party disperses. Koler drives back in his official car and Klocker, who had hoped to take along a certain Signor Capsoni, leaves with his coffin empty.

Of course, rumours abound. It is said that the body has burn marks on its back and an open wound on its head, even that the dead man was in fetters. Doubts have arisen about the Capsoni identification. Back at the police station, Schöpf first deals with the axe, depositing it in the air-raid shelter. Then he sits down to type his report:

Alpine incident: body discovered at Hauslabjoch (Niederjochferner) – supplement. On 20.9.1991 District Inspector Koler of the Imst police was flown to the Hauslabjoch by the helicopter of the Federal Ministry of the Interior in order to effect the recovery of the body. Because of unfavourable weather conditions the recovery operation had to be cut short. It was only possible to free the corpse from the ice up to the region of its hips. The dead man's identity has not as yet been established. On the strength of the articles found near the body it may be assumed that the accident happened as long ago as the nineteenth century. According to information from Alois Pirpamer, the mountain rescue chief from Vent, the body cannot be Capsoni's as this person's body was recovered some years ago. Given appropriate weather conditions recovery will be continued and/or completed. When further details become known a report will again be submitted.

At 3.52 p.m. this report on the latest findings is transmitted to Imst. It, too, will appear on Rainer Henn's desk three days later, after the weekend. Schöpf continues with his duties. At 4.10 p.m. he reports by phone to State Prosecutor Dr Robert Wallner of the Provincial Court in Innsbruck, and a criminal investigation is set in motion. In accordance with the Code of Criminal Procedure a file is started under the number ST 13 UT 6407/91. The criminal action is directed against U.T. (*unbekannte Täter*, persons unknown). The case is assigned to the examining judge responsible for the letter U, Dr Günther Böhler. For the moment, however, nothing happens at the provincial court. It is Friday evening, the beginning of the weekend.

Late in the afternoon Alois Pirpamer makes the ascent to the Similaunhütte. He wants to hear his son's account of events and to see the body for himself the following morning. An exciting day is nearing its end. Meanwhile the world-famous South Tyrolean mountaineer Reinhold Messner, along with his climbing companion Hans Kammerlander, is having dinner at the Weisskogel refuge. The next stage of their journey will take them over the Hauslabjoch to the Similaunhütte.

3 Saturday, 21 September 1991

Even before the axe disappeared behind the heavy steel doors of the air-raid shelter below the police station in Sölden, it had been seen by many people. Wooden hafts of that kind have been found in many prehistoric pile-dwellings of the circum-Alpine region. Similar picks with the characteristic elbow-shafting are known, especially in Austria, from the Iron Age salt mines of Hallstatt and Hallein. They should be familiar to people outside the narrow circle of specialist academics. But never once did it occur to anyone who had seen it that the Hauslabjoch axe might be a prehistoric artefact which should, in the first instance, have been reported to the Ancient Monuments Office.

After Schöpf had made his report, the public prosecutor issued instructions to the police in Sölden 'to transfer the body after recovery (including the items found with it) to the Institute of Forensic Medicine for it to be examined there. An autopsy order has not been issued.'

In the mountains everyone helps everybody else. Hence the mountain rescue service also felt involved.

The refuge comes to life even before dawn. The two Pirpamers, father and son, have a quick breakfast together. During the season they don't see much of each other. Markus Pirpamer recounts the previous day's events, but today, he stays behind. Shortly after 7 a.m. Alois Pirpamer prepares for the ascent. He has spent his life in the mountains. His fine narrow face is framed by thick snow-white hair. With a sure, easy step he sets out for the Hauslabjoch.

Everything is just as his son described it. He sees the leather-brown naked corpse lying on its stomach in the ice. The water has frozen overnight. Leaning against the rock is a brown trimmed stave. He tries to pull it free, but fails. He picks up the other remains of fur and bits of wood which are scattered about, examines them with interest and replaces them. After staying for something like half an hour he heads back down to the Similaunhütte.

Alois Pirpamer seems to remember that that Saturday, 21 September 1991, a day climber had shown around photographs in the lodge – pictures taken with a Polaroid the previous day. Afterwards the man descended on the South Tyrolean side to the Schnalstal, the Senales valley, on the Italian side of the frontier. We are not sure whether this man is one of the visitors we managed to trace. If not, we appeal to him in these pages to contact us and, if possible, to let us see his photographs.

Later in the morning a call is received at the Similaunhütte from the Sölden police. The recovery operation cannot continue that day. Over the weekend all the helicopters are needed for more urgent work with mountain and road accident victims. 'Besides,' the pilot Anton Prodinger of the Air Operations Centre at Innsbruck airport told us later, 'there was no urgency. After all, everyone thought that this was a normal glacier corpse. No urgent instruction was received from any quarter to push ahead with the operation.'

In view of this, Markus Pirpamer and Blaz Kulis once more climb to the Hauslabjoch. They intend to cover the body to protect it from the prying eyes – and hands – of weekend tourists. After all, it is lying fairly close to the path. Although mountain hikers are, for obvious reasons, always urged never to stray from the marked routes, many – like the Simons – disregard the warning.

As a result of the high temperatures the meltwater lake around the dead body has risen again. The two hack at the ice around the corpse in order to drain off the water. Pirpamer covers the body with a cut-open rubbish bag weighed down with snow to prevent the body from rotting in the heat generated by the plastic. The whole operation takes only ten minutes.

On their way back they find a piece of birch bark in the next ice-free hollow, some 20 metres from the corpse. It is a little larger than a human palm and has holes along the edge. Markus Pirpamer puts it in his pocket and takes it along to the refuge.

Blaz Kulis has a colleague, a trainee cook from Slovakia, Osvald Lubomir. Kulis tells him about the corpse and wants to show it to him. So they climb up. They find everything untouched. But for the young cook the sight of the snow-covered plastic bag is enough, and the pair start back at once.

Shortly afterwards, at 11.45 a.m., Brigadier Herbert Niederkofler of the Italian police in Senales telephones his colleagues in Sölden. He confirms Alois Pirpamer's statement that Capsoni's body was recovered as long ago as 1952. The music professor from

Verona is definitely out of the running as far as an identification for the Hauslabjoch body is concerned.

Carlo Capsoni was born on 11 March 1903 in Verona, the son of Giovanni and Maraja Ermengilda Capsoni. Before his death he lived in Piacenza. On 25 August 1941 he made the ascent to the Hochjoch, having left word that he intended to walk on his own via the Schöne Aussicht to the Similaunhütte. Nothing more was heard of him. On 8 August 1952, towards 5 p.m., mountaineers found his body in the glacier ice on the Hochjoch near the Schöne Aussicht refuge. His mortal remains were buried at the cemetery of Unserfrau im Schnalstal. That the Simons should have set up their base in the same village, and from there discovered a glacier body many thousands of years older, is pure coincidence. But Capsoni had lain in the ice for only eleven years.

Meanwhile Lubomir is tormented by curiosity. Towards 3 p.m. he sets out for the Hauslabjoch on his own. This time he is braver. He lifts the cover a little – yes, there is a dead man under there. Shaken, he leaves the scene some ten minutes later.

At the same time Messner and Kammerlander, along with their guide Kurt Fritz, arrive at the Similaunhütte as planned. There they have arranged to meet the Ötztal ethnographer Dr Hans Haid and his wife Gerlinde, a folk-music researcher. Hans Haid intended to talk to Messner about legendary cult centres in the Ötztal, but the only topic of conversation in the taproom is the man in the ice. Markus Pirpamer makes a sketch from memory of the axe which was taken to Sölden the day before. Messner repeats: 'The axe is iron and doesn't have a hole?' Markus Pirpamer says yes. 'Then it is at least five hundred years old, maybe three thousand, and an important archaeological find.' He was to be proved right.

The metal analyses later conducted on the axe showed that the blade consists of almost pure copper. The original guess that it was iron is understandable. Objects made from copper, as from bronze, if left lying in a damp environment for any length of time develop a so-called water patina. As a rule, this is a wafer-thin layer of brownish, occasionally rust-coloured, rarely greenish, hue on the metal. But prehistoric finds in water are not very frequent. A prehistoric axe found in glacier ice is unprecedented. Infinitely more common are articles of copper or its alloys found in mineral soils. Generally these have a thick covering of verdigris, sometimes called 'noble patina', which has become the common mental picture of prehistoric artefacts. The finds at the Hauslabjoch were not strictly compatible with any of the previous theories, which was one reason why the authorities decided to exercise extreme caution. Messner alone had come close to the truth.

Markus Pirpamer then exhibits the piece of birch bark. Discussion is brief. They all want to visit the site of the mystery. Together the three mountaineers and the Haids set out. Messner, Kammerlander and Fritz choose a short-cut over some scree and only take half an hour. Hans Haid arrives some fifteen minutes later. Gerlinde Haid finds herself in quite dangerous fog and takes an additional half hour. The old mountaineers' rule that the pace should be set by the weakest member of the party seems to have been somewhat modified in this instance.

Messner and Kammerlander pull aside the plastic sheet. The dead man looms above the ice and water. Through the meltwater they see that he is wearing something rather like trousers on his legs. For that reason, and in order to look for more pieces of equipment, they hack open the ice around the man's buttocks and along his thighs. Messner uses a ski-pole. Kammerlander has nothing suitable to hand and so picks up a piece of wood lying nearby. Not until later does he realize that this was the thinner strut of the frame of the back pannier on which, 5,000 years ago, the man had dragged his possessions up to the Hauslabjoch. The hazel switch is still serviceable. The birch-bark vessel and its contents have by now been crushed underfoot. Nevertheless the mountain guide Kurt Fritz notices 'wilted leaves whose veining was still clearly recognizable'.

These leaves were subsequently identified by our botanists as those of the Norway maple (*Acer platanoides*). They also noticed that numerous flakes of charcoal were adhering to the leaves, a fact which provided vital proof that the function of the birch-bark container had been to carry embers. Hunter-gatherer peoples usually took some embers along from the hearth-fires of their last camp, as nomads do today. The charcoal fragments were wrapped in fresh leaves or grass, and the lot placed in a container of wood or ceramic material. A leather pouch, a basket, or, as in our case, a bark vessel could serve the same purpose. This keeps the embers live for a long time. Even today Austrian housewives will wrap a briquette in damp newspaper and shove it in a tile stove. This helps to kindle the fire quickly the following morning.

The visitors to the Hauslabjoch endeavour to pull what was later found to be a bow clear of the ice. A conical hole about 20 centimetres deep is hacked around the lower end of the stave. But the glacier refuses to give it up. The attempt is abandoned, especially as meltwater continues to flow in.

Attention is also focused on the body itself. The leatherlike leggings are inspected, and fine seams are detected. Fritz reaches into the water and picks up a piece of leather. One of the man's feet is likewise wrapped in leather, with grass spilling out. It reminds Messner of the footwear of the Lapps. Then Fritz raises the dead man's head, so that his crumpled features are visible for the first time. A fibrous 'mat' is revealed frozen to the boulder against which the man's torso is resting.

This matted grass was found again during the second archaeological examination at the beginning of October 1991. Only then was the exact position of the body at the point of discovery accurately determined. The so-called mat formed part of a straw, or rather grass, cloak in which the man had wrapped himself against the bad weather. As recently as this century pastoral peoples used the same kind of straw cloak as protection against cold and rain.

A piece of wood barely 20 centimetres long is found floating near the body in the meltwater pool. It has holes at regular intervals – a flute springs to mind. However, Gerlinde Haid, the folk-music expert, disagrees. In fact, it is a broken-off piece of the quiver stiffening. The fact that this fragment was found not near the quiver, but right next to the dead man, is an important clue to the drama which, over 5,000 years ago, occurred near the main ridge of the Alps. But more of this later.

A little way behind the rocky ledge against which the bow rests Gerlinde Haid finds numerous pieces of birch bark. This spot is about 5 metres from the corpse. As their presence at this altitude strikes her as odd she pockets the scraps and takes them away.

A few weeks later Gerlinde Haid handed the birch-bark fragments to us. Without a doubt they too belong to a container. Some of the pieces even show clear signs of stitch-holes. She also described to us their precise location. Subsequent examination showed that the fragments were all natural-coloured on the inside; the birch bark found near the body, on the other hand, was all stained black, which was one of the main arguments in favour of its identification as a cinder carrier. The pieces collected by Gerlinde Haid evidently came from a second container which the man had been carrying. These details show how important it was to question every visitor to the Hauslabjoch site minutely about what they had seen. The piece which Markus Pirpamer picked up on the morning of the same day, some 20 metres away, is the base of the second container. Obviously the wind must have moved some of the lighter items once they were released from the ice, and some of them will probably never be found.

The visitors again carefully cover up the corpse. They place lumps of snow and ice and a few stones on the black plastic sheet before, about half an hour later, they leave the Hauslabjoch.

Meanwhile Michael Tschöll, the guide who will accompany Messner on his next day's climb, has arrived at the Similaunhütte. Paul Hanny, the mountaineers' manager, has also joined the party. In the refuge the discovery of the corpse is still the main topic of conversation. Hanny sends the mountain guide up to the Hauslabjoch to take photographs. He sets out at about 5 p.m., but the Messner party, by then on their descent, do not meet him. Tschöll removes the plastic cover. Hurriedly, because dusk is falling, he takes a few photographs using flash. After five minutes he leaves the Hauslabjoch.

In the evening they all sit cosily together in the Similaunhütte. The news of the day is exhaustively discussed. Messner and Kammerlander, guided by Tschöll, have to reach the Zwickauer Hütte, fifteen peaks away, the following day. They put out a press release, as they have done every evening during their circular tour of South Tyrol. Messner sticks to his original statement: 'At least five hundred, maybe even three thousand, years old – because of the iron axe.' But Ezio Danieli, the correspondent of *Alto Adige*, an Italian-language South Tyrol paper, thinks three thousand is a bit rash, and he'd better stick to five hundred. And so the next few days give rise to the myth of a mercenary of Frederick Empty-Purse (1403–39, Count of Tyrol), the very story which originally drew my attention to the discovery.

A few weeks later Messner and I appeared together on a television programme. I was impressed to meet him. He seemed quite different from the way he is often presented by the media – shorter than one might imagine, hardly taller than the Iceman who was the subject of the programme. They share the typical mountaineer's stature. Messner doesn't seem particularly wiry or muscular, as one might expect of a man who has done what he has; he has the figure of someone who keeps in good shape. It is said that he spends hours practising pull-ups with his fingertips. His handshake is firm and confident, his laughter open and hearty. When he thinks no one is watching him, his face seems almost sad. He worries about civilization's desecration of the wild, and in particular of his beloved mountains. When challenged by nature, he has unbelievable reserves of energy.

As so often in cases of this kind, the reporters are quicker off the mark than the authorities. They have a nose for a big story – which, after all, is what makes a successful journalist. On the midday radio news that Saturday the press photographer Max Scherer first hears about the glacier corpse. He telephones his colleagues – 'I have a hunch that this isn't just any dead body.' He even telephones a helicopter company. But a directive has been issueed by the Tyrol Provincial Government saying that, for reasons of environmental protection, no landing permission is being given above an altitude of 1,700 metres. For the moment Scherer must be content with that ruling.

4 Sunday, 22 September 1991

On Sunday, 22 September 1991, Alois Pirpamer and Franz Gurschler, a retired joiner who helps with the maintenance of the goods lift to the Similaunhütte, set out at sunrise for the Hauslabjoch. The time is 7 a.m. In his capacity as mountain rescue chief Alois Pirpamer intends to free the corpse entirely from the glacier ice, ready for its official recovery which is planned for Monday. The ascent takes about an hour. The corpse lies before them, exposed. The plastic sheet, presumably lifted off by the wind, now lies near the bow. Overnight an ice crust roughly two centimetres thick has formed on the meltwater pool surrounding the dead man's torso. Systematically they set about freeing the body with their ice-picks.

There are no longer any clothes to be seen on the man's back, buttocks or thighs. Only below the knees can they still detect remnants of his trousers. His left foot can only be freed with great difficulty, as meltwater keeps running in. The right arm cannot be freed at all. It sticks down diagonally into the ice. They assume that the dead man might be holding something that obstructs movement. As they try to lift the body, remains of fur stay clinging to the ice. It is not impossible that the lesions in the genital area, which will be noted in the forensic examination, were caused then. Once more they cover the body with the plastic sheet. Once more the sheet is weighed down with chunks of ice and stone slabs.

Next the two men collect all the loose objects. Into a plastic refuse bag they place the pieces of wood, the rod with the holes, the remains of cord and the scraps of fur, which are lying on or beside the rocky shelf. Another attempt is made to hack the stave out of the ice, but again it fails. After two hours' work Pirpamer slings the bag over his shoulder and they leave the Hauslabjoch. That same afternoon Pirpamer takes the bag with the finds down to his hotel in Vent.

A short while later Messner's film crew – cameraman Fulvio Mariani and sound recordist Nicolas de la Cruz – arrive at the Hauslabjoch. The two had spent the night at the Similaunhütte and heard the talk about the man in the ice. Messner wanted shots of the finds and the spot where they lie. They approach the covered body, but think it's a big fuss over nothing: 'Probably a soldier from the First World War.' They return to the valley without having seen the body.

Otherwise there is only some more telephoning. Klocker, the undertaker, asks the police in Sölden to notify him of the time of the recovery flight. Alois Pirpamer informs the same police station that the body is ready for collection. The *Kronenzeitung* newspaper rings Max Scherer to express interest in any photographs he might take. Scherer himself contacts ÖRF, Austrian Radio and Television, to get them to use their influence to secure a landing permit for the following day. Rainer Henn, the forensic expert, spends the day at his home on the Starnberger See, devoting himself to his collection of precious Russian icons – totally unaware of the excitement ahead.

At about 5 p.m. ÖRF reporter Rainer Hölzl returns to the Rennweg studio in Innsbruck from a filming assignment. His colleague Georg Laich tells him of a press release from the Security Directorate which has just arrived. After a telephone call to Markus Pirpamer, the refuge keeper, who assures him that the body is at least 150 years old, Laich broadcasts the story on the radio. Hölzl, the television man, is very excited, and gets in touch with the Air Operations Centre at the airport to find out when the recovery is scheduled. But the people on duty know nothing about it and tell him to wait until Monday.

In its Sunday edition *Alto Adige* runs a big story on the sensational find at the Hauslabjoch below the Finailspitze at an altitude of 3,200 metres. The great mountaineer Reinhold Messner has seen an early hunter, probably 500 years old at least. Frederick Empty-Purse's retreat after a lost skirmish, through the Senales valley into the Ötztal, is also mentioned. The corpse seems to be very well preserved. In his hand the dead man holds an axe. On his back are burn marks or lash-weals. On his feet is the kind of footwear worn by the Eskimos. His head has been pierced by a sharp object or shattered by crashing against the rocks. Messner, the article said, was speechless with amazement. As the find was made on Austrian territory, an Austrian university would conduct the investigations; the authorities had already been notified. There was a need for urgent action. The paper published the first, hand-drawn, picture of the Iceman – an unkempt figure with puttees on his legs, stretched out on the blank ice, his left hand clutching a battle axe, or rather an executioner's axe. In the background the mountains are outlined against the sky. Two walkers, presumably Messner and Kammerlander, are approaching...

An article like this would make any self-respecting archaeologist leap out of his chair. But unfortunately on that Sunday no South Tyrolean prehistorian read the morning edition of *Alto Adige*.

5 Monday, 23 September 1991

The day of the official recovery dawns. It is Monday, 23 September 1991. At first light, about 7.30 a.m., Hölzl rings the Air Operations Centre. The air rescue operator Peter Strasser tells him that yes, recovery is planned for that day, but the flight time will be decided at short notice, since the weather is uncertain. Hölzl stays near the phone. Eventually, at 10.30 a.m., he is informed that the recovery flight is about to take off. Scherer has already arrived at the ÖRF studios. Hurriedly the administration department obtains permission to land at the Hauslabjoch, and charters a helicopter. Swiftly the reporter assembles a camera team. Everyone is tense and in a rush. They drive to the airport, and 11.31 a.m. they take

The first drawing of the Iceman, made on the basis of telephoned information from Reinhold Messner and published in *Alto Adige* on Sunday 22 September 1991.

off in the direction of the Hauslabjoch. On board are the pilot Gilbert Habringer, flight engineer Hannes Gehwolf, the cameraman Anton Matthis, Scherer and Hölzl. As the cabin is full, the sound recordist Thomas Eber has gone ahead by car to Vent. But he will arrive too late and never see the glacier corpse.

The pilot tries to find the spot described to him by Strasser, but last night's snowfall makes it impossible. They fly back to the Similaunhütte and invite Markus Pirpamer on board. Hölzl offers him his seat and stays behind. This time the refuge-keeper points out the rocky gully from the helicopter and Halbringer gets his bearings. They return to the refuge and seats are once more exchanged. Eventually a landing is made at the Hauslabjoch at 12.29 p.m. Recovery has not even started.

Fresh snow has covered all previous traces. They fan out, and Gehwolf finally finds the spot. They scrape off the snow, lift the sheet a little and assure themselves that the body is still there. Everyone is highly disciplined. At 12.37 p.m. the helicopter from the Ministry of the Interior appears. The ÖRF team have won the race, and the official recovery takes place with the cameras rolling – film that will be seen around the globe.

Rainer Henn usually arrives at his office in the Forensic Medicine Institute very early. This morning he finds two fax messages waiting on his desk. A body has been found in a glacier at the Hauslabjoch – nothing out of the ordinary this year. He arranges a flight because he wants to be present at the recovery. He also gets in touch with the public prosecutor. They agree to meet as soon as the body has been brought to Innsbruck. At about 8.30 a.m. a professor of archaeology named Spindler telephones, claiming that the find could be exceedingly old and of historical interest. He even wants to fly to the site himself, but Henn dissuades him.

With a slight improvement in the weather the helicopter takes off at 11.40 a.m. The pilot Anton Prodinger, the air rescue operator Roman Lukasser and Henn first fly to Vent for an operational briefing. At 12.05 p.m. the machine touches down. A curious crowd has gathered, as well as the police from Sölden and a few journalists. Alois Pirpamer reports that the dead man is already clear of the ice and so stretcher, ice-picks and shovels are off-loaded as useless ballast. The press photographer Werner Nosko and an unknown individual press their cameras into the hands of Prodinger and Lukasser, begging them to take pictures up in the mountains. Fifteen minutes later they take off again, circle over the gully and touch down at the Hauslabjoch at 12.37 p.m.

Much to his surprise, Henn is welcomed by a freezing camera team. (As the recovery that followed was filmed from first to last, it is exceptionally well documented.) Henn trudges through the snow to the site of the find. Lukasser walks ahead, dragging the body bag. They have to clamber over a few boulders. Henn carries his black bag with his special instruments and his camera round his neck. Then he stands in front of the plastic sheet. He pulls it aside a little to reveal the dead man. Lukasser doesn't want to touch the corpse with his bare hands, so Henn hands him some rubber gloves. Then he spreads out the zipped transparent-plastic body bag, while Lukasser removes the black sheet completely. Henn tries to move the dead man's head. But something unexpected has happened – overnight the body has once more frozen into its base and can be neither moved nor shaken free. What's to be done? No ice-picks, no shovels. Henn takes photographs, Lukasser considers bringing up the tools. Henn turns up the collar of his steel-grey anorak. It is bitterly cold. The camera is still rolling.

Hölzl seizes the opportunity for an interview. He tries to get Henn to make a guess at the date of the corpse, but the forensic expert refuses to speculate before he has seen the results of scientific examinations. Only when he is asked how it is at all possible to date a corpse does he give a precise answer. He explains that if very long periods of time are involved, the carbon-14 method can be used. Conclusions could also be drawn from the nature and degree of conversion into grave-wax (see p. 278). If the location of the discovery is taken into account, deductions could be made, with the help of glaciologists, from the speed of flow of the glacier ice.

Fortunately a mountaineer, the South Tyrolean Markus Wiegele, happens to arrive at the site. He carries an ice-pick and a ski-pole. Of course he did not simply turn up out of the blue. The previous evening, at the Schöne Aussicht refuge, the conversation had turned to Messner, who was said to have discovered a body near the Similaunhütte the previous day. Wiegele is resting on the crest when he sees the helicopters circling. Immediately he connects them with the story he heard the night before and makes his way down. He willingly offers his pick and ski-pole and enthusiastically lends a hand.

While Lukasser works with the ice-pick, Henn kneels down and recovers tatters of leather, stacking them up beside the corpse. Meltwater keeps running in. The tools ring on the ice. Gradually the body is loosened. Lukasser pulls at the corpse's arm with the ice-pick; Wiegele helps with the ski-pole. Henn takes hold of the feet. And so they ease the body out of the ice and lay it on its back by the hole. It is a shocking sight. Everyone stands silent around the dead man. Henn is the first to speak, and takes a businesslike approach: 'What matters now is that we bring out a few more things for the archaeologists.' Items of equipment are duly fished out of the trough in which the body has lain.

Hölzl conducts a second interview: 'Herr Professor, what can you say now that you can see the body as a whole?' Henn sums up: 'Teeth worn down from chewing, rather ground down. Partially mummified. Must have been exposed to the air for some time before he got into the ice. Clothes? Unfortunately nothing left. That's all we can say at this point.'

Henn turns the body back onto its stomach. Then follows its insertion into the body bag. Henn takes the feet, Lukasser the shoulders. The zip is fastened. Lukasser keeps hacking with the ice-pick in order to recover more items of equipment, while Henn wipes his hands with snow. The ÖRF helicopter starts up its engine. Markus Pirpamer, the refuge keeper, is brought along to the Hauslabjoch for an interview. He reports how, having been notified of the discovery by the Simons, he had informed both the Italian and Austrian authorities. He describes the location of the find, his observations at the spot and the equipment of the dead man. He too has no idea of the age of the corpse, but thinks it can hardly be more than a hundred years old.

The helicopter which has brought the refuge keeper now brings further tools, including a shovel. Henn uses it to make a channel in the snow to let the meltwater drain away from the ice trough. Lukasser continues to dig for further items of equipment. A large scrap of leather or hide appears. Then Henn joins in, probing with the ski-pole. Meanwhile Wiegele has taken over the shovel to drive the water through the drainage channel. Lukasser works untiringly with the pick, throwing bits and pieces on a pile. Then Henn bends down and frees a small oblong article from the ice, roughly at the spot where the dead man's hip or belly had lain. Henn lifts the object, and part of it falls away. It is the Iceman's miniature dagger with its short flint blade and wooden handle. Only the scabbard drops back into the water. Henn shows the find to Lukasser. Then he places it with the rest of the things on the heap. The others also scrutinize the object and Matthis takes a close-up of it. The television people have now finished and they fly back down to the valley.

Those left behind pack their things. The implements found are also stowed away in the body bag. It is not zipped up completely because the corpse has begun to smell. Wiegele meanwhile makes a last attempt to pull the bow out of the ice – in vain. It is therefore broken off and stuffed into the body bag. The end left in the ice will not be salvaged until the following year. The bundle is dragged to the helicopter and shoved into the orange recovery bag. The karabiner on the transport rope of the recovery bag clicks into the underside of the helicopter. The crew boards and at 3 p.m. they lift off in the direction of Vent. Only Markus Pirpamer leaves the spot on foot.

Afterwards, especially when the television pictures were transmitted around the world, fierce accusations were levelled against the forensic examiner. But these accusations only began to circulate once the find had been classified as prehistoric and assigned an age of at least 4,000 years. It is easy to be wise after the event. The Viennese, in particular, voiced loud accusations without knowing the circumstances. Naturally enough, traditional rivalries between Vienna and the Tyrol played their part. At least, it was argued, a prehistorian should have been brought in on the recovery, which in any case had been performed in a most unprofessional manner. However, prior to the recovery there was no real contact between Henn and myself. Neither he nor I expected a prehistoric find, nor could we have anticipated its sensational nature. According to the press reports I have seen, Modern Archaeology should at least have been involved – but this is a young discipline often still not taken seriously in professional circles. The fact that Henn flew off without me didn't worry me in the least.

And then there is Henn. Rescue operations in Alpine glacier regions, at an altitude of 3,200 metres, at which helicopters fly with difficulty, are rather different from recovery operations in traditional archaeological settlement areas. It is significant that none of the principal critics ever took the trouble, then or later, to climb to the site in person. Such critics usually voice their opinions from the comfort of their own offices... The police report merely stated that the Alpine incident could have happened as early as the nineteenth century. To bring in a prehistorian on that basis would have been absurd.

As for the Hauslabjoch site, not a single person who had attempted to free the body during the preceding days, or who had been otherwise involved, was present during Henn's recovery operation. Not one of them informed anybody that all the objects scattered about the point of discovery had already been removed; most significantly, that the axe was safely stowed away in the air-raid shelter of the Sölden police station. All Henn knew was that the body was ready to be transported – which was why no recovery equipment was taken along.

Even when the forensic examiner approached the site and saw the plastic sheet scraped clear by the reporters, it was by no means obvious that the sheet had only been there for a couple of days and was quite unconnected with the tragic incident. Under the sheet all that could be seen was the uncovered head and back of the corpse, without the slightest trace of any equipment. And although pick-axes are common excavation tools, Henn was immediately accused of being a 'ski-pole archaeologist'. An unfair criticism and quite unwarranted, it was part of a no-holds-barred campaign of defamation against the Innsbruck academics, the sole aim of which was to have the valuable find removed from their care for study elsewhere.

Moreover, all our informants told us without any prompting that throughout the clearing work Henn continually urged them to take the greatest possible care. Later, during one of my many talks with Henn, he told me his first impressions: 'When I discovered the stone knife I wondered whether this might not be an escaped convict who had made the thing himself. Many of the seams on the scraps of leather were as straight as if they had been made on a sewing machine.' That evening, when Henn first saw the remaining items of equipment which had been brought to the Institute of Forensic Medicine, he immediately got in touch with the Archaeological Institute. At that point he realized – and he later confirmed this to us – that the find was no longer a case for the dissecting room. Henn died on 25 July 1992, barely sixty-four years old. He was killed at St Stefan an der Gail, in south Austria, an innocent victim of a dreadful road accident, while on his way to give a lecture on the 'Man in the Ice'.

At 2.03 p.m. the helicopter lands in Vent. The orange recovery bag is unhooked and opened. Klocker drives up in his hearse. The coffin is pulled out and its lid opened. The dead man's left arm is still sticking out. When it is straightened, there is an audible crack; subsequently the medics will diagnose a fracture of the left humerus. Alois Pirpamer brings the refuse bag with the finds retrieved the previous day. It too is placed in the coffin, then the lid is replaced and screwed down. The undertaker drives off towards Innsbruck. En route he stops off at Sölden to pick up the axe, which the police have carefully packed in a carton.

Not entirely by accident the Ötztal ethnographer Hans Haid is also at the Sölden police station. His talk with Messner at the Hauslabjoch on Saturday has been preying on his mind. He has tried to mobilize the authorities – almost an impossibility over a weekend. Not until Monday morning, shortly before the recovery operation, did he manage to get hold of the Tyrolean Provincial

Museum. Its director, Professor Gert Ammann, took the call. But the appropriate curator was not there, and the find did not sound like it would be of interest to the museum. Ethnography in action: Haid lives on a mountain farm idyllically situated between Vent and Sölden; he himself scythes the grass and shears the sheep. But from such a base it's not easy to get things going. On the following day, when it is already too late, he tries to mobilize the ÖRF. By then, television has long had the pictures on film.

'The dead man dates from the Hallstatt period, or shortly before or after. That much is indicated by the tools. You did not include the axe in your investigation. The man was looking for minerals, like the legendary "Venedigermandl". The hole in his occiput is from trepanning, which was a common practice in prehistoric times, mostly for medical reasons. A rough dating between 900 and 500 BC is indicated.' Haid, too, came fairly close to the truth.

When Klocker arrives at the Sölden police station, Hans Haid asks for another look at the dead man. Briefly, the undertaker opens the coffin. Even though the zip of the body bag has not been completely closed there is an unpleasant smell. The corpse's skin is already getting darker. Klocker skips lunch and drives to Innsbruck in a hurry.

The Hotel Post is immediately next to the helicopter pad in Vent. Henn drinks another cup of coffee before he, too, drives off to Innsbruck. The hearse has arrived before him.

At 4.08 p.m. Klocker drives up to the Forensic Medicine Institute where he hands the body over to Dr Unterdorfer. Together they bring the coffin and the carton to the dissection room on a trolley. Official representatives are also present. Inspector Konrad Klotz of the Tyrol Police Command is waiting. The coffin lid is unscrewed, then Klocker and Unterdorfer lift the body, complete with plastic bag, onto the dissection table, unzipping it to allow the humidity to escape. Unterdorfer throws all the windows open. The temperature in the room is 18° centigrade. Then Henn joins the group. Unterdorfer makes the first diagnosis: 'This is presumably an original inhabitant of the Ötztal.'

Next Klocker goes to fetch his camera from the car. He has already taken lots of snaps at the landing pad in Vent and now he uses up the rest of his film. Eventually, hunger forces him to drive home. As he is leaving the institute, his path is blocked by a young man with the letters APA (Austria Press Agency) embroidered on the back of his jacket. He offers Klocker 20,000 Schillings for his film. Klocker sends him packing – the photos are for his own use. Subsequently, that reporter will be reluctant to recall his offer.

There is no question but that the case is taking on exceptional proportions. The man from the Hauslabjoch is the 619th body which fate has placed on the forensic dissection table in 1991, so it is given the reference number 91/619. Public Prosecutor Wallner and Examining Magistrate Böhler arrive after being summoned by telephone. They bring along a junior lawyer, Marlene Possik, to act as secretary. The refuse bag with the secondary finds lies on a second dissection table; the items have yet to be tipped out. The top of the bag is merely rolled down a little to allow a view of the contents.

Böhler dictates:

Public prosecutor Dr Wallner on 23.9.1991 reported the following state of affairs: In the Ötztal, near the Finailspitze, a body was found. According to rumour the body is said to have been fettered and to have burn marks on its back. Possibly it is a very ancient corpse. Upon application of public prosecutor Dr Wallner I ordered a forensic examination of the body with a view to establishing its identity and the possibility of culpability by a third party. This instruction was passed on by Dr Wallner to the security authorities. In the afternoon I participated jointly with Dr Wallner and Magistrate Böhler in the forensic examination. Purely external examination of the mummified corpse revealed that it is evidently several hundred years old. The alleged fetters are remnants of clothes or footwear. No burn marks could be identified on the back. With the corpse some primitively wrought tools (knife with stone blade and wooden handle, a kind of 'ice-pick' with stone point) were found, as well as hand-sewn clothing (leather parts crudely sewn), so that it may be assumed that the corpse had been lying at the place of discovery for several hundred years. The reason why there was no grave wax formation is probably that the body was already mummified before being frozen into the ice. In the opinion of Dr Unterdorfer the body discovered is historical. I have given instructions that the body should be handled with care and that, after examination, it should be made available to the historians, seeing that there are no indications of third-party culpability or of a culprit still pursuable. I have raised no objections to the corpse being photographed by the press, etc. Innsbruck, 23.9.1991.

Henn telephones the Institute for Pre- and Protohistory. After all, we had spoken on the phone earlier that morning. But it is outside office hours and I am on my way home. So the involvement of the 'historian' is postponed to the following day. The material of the axe, described by the examining magistrate as 'a kind of ice-pick', is carefully inspected and the blade scraped with a scalpel; from its colour and softness the material is thought to be copper.

Unterdorfer dictates the following report to his secretary, Uta Halper:

Name XY [the pre-printed boxes for Sex, Date of birth, Occupation and Address are left empty].
Identity: not yet established; unknown mummified mountain corpse from the Niederjochferner, Ötztal.
External diagnosis:

1 The body, along with scraps of greyish-black sodden material, was delivered in a colourless body bag and placed on the table by the undertaker Klocker from Längenfeld. The other items brought with it are contained in a black bag.
2 Sodden body of a largely mummified, externally ochre-coloured to brownish-black person. The body displays local blackish-grey sandy encrustations, along with remnants of interspersed fur-like patches. A clump of ice adheres to the area of the crotch and the buttocks. In the area of the back there are likewise similar, partially ice-impregnated, deposits.
3 As far as can be judged, no head-hair, body-hair or pubic hair; evidently the epidermis is no longer present; as far as can be seen, there are no nails left either.
4 In the area of the right foot and ankle joint: remnants of evidently simple footwear. As far as can be judged, this is dried straw and grass with leather-like footwear wrapped around it. This in turn is tied with a twisted cord; some interlacing recognizable.
5 The body relatively light, its weight is provisionally estimated at between 20 and 30 kg. Its length is given at 153 cm along the left leg.
6 With the body lying on its back, the left arm is extended from the shoulder joint at an angle upwards to the right. The right arm, with the shoulder area angled upwards, is extended outwards at an angle of approximately 60 degrees, the right hand held as though gripping a round object.
7 The body is lightly rinsed down with cold water; it is evidently a very, very old corpse.
8 In the eye sockets the dried-up eyeballs are still recognizable, with major damage; the chewing surfaces are clearly diagonally worn down by chewing.
9 The front and side of the neck without outwardly visible indications of fatal injuries.

10 Chest and abdomen sunken, the skeletal parts clearly identifiable; as far as can be judged from the outside, without injuries.
11 The lump of ice between the legs detaches itself; in it are thong-like leathery materials.
12 The external genitals foliated, as far as can be judged most probably male, desiccated. In the region of the left thigh and hip extensive damage. As far as can be seen, the muscle tissue fibrously transformed, torn out in hide-like scraps: one-third of the left thigh-bone (the third nearest to the trunk) sticks out, the head of the bone without identifiable degenerative changes, in the sciatic area of the back some degree of damage noted, evidently as a result of being eaten by animals.
13 In the area of the outer and rear side of the left leg superficial soft-tissue damage, evidently as a result of being eaten by animals.
14 The body is carefully turned to lie on its stomach and cleansed with cold water of the adhering blackish-grey sandy encrustations.
15 In the occipital area there is damage about 3 × 4 cm to the scalp, evidently made by a carrion-eating animal (bird of prey?), jagged along the edges, towards the left scratch-like superficial defects. The cranium, as far as can be seen, does not exhibit a corresponding fracture. The area of the nape without external indications of injury of an especially fatal nature.
16 In the left-hand lower dorsal region there are four groups of longitudinal, blackish-grey linear discolorations of the skin. The individual lines are arranged in parallel series, in four groups slightly below each other. They show discolorations of the skin. The individual lines have a length of about 2.8 cm to 3.0 cm, and a width between 2 mm and 3 mm. From top to bottom, first a group of four, then at closer intervals two groups of three, and finally, faintly identifiable, another group of four. This one, considerably fainter, localized in the area of the left sacro-iliac joint.
17 In the area of the sacro-coccygeal region and the left rear half of the pelvis, as well as in the adjoining left thigh region, there is extensive damage to the soft tissue suggesting that the flesh has been eaten by animals (dog-like predator). Ice still at this spot. As far as observable, the skeletal parts are also affected, exhibiting something like clipped marginal areas. More detailed assessment not possible because the deeply ingrained ice has not yet melted.
18 In the area of the hollow of the right knee a cross with a length of bars of 2.8 cm and 2.7 cm and a bar width of between 3 mm and 4 mm dyed-in or tattooed in blackish colour.
19 In the area of the lower extremities no indication of major injuries.
20 The hay-upholstered footwear remains on the right foot are left in place.

21 In the area of the right lower dorsal region, more feebly, another group of four lines in the shape of longitudinal blackish parallel skin discolorations as on the left side.
22 In the area of the back of the left hand, near the wrist, two transverse parallel linear skin discolorations, 4.2 cm in length at a distance of 7 mm to 8 mm.
23 The Institute of Forensic Medicine is informed that, on the strength of its external appearance, independently of the accompanying articles, this is a corpse at least many hundreds of years old. Following consultation with Professor Henn the accompanying items are laid out on the second dissection table and the entire find left as it is to await the arrival, after notification, of Professor Spindler of the University Institute for Pre- and Protohistory.
24 In view of the circumstances Dr Böhler, the examining magistrate, issues no instruction for an autopsy. After further telephone consultation, no prohibition of photography is issued.

Now it is the turn of the police photographer, District Inspector Roger Teissl, of the Criminal Investigation Branch of the Provincial Police Command for the Tyrol. His photographs are used for the police report, and are sent also to the press office of the Security Directorate. One of these pictures, showing some of the objects found, will shake me wide awake the following morning, more violently even than the ringing of my alarm clock.

Before we follow subsequent developments, the course of the criminal investigation against persons unknown and its conclusion can be summed up. The file has to be entered in the registry and is passed on, for information, to the public prosecutor's office. The next day it goes back to the examining magistrate with the request to 'pass it on for further action once the results of the forensic examination are available'.

This is done by enlarging the batch of files with the findings of the next few days:

(1) According to press reports the body discovered ('Similaun man') is said to be a 4,000-year-old corpse from the Bronze Age; (2) File cover and registration to be amplified by the words 'Similaun man' (because of historical interest in the assessment of file indexing).

The juridical fate of the corpse undergoes the following change:

As the body has been released by the courts in accordance with the Criminal Procedure Code, the body will, in accordance with Article 31 of the Tyrol Municipal Public Health Act, be impounded by the public-health police on 25.9.1991. Provincial Public Health Director Hofrat Dr Christoph Neuner is formally instructing the competent District Administration in Imst to arrange for a forensic autopsy to establish the cause of death (Dr Unterdorfer).

The files do not reveal that this notice of impounding has ever been rescinded. It exists to this day, probably only because of an oversight. The notice continues:

On 26.9.1991 at 09.05 the right shoe-like item of clothing is preserved and taken over under the direction of Dozent Egg. Prior to its removal the footwear is viewed by the Innsbruck master shoemaker Walter Reithofer.

Reithofer is the Forensic Medicine Institute's footwear expert. A retired shoemaker, he can determine the approximate date of any shoe at a glance. When unidentified bodies are found, his knowledge helps determine the date of death. But the Man in the Ice defies even his experience – the first time it has ever happened. The police report is added to the file. Eventually, on 14 October 1991, the criminal procedure is suspended. Significantly, the examining magistrate rules that a label is to be attached to the file stating that, in view of its historical interest, it is to be preserved in perpetuity.

The man from the Hauslabjoch remains on his slab in the forensic institute. Recovery and initial examinations are complete. It is late; the officials go home. This is to be our last quiet night for weeks to come.

6 Tuesday, 24 September 1991

The man from the ice is stored in the refrigerated chamber of the Forensic Medicine Institute at a temperature of 0° Centigrade. It is daybreak on Tuesday. Unterdorfer turns in his bed, unable to sleep. He cannot contain his excitement any longer. This man who, as he is so fond of saying, has carved up 'an entire small town', performing between fifteen and twenty thousand autopsies, loses his customary calm. At 5 a.m., when it is still dark, he walks to the institute. He pays a visit to his primeval Ötztal inhabitant and gets everything ready for the prehistorian. Once more the dead man is wheeled into the dissection room, and the more interesting of the accompanying objects are displayed to effect on a green cloth. Before 8 a.m. is too early to phone anyone. And then the moment arrives. The short-sighted archaeologist with his metal-rimmed spectacles bends over the table and gives his verdict: 'At least four thousand years old.'

Briefly I explain the dating clues – the flint implements suggest the Neolithic, but similar items are still to be found in the Early Bronze Age. The metal blade of the axe – it is still unclear whether it is copper or bronze – has a shape which is also common in the Early Bronze Age, but exceedingly rare in the Neolithic. All the other finds fit very nearly into that framework, but cannot as yet be dated more accurately. There is still disbelief. Then I ask if the Ancient Monuments Office has been notified. No one shows any embarrassment when they say no. The finds exhibit the early signs of desiccation, so I make arrangements for them to be carefully sprinkled with distilled water. A few minutes later I return to our institute where, apart from Julia Tschugg, no one else has arrived. I ask her to round up everybody. The way the objects are lying about in Forensic, they won't be on display for long. Besides, I must get through at once to the Roman-Germanic Central Museum in Mainz, Germany, and to the Federal Ancient Monuments Office and the Ministry, both of them in Vienna.

I get through first to the Ancient Monuments Office. Dr Christa Farka takes my call. In a few words I try to explain the situation. She understands at once and gives me a completely free hand – I am to do whatever I consider right. Shortly afterwards Mainz is on the line. My colleagues there realize the urgency of the situation and promise to come the following day. A load drops off my mind when, quite spontaneously and without any conditions, they promise to restore and conserve the objects free of charge. I also ask about the corpse. But Dr Ulrich Schaaf, the director, declines to take the body as well: they have no expertise in that field.

Meanwhile the working day has also begun in the Ministry of Science and Research. Dr Johann Popelak is sitting at his desk in his plain, businesslike office. At first he doesn't know what to do with the excited professor from Innsbruck on the other end of the line. He should authorize a helicopter? Where to, the Hauslabjoch? Never heard of it. So late in the year there's no money available anywhere. I explain the situation again. I really ought to be more concise, the ministry shouldn't think that government money is being wasted on endless long-distance calls. But I am past caring. And then the ministerial adviser cuts me short: 'Go ahead, fly out! I'll take responsibility.'

'We also have to excavate and conduct archaeological investigations at the location of the find.'

'No problem. I'll chase up the necessary finance.' I doubt if such an answer has ever been given by that ministry before or since.

For the moment, no one else in Vienna knows what is happening. The first television report last night only went out on the Tyrolean regional programme. All this will change dramatically during the next few hours. Anyway, I didn't even see the brief television clip on the day's news, because we don't have a TV at home.

By now my colleagues and some of the students have assembled at the institute. We walk across to Forensic. A few journalists are also hovering around. They must have noses as sensitive as butterflies, which can detect a single molecule of their partner's pheromone. The reporters are eagerly listening to the shoptalk between prehistorians and forensic medics. While archaeologists are agreed on the great age of the finds, the men in white continue to have their doubts. The humanities do not command much credibility among scientists and medical men.

One of the main reasons for the initial disbelief at such an early dating is the argument that a dead body could not possibly stay within the glacier ice for so long. Even in extremely long and gently sloping glaciers, which in consequence have a very slow rate of flow, the period of deposition, it was thought, could not exceed a thousand years. In point of fact, the oldest glacier corpses previously known in the Alps had an age of 'only' 400 years. These are the so-called 'Theodul Pass mercenary' and the 'Porchabella glacier shepherdess'.

Between 1985 and 1990 the remains of a body emerged from the ice of the upper Theodul glacier near Zermatt, at an altitude of about 3,000 metres above sea level. The sheering and grinding forces of the glacier had dismembered it into small pieces.

The gently inclined snowfield is very popular with beginners on skis. An instructor, Annemarie Julen-Lehner, noticed an obstacle and decided to remove it. 'How on earth did a coconut get here?' she wondered. What she picked up, with a shock, was a human skull covered with matted hair.

After that she and her husband Peter Lehner examined the edge of the retreating glacier tongue until the site was exhausted. It was a colossal yield. In addition to further fragments of human bone there also appeared some from two mules, complete with horseshoes, all scattered within an area about 75 metres long and 10 metres wide. Then leather footwear was discovered, parts of a silk shirt, a horseman's épée 125 centimetres long, a dagger with pommel and parrying bar, a wheel-lock pistol, several knives and an iron shoehorn, as well as two silver pendants, on one of which the initials HN were engraved, surely the monogram of the unidentified dead man. Heaps of coins were scattered about – more than two hundred pieces of silver altogether. They enabled the date of the Alpine accident to be determined very accurately at 1595 or thereabouts.

The discovery at the Hauslabjoch is totally different. It seems imperative to consult the glaciologists and there is an institute for High Mountain Research at the University of Innsbruck. We have collaborated for years with its director, Professor Gernot Patzelt, and I make an appointment with him for the following day. He is delighted at the news and ready to go up into the glacier world with his assistants at once. All of them, if only because of their profession, are excellent mountaineers, quite at home among rocks and ice. Unfortunately, the same cannot be said of me.

Back at our institute, we make preparations for the archaeological expedition into the mountains. Only those with a fair amount of mountaineering experience can be included. Henn and I charter a helicopter for the next day. I mention this because a colleague will subsequently claim in his book that 'evidently they were not expecting any further finds at the Tisenjoch (i.e. Hauslabjoch), since the prehistorians in Innsbruck failed to visit the location in spite of superb weather.' Above all, I am anxious to show what bad weather we had to contend with during the next few days and what dangers the mountains hold for the careless.

But this Tuesday there is still a lot to be done. The pressure from reporters is continually increasing. It is almost impossible to move freely. We promise them a press conference at 5 p.m. in the lecture theatre of the Forensic Medicine Institute simply to get them out of our hair. But it does not help much. Henn locks up his institute. For us, in a building of the Faculty of Humanities, this is not so easy. There are several departments on every floor and the passages are designated fire exits in the event of an emergency. We were to have a taste of this a couple of weeks later when the ten-storey building had to be evacuated because of a bomb alert. It is impossible simply to lock ourselves in. Swarms of reporters and camera crews crowd the passages, all the way down the staircases. Our telephone system is hopelessly overloaded. Nothing is working properly any more.

At the university only full professors are allowed to make direct external calls. Evidently it is assumed that all other staff would make too many private calls; they must, therefore, book their calls through the switchboard. But now the switchboard, too, breaks down under the load. In this situation our secretary reveals an unsuspected talent for organization. She gets on to the house technicians. We need a lot more external telephones, and we need them now. A miracle happens. Ten minutes later the technicians arrive. In blue overalls, with stepladders over their shoulders, rolls of cable and toolboxes in their hands. A ripple runs through the crowd of reporters. There is some 'action' in there. While the workmen drill holes through the walls and clip on strands of cable, the scene is lit by spotlights. Solemnly the TV crews film every hammer-blow.

Our technical equipment proves hopelessly inadequate under the onslaught. Julia Tschugg speaks to the finance people and demands fax machines, desktop copiers, etc. In the Institute for Pre- and Protohistory all hell has been let loose by the Iceman. Then a second miracle happens. Within an hour the equipment is actually in place. The telephone service is organized: Christiane Ganner takes the Italian calls, Nadja Riedmann the English ones and Elisabeth Zissernig the French ones. The rest brush up their English. The atmosphere is fantastic. No one dreams of taking a break or knocking off for the day. Meanwhile I am offered free flights to the television studios in Mainz, Hamburg and London for live interviews; I am to board the plane at once; everything has been booked. I decline with thanks. There is more urgent work to be done.

Our main concern is the corpse. None of us has any experience of permafrost mummies. We hope it will not have disintegrated into dust by tomorrow. Are there any other historical or prehistoric permafrost corpses anywhere in Europe? No. I think of the mummies of those Scythian nomad rulers from the Siberian Altai, who had been frozen into an ice core below their burial mounds. Sergey I. Rudenko excavated them in the years before and after the Second World War. The impressive tattoos of animal forms on the skin of one man in particular caused a great deal of excitement. How had those bodies been preserved? We can't get through to the Hermitage in Leningrad (soon to be called St Petersburg again). Russia is in turmoil. The Soviet regime is collapsing. Gorbachev switches to reform. It is hopeless. Henn and I confer; 'Platzer,' he suggests.

Professor Werner Platzer is the director of the Anatomical Institute of Innsbruck University, Henn's 'neighbour institute'. Whenever they have too many bodies in Forensic, Platzer helps out with his refrigeration cabinets. All in all, the people in Anatomy have a good deal of experience in preparing human bodies. But one thing is becoming increasingly obvious: this valuable human find has to be preserved in perpetuity so that it is freely available to future generations of researchers. The usual chemical methods of conservation must be avoided. We agree that no positive guidelines exist for the long-term preservation of human ice mummies. And so an idea takes shape: the man from the Ötztal Alps should be stored under glacier-like conditions. If he has survived in such a good state for thousands of years in a glacier, then he ought to last thousands more if the conditions are simulated.

In the meantime the corpse has thawed out to the normal indoor temperature of the dissection room, 18° Centigrade. So we discuss whether a refreezing to the mean annual Hauslabjoch temperature of minus 6° Centigrade could damage it. Considering that during the fair weather before and after its discovery the body had regularly thawed out, at least partially, by day and frozen again overnight, we reckon that the possible damage factor in the event of careful refreezing is comparatively slight. At any rate, we prefer this method to the alternatives. There is no lack of good advice. It ranges from shock freezing via storage in liquid nitrogen to the complete desiccation of the corpse, and includes various recipes for preserving fluids. A Viennese human biologist actually recommends placing the body into a domestic deep-freeze as if it were a pack of sausages.

During the press conference some journalists believe they can see a slight but spreading discoloration of the body's skin. Immediately there is talk of rampant fungi. The men in Innsbruck are letting the corpse go mouldy, the Viennese rant. Actually, these are, first, only slight condensation phenomena, due to the fact that the room temperature is higher than that inside the body. They are harmless. Second, flat bluish to blackish pustule-like spots appear. Similar phenomena have been seen in the past, for instance, with Siberian mammoths. These tiny crystals were identified as vivianite, a complex iron phosphate. They occur where the surface of the corpse has made contact with ferrous material – in our case the points of contact with the sediment on which the body rested at the Hauslabjoch. These mineral impregnations, too, are harmless.

Nevertheless we take certain precautions to protect our precious corpse with its irreplaceable equipment against further contamination until it is definitively preserved, and several dishes of diluted phenol are placed on the two dissection tables under the plastic sheets. The forensic experts swab the body with the same solution. This measure does not impair future examination, such as scientific dating.

By late afternoon of this exciting day everything necessary to preserve the body in the Anatomical Institute is in place. On the evening of 24 September 1991 Platzer takes possession of the corpse. Over the next few days the atmosphere in the refrigeration chamber was further improved with regard to keeping the air humidity constant. This method of storing and preserving the Hauslabjoch man will prove highly successful in years to come. Admittedly it is not cheap: it costs nearly 150,000 Schillings per month.

Fortunately the journalists don't notice that the body has been moved, because the Institute of Forensic Medicine and the Anatomical Institute are connected by an underground passage. So the corpse, unobserved by the public, can be transferred to its refrigeration chamber. Meanwhile we have virtually every country in Europe on the line. Later in the evening the American institutions call in. In the dark I am driven to Studio Tyrol on the Rennweg. The interview is broadcast by monitor, the final question, What could have driven the Iceman into the mountains? 'Perhaps he wanted to visit his girlfriend in the next valley,' I reply. A little after midnight – I am still not home – our telephone rings. Sleepily my wife picks it up. 'This is New Zealand.' She thinks it's a joke and puts down the receiver.

7 Wednesday, 25 September 1991

Gernot Patzelt walks in. It is Wednesday, 25 September 1991, and the time is 8 a.m. I have asked for the telephone on my desk to be disconnected, but to no avail. The reporters get through, perhaps by bribing the girl on the switchboard. But the glaciologist, with his cheerful weatherbeaten face, is above bribery. What he says makes sense. Briefly we discuss the problem. Because of the predictable worsening of the weather his three colleagues, Dr Ekkehard Dreiseitl and Gerhard Markl of the Institute of Meteorology and Geophysics and Dr Heralt Schneider of the Mathematical Institute, have already set out from Innsbruck for the Ötztal. They left at 6.30 a.m. and will reach the Hauslabjoch at 10.30 a.m. Patzelt will arrive somewhat later, at about 1.30 p.m. As soon as they reach the spot Schneider discovers the quiver. It has frozen on to a stone slab, with its upper part sticking out of the ice by about 20 centimetres. It contains at least ten arrows. Everything is left untouched in anticipation of our arrival.

Shortly before 9 a.m. Henn and I drive to the airport in his Mercedes. On the way he tells me a little about himself. Born in Mannheim, he studied in Heidelberg and then became a scientist at the Max Planck Institute. From there he switched to the University Hospital in Würzburg. After that he continued his career at the Institute for Juridical Medicine in Freiburg-im-Breisgau. His next stop was Munich. Finally, he was invited to take over the directorship of the Forensic Medicine Institute in Innsbruck. His hobby is gliding. Nothing, he says, is more impressive than floating soundlessly and weightlessly above the Alpine peaks.

The helicopter takes off and leaves Innsbruck behind. Below us is the loop of the Inn. We fly upstream, at first in a westerly direction. On our right and ahead is the Nordkette with the Lechtal Alps. On our left, far to the south, are the peaks of the main ridge of the Alps, veiled in sheets of cloud. The pilot looks worried. He is continually talking to the weather men through his helmet microphone. After a few minutes the valley mouth of the Ötztal opens up on our left. The machine veers to the south and flies between the steep flanks of the valley.

With a north-south extension of nearly 60 kilometres the Ötztal is one of the longest transverse valleys slicing from the Inn into the chains of the northern Alps, in the direction of the main ridge. That is why since ancient times it has served as a transit route to the south and in the opposite direction. Small passes facilitate the crossing of the main ridge – the Hochjoch, the Finailjoch, the Hauslabjoch, the Tisenjoch, the Eisjoch, the Niederjoch and the Similaunjoch. From the South Tyrol side, from the Vinschgau (in Italian the Val Venosta), the Schnalstal (Val di Senales) pushes towards the ridge. To this day the Val Venosta peasants have grazing areas in the Ötztal. Every summer they drive their flocks of sheep and herds of goats northwards, up to the mountain pastures, and bring them back down again in the autumn. The Iceman, too, must have been familiar with these crossings.

We have passed through the narrow mouth of the valley. The steep slopes recede. The valley opens out: below us lies the village of Ötz, from which the valley takes its name. It is thought to derive from the Old German word *Atzen* (to eat, to graze), and indicates the area's ancient grazing economy. The helicopter climbs and overflies the next ravine down which the Ötztaler Ache, the valley's stream, rushes. The road resorts to hairpin bends. Then the valley opens out again. Small villages lie below us. Minute mountain farmsteads cling to the hillsides. In the next basin is Längenfeld, where Klocker builds his coffins. The mountain flanks are hemmed with dark pine forests. Above them, beyond the treeline, are the pastures. We are approaching Sölden, where the valley divides. To the left it climbs towards the Timmelsjoch. We turn slightly to the right, in the direction of Vent. Here the road ends.

Now the landscape changes. We have left the forest zone. But mountain meadows still follow the valley floor. From all sides descend the lifeless screes of glacial moraine. It is a mystery to me how the pilot manages to find his way. Then we are above the glacier. Clouds pile up ahead of us, presaging the föhn, a warm dry Alpine wind. The helicopter avoids them, skips over a ridge, rises upwards. 'We are quite close. Down there is the Martin Busch Hütte,' the pilot shouts into his microphone. 'Those damned clouds.' He tries to dip under them. Huge crevasses yawn at us like hungry jaws. Only a few hundred metres, at most, to the Hauslabjoch. Just then an impenetrable fog bank rolls over the ridge. It is hopeless. The pilot decides to go back. For a whole hour we wait at the Martin Busch refuge, but the weather does not improve. We are forced to fly back to Innsbruck.

The glaciologists at the site wait for us in vain. They hear the helicopter's rotors. Then the noise recedes. The machine is turning back. It is too dangerous to attempt a landing in these conditions. What should they do about the quiver? It is not entirely impossible that someone else might chance by. The site is too close to the Hauslabjoch. Besides, it might snow. Then it would be lost for ever. It cannot be expected to emerge again from the ice this year.

The glaciologists make the only correct decision. They recover the quiver. Carefully they detach it from the ice and from the slab of rock without once touching it with a tool, by ceaselessly pouring meltwater over it from a picnic bag – a procedure that takes roughly two hours. Then they wrap the valuable object in a quilted jacket. They are just trying to splint the bundle with ski-poles when a figure emerges from the fog. It is Vice-Brigadier Silvano Dal Ben, an officer of the Italian Financial Guard. Two members of the team keep him talking while the others hastily stow the quiver in a rucksack and then disappear into the fog.

Only Dal Ben is left behind. From the hollow in the ice, where the body once lay, he collects numerous scraps of fur and leather. In the evening he delivers them to the South Tyrol Ancient Monuments Office in Bolzano. A transfer protocol is signed. It was to take approximately nine months and considerable red tape before those finds could be united with the rest of the Iceman's equipment for restoration in Mainz.

A group of archaeologists also set out from Innsbruck at 9 a.m. that day, taking their excavation equipment with them in the car. We all intended to meet at the Hauslabjoch. Other people are on their way in a second helicopter. They are to be disembarked at the site of the find and the rest of the team will then be brought up from Vent. Josef Ullmann, a student of prehistory, is already on board. With him are the ÖRF reporters Hölzl and Matthis, who are financing this flight. Dr Gerhard Tomedi is waiting at the landing pad with Julia Tschugg and Christiane Ganner, a student. But the machine returns fully laden. Just below the Hauslabjoch they have come up against the by then full-blown föhn front. Any continuation of the flight, let alone a landing, is out of the question. The ascent on foot from Vent to the Hauslabjoch, given normal mountaineering equipment, takes about four and a half hours. There is nothing for it. The expedition has to be cancelled for the day.

As Henn and I return from the airport the party from Mainz arrives. Dr Markus Egg was born in Innsbruck. He completed his studies at our institute and then began work at the Roman-Germanic Central Museum. A few years later he qualified as an assistant professor at our institute. The situation, which will become somewhat tense over the next few days, is made easier by the fact that a native Tyrolean looks after a Tyrolean matter abroad. Egg is accompanied by Roswitha Goedecker-Ciolek. She has a reputation as an outstanding specialist restorer, particularly where the preservation of organic finds is concerned.

The afternoon passes with preliminary restoration measures to safeguard the finds. Tomedi, who has meanwhile returned from Vent, assists. In particular, the Iceman's equipment has to be made ready for transportation to Mainz. We make our first enquiries concerning the formalities of an export licence. There are problems. Why, they ask in Vienna, can the objects not be restored in Austria?

The weather still rules out any helicopter flights, and no change is forecast for the next few days. We shall have to risk it on foot. It's a difficult decision. Only archaeologists with experience in the mountains can undertake the initial research on the spot. The Tyrolean Provincial Museum comes to our aid, and Liselotte Zemmer-Plank, the curator in charge of prehistoric artefacts, makes two of her staff available. The expedition has to be prepared with extreme care. Tomorrow the two will set out. Meanwhile, Wednesday is drawing to a close. We ask for a few rolls to be sent up from the cafeteria. We've been drinking coffee all day.

8 Thursday, 26 September 1991

In the early hours of 26 September, at 2.30 a.m., the glaciologists eventually return. At 8 a.m. sharp Patzelt is back at the institute, and hands us the quiver. The Mainz restorers are already at work and take it in charge.

Patzelt reports on their observations at the Hauslabjoch. The location of the find is at 3,210 metres above sea level, close to the summer track from the Similaunhütte to the Hauslabjoch, some seventy-three vertical metres below the saddle. The nearest geographic point on official maps is the Hauslabjoch. It therefore lends its name for the scientific labelling of the Iceman. The surrounding terrain is only slightly inclined. It is part of the accumulation area of a small unnamed glacier which once flowed eastwards into the Niederjochferner. Now, however, they are totally separate. Because of exposure to the wind the thickness of the snow and ice at that spot is likely to have been slight under post-glacial conditions.

Melting had increased during the past few warm summers, reaching an exceptional level in 1991. From January to mid-April the glacial winter was characterized by a relative dearth of snow, i.e. by drought. In a very mild late winter, with an excessively warm March, registering a mean temperature around 3.5° Centigrade, the slight snow cover began to melt before Easter even at high altitudes. The vegetation period began very early. A return to cold weather on 17 April admittedly meant a return to winter conditions with new snow all the way down to the valleys and night frosts. The crucial cause of glacier recession, however, had been a heavy fall of Saharan dust over a wide area. This occurred between 5 and 8 March with a high-level southerly air current. It had darkened the sky and coloured the snowfield and icefields yellowish brown.

The six months of summer began like winter. May was too cold by almost 4° Centigrade. Regionally, it brought two to three times the normal amounts of precipitation, which in the mountains mostly fell as snow. Thus at the end of May there was an above-average amount of snow lying on the glaciers. The Sonnblick observatory in the Province of Salzburg measured a maximum snow-cover of 810 centimetres. Exceptionally big avalanches from the mountain flanks carried additional quantities of snow into the basins. The mean June temperatures were likewise unseasonably low. By the end of that month there were numerous fresh falls of snow in the mountains, delaying the reduction of the snowcover. Then the situation changed radically. July, August and September were each of above-average temperature by 2° Centigrade and marked by prolonged periods of fine weather. The short-lived cold spells around 17 July and 29 August produced only small quantities of new snow and scarcely interrupted glacier melting. By the second half of July the Saharan dust reappeared. Then an incredibly rapid melting of ice began. The sunlight was no longer, as usual, reflected by the white snowfields; instead, the dark-yellow layer of dust absorbed the radiation and accelerated the reduction of the snow-cover and glacier melting. Even after fresh falls of snow from 7 to 10 September, and during the night of 22/23 September, the dust-contaminated snow and icefields continued to experience melt conditions, even at high altitudes. The fact that the Iceman began to emerge from the ice in mid-September is therefore ultimately due to a desert storm in North Africa.

The spot where the body was found is in an almost horizontal gully running from west-southwest to east-northeast. This hollow in the shape of a ship's hull lies some 2 to 3 metres below the ribs of rock bounding it. It is up to 5 metres wide and approximately 50 metres long. When the glaciologists arrived its base was still filled with ice, and the meltwater was up to 30 centimetres deep. The gully drains towards the northeast, towards the Niederjochferner, though its lowest part does not drain at all. In this gully, therefore, the ice, even when very thick and with an annual mean temperature of minus 6° Centigrade, was virtually motionless. In the event of extreme accumulations the glacier from the main ridge flowed down, over the ice core inside the gully, to the Niederjochferner. On the gently inclined terrain above the gully the ice can have had no more than a moderate speed of flow. And as the fall line of the glacier is at right angles to the gully, it did not move the ground ice inside the gully along with it. Given the flat wind-exposed position near the saddle, the snow and ice cover, even at maximum glaciation, is unlikely to have been deeper than 20 to 25 metres.

The nature of the terrain below the Hauslabjoch explains why the objects trapped in the ice since prehistoric times were not subjected to the grinding and sheering effects of flowing ice and why, in consequence, they were preserved at their original place of deposition.

Closely connected with these facts, which tie up several loose ends, is the problem of whether, in the course of his more than 5,000 years in the glacier, the Iceman has once before, or indeed more than once, emerged from the ice. In other words, was September 1991 the first time that he found himself in melt conditions? According to Patzelt's knowledge of the glacial and climatic history of the Ötztal Alps, such a possibility might have existed during the warm phases of the Middle Ages or the period of the Roman emperors. One might speculate whether the skin lesion on the back of the head (the highest point of the body as it lay in the ice) might be due to that part of the body having once before become desiccated and hence been subject to decay.

On the strength of glaciologists' experiences of dead animals ejected by glaciers it seems extremely unlikely that the Iceman had ever before emerged from the ice to the extent that he had when the Simons discovered him. The resulting desiccation and re-moistening, the daily change from frost to solar irradiation, which would surely have heated the surface of his skin to over 20° Centigrade, would have destroyed the soft tissue in no time. Predators and carrion birds would have seen to the rest.

From these considerations, and bearing in mind the topography of the site, we may conclude that in the Iceman's lifetime climatic conditions were similar to those prevailing at present. The gully in the rock must then, as in September 1991, have been wholly or largely clear of snow and ice, with mean annual temperatures of about 0.5° Centigrade above the longterm average. During the known warm intervals of the Roman period from the third to the fourth century, or those of the Middle Ages from the ninth to the tenth century, the ice at the Hauslabjoch certainly did not melt to the present extent. Either precipitation was less then or it was less warm than today.

The state of preservation of the corpse is so good that the man must have been covered by snow while he was dying or immediately after his death, without subsequently ever resurfacing. If the snow remained cold and dry for some time, and hence permeable to air, temporary dry-freezing would be conceivable given the terrain's exposure to wind. Of course, the body did not dehydrate entirely. That some of the body fluid remained was effectively demonstrated when the corpse thawed out completely in the dissection room of the Forensic Medicine Institute and could again be manipulated. Totally desiccated mummified bodies, by contrast, are feather-light and break if one tries to bend them.

Patzelt's expert statements on the prehistoric climate show that glacier recession and warming-up occurred in the Holocene period of geological history quite independently of the man-made greenhouse effect. Although present-day climatic conditions show a persistent warm phase, they keep entirely within the post-glacial fluctuation range.

I am fascinated by Patzelt's exposition. But the glaciologist is still ill at ease. Then he comes out with it: 'I'm afraid that the site of the find is on Italian territory.' Although I have visions of impending complications, regardless of what may happen it is imperative that the site is immediately closed to everyone and guarded. We contact the Alpine police, who presently take up their positions and in an exemplary manner remain at the Hauslabjoch despite occasional severe cold, until the increasing snow-cover makes any additional protection unnecessary.

Wolfgang Sölder and Gerhard Lochbihler receive the first two access passes. Sölder is a scientific assistant at the Tyrolean Provincial Museum Ferdinandeum in Innsbruck who has nearly completed his study of prehistory and early history at our institute. Lochbihler is in charge of the prehistoric restoration workshops at the same museum. Both are experienced excavators and also trained mountaineers – an ideal team for the planned expedition.

Sölder and Lochbihler set out from Innsbruck in the early afternoon and drive as far as Vent. It is completely dark by the time they reach the Similaunhütte at 8.25 p.m. after a four-hour ascent over the glacier which has been made icy by the rain. Apart from Markus Pirpamer there is no one there.

While the two Alpine archaeologists are on their way, the group from Mainz provisionally safeguard the finds and complete their initial conservation measures. Just then a directive arrives from the Ministry of Science and Research in Vienna: the export permit to Mainz has not been granted for the time being. The Vice-Chancellor wishes to know why the items are not kept and preserved in Austria. Meetings are scheduled and commissions appointed. All the potential restorers in Austria are invited to Innsbruck for the following day.

9 Friday, 27 September 1991

On Friday, 27 September 1991, the weather is still against us. At 7 a.m. the day watch begins at the Similaunhütte. A short while later Sölder and Lochbihler start their ascent to the Hauslabjoch led by Markus Pirpamer. They reach the spot in thick fog at 9.15 a.m. Meanwhile a heavy blizzard has started up, and they can only just make out the objects frozen to the ice. A round piece of wood, a sloe, some bundles of grass and scraps of leather are recovered before the snow blankets everything again. At 12 noon the weather forces them to go back. To make matters even worse, a thunderstorm approaches over the ridge during their descent. The air is charged with electricity. The metal parts of their excavation equipment begin to sing. Two ice-picks and one folding spade have to be left behind under an overhanging slab of stone because of the danger from lightning. It takes the scientists three times longer than usual to battle their way down to the Similaunhütte.

That morning the Austrian restorers arrive. First they inspect the finds. It is immediately clear that this collection, consisting as it does of numerous, diverse, and mostly extremely vulnerable materials, calls for new methods of restoration. One of the initial tasks is the cataloguing of the finds and the preparation of a detailed record. These jobs must be done before preservation can start. It means that not only photographs, but also computer tomographs, X-ray pictures, and, if necessary, sketches have to be made.

The second stage will require the taking of samples and cleaning. Samples are needed to determine the materials; with sophisticated present-day techniques samples in milligrams or even micrograms are usually sufficient. It is important that all these samples are taken before any preservation treatment begins because restoration will involve impregnating the finds with a variety of chemical substances that could affect the test results. It is especially important to take samples of uncontaminated organic particles for the scientific dating of both the body and the accompanying objects. Cleaning is done using distilled water only, which is passed three times through increasingly fine-meshed sieves and finally filtered as well. In this way even the smallest particles are retained, down to grains of pollen adhering to the clothing and the implements.

In the third stage the objects must be preserved. The grasses will be dry-frozen following vacuum soaking with Luviskol, Lutensit and polyethylene glycol. The leather and fur materials have to be re-greased and likewise dehydrated in the dry-freeze equipment. The pieces of wood present no particular problems as there is a good deal of experience of preserving such items. First they are cleaned in a six-month or twelve-month bath in de-ionized water, then they are allowed to dry out very slowly under continual observation. In view of the excellent state of preservation even of the cell membranes, treatment with Lyofix, polyethylene glycol or other stabilizers does not seem necessary.

Particular difficulties, however, arise from the fact that several items consist of a variety of materials. In the quiver, for instance, a hazel switch was used as a reinforcing brace and fur for the container. The axe consists of a wooden helve, while the binding is of lashed thongs of leather or hide. Construction of the arrows was particularly complex. In addition to the flint points, examination has shown the presence of feathers, a black adhesive, fine textile threads and, again, wood for the shafts. Their restoration by different methods has to follow a logically structured timetable.

The crisis meeting is held at the Institute for Pre- and Protohistory where archaeologists and restorers jointly discuss the problems. They unanimously agree to recommend to the minister that the collection of finds be entrusted to the workshops of the Roman-Germanic Central Museum in Mainz. There the necessary apparatus is available, as well as the greatest wealth of experience in the world.

Rumours are now rife that, contrary to the assumption of the first few days, the site of the finds may after all be on Italian territory. The press eagerly seizes on this aspect and blows it up. After the First World War, which she lost, Austria had to cede a part of ancient Tyrol to Italy. To this day many South Tyroleans feel oppressed by what to them is alien rule and more than once over the past decades extremist groups have drawn attention to their grievances. In the normal run of things, however, no one bothers about the precise line of the frontier up in the mountains. But the Iceman breathes new life into the arguments. The uncertainties derive mainly from the fact that, under the terms of the treaty of St Germain-en-Laye drawn up between the Austrian Republic and the Allied and associated powers on 10 September 1919, the frontier between Italy and Austria was to run along the watershed between the catchment areas of the Inn in the north and the Adige in the south. The place where the body was found is unquestionably northwest of the main ridge of the Alps, which makes a small swerve at the Hauslabjoch and then runs south. This area drains into the Inn.

On the other hand, the commission responsible at the time for the definitive drawing up of the frontier was authorized to deviate from the actual watershed line where practicalities made it expedient. This option had to be resorted to at the Hauslabjoch because the main ridge of the Alps was not then visible owing to glacier cover. Nevertheless, once established, the frontier line retains its validity under international law even if topographical conditions change.

And so, that afternoon, four Alpine police officers arrive at the discovery point. As they do not have any proper surveying instruments with them for checking the frontier line marked out with stones, they use their climbing ropes. They find that the spot is on Italian territory by approximately three rope lengths - roughly 120 metres.

Now the first lawyers demand to be heard. Meanwhile post is brought to the institute by the sackful, most of it to remain unread. No letters are answered. Letters from lawyers are opened, just to be on the safe side. We learn that the catchment area of the Upper Ötztal up to the main ridge of the Alps is politically part of the municipality of Sölden. A lawyer from Innsbruck, Dr Andreas Brugger, informs us that he represents that municipality. He writes to me as follows:

My client takes the view that the right to decide what is to be done about the body found on the Similaun glacier and the objects which it evidently had on its person, etc., belongs exclusively to the municipality of Sölden. It follows, therefore, among other things, that it has always been the unchallenged task and duty of the municipality of Sölden to take the necessary steps with regard to corpses found on its municipal territory. Besides, the person found had without any doubt lived within the municipal territory of my client, for which reason the inhabitants of the municipality of Sölden also enjoy the closest relationship with the deceased. On the grounds of the above, my client insists that, without its express consent, no dispositions whatsoever may be taken with regard to the above-mentioned body and the objects found with it. As my client has learned from the media that it is alleged that transportation of the corpse to a foreign country is being considered – albeit for restoration purposes – I am requested, on behalf of my client, specifically to forbid you to transfer the body, along with the objects, to another place, let alone to a foreign country. I am, moreover, instructed to bring to your notice that my client does not at present agree to any interventions or examinations. Regarding any measures that are absolutely necessary for the preservation of the present condition of the corpse and the objects found with it, my client's consent must be obtained....

The forensic people receive the same letter. But they are much more businesslike than we are. Unterdorfer replies at once:

> ... as the cause of death has not been established and a death certificate cannot therefore as yet be issued, the body is still within the jurisdiction of the Provincial Public Health authority. The Imst District Administration has been authorized, acting for the Provincial Public Health Directorate, to perform the public-health police seizure of the corpse and the forensic autopsy thereof was ordered for the purpose of establishing the cause of death. Only subsequent to the determination of the cause of death and the possible clarification of identity can the papers concerning the body be issued.

He concludes on a conciliatory note:

> In this connection we would request you to consult your client, the municipality of Sölden, about whether, during the period of the scientific measures now proceeding until a possible determination of its identity, the corpse may be called the 'Similaun corpse'. We would take pleasure in this and would not oppose a solemn naming ceremony on the initiative and with the participation of your client.

In view of the uncertain legal situation, and regardless of all other motives, immediate protection of the find now seems essential. Under its Ancient Monuments legislation, the Republic of Austria is entitled to effect this even though the question of ownership remains to be clarified. In such a case it would regard itself as the trustee. Indeed, this implies the duty of particular care for the safeguarding of the monument. But here, too, legal uncertainties – to put it mildly – are beginning to surface. The letter of the law does not provide for a human body to qualify as an ancient monument. Monuments are solely objects created by men, immovable or movable, of historical, artistic or cultural importance. Some bright sparks are suggesting that the tattooing on the body's back should be recognized as a work of art in order to extend the protection of the law to the body.

There is no question but that all these matters will have to be decided at a higher level. There is a need for speed, because administrative offices tend to be deserted after midday on Fridays. The minister appoints an 'Iceman Commission' with Platzer as chairman. We are summoned to Vienna for Monday.

That evening there is a lot of activity at the Similaunhütte. The four Austrian Alpine police, who have taken turns standing double sentry duty at the Hauslabjoch all day, are warming themselves. Twelve Italian police officers are already seated at the tables in the restaurant. But the two groups don't communicate, and not only for linguistic reasons. Our colleague Lippert has finished his field trip through the High Tauern range according to plan and is now on his way home. The high point of the trip is to be a visit to the Hauslabjoch. He too has arrived at the refuge with his seven students. Sölder and Lochbihler are also staying on. The lodge-keeper does a brisk trade.

10 The Weekend of 28–29 September 1991

The forecast for Saturday, 28 September 1991, at long last promises a slight improvement in the weather. Liselotte Zemmer-Plank and I fly to the Hauslabjoch. The pilot has no difficulty finding the spot. Towards 9.30 a.m. we touch down below the location of the find, on the ice of the Niederjochferner. Everything is deep in snow. There is an icy wind. I am not wearing proper mountain clothes, only jeans and an old leather jacket. I forgot to bring a cap. True, my colleague from the Provincial Museum has brought her husband's hunting trousers along for me. They have zips on the inner and outer seams, so that they can be put on and taken off without removing one's boots. But the quilted trousers are much too small and just dangle in front of me, legs unzipped, like a leather apron. In a few minutes I am chilled to the bone.

As a punishment for my lack of forethought I develop a cough over the next few days – not just my normal smoker's bronchitis but a bad sore throat as well.

Over a scree slope of ice-covered boulders we climb – with me on all fours – up to the gully in the rock. So far I have only seen it in photographs. Then we stand on the rock shelf and look down into the gully. The snow covers the jagged terrain with a soft blanket. The boulder where the corpse rested is a gently rounded feature. A picture of peace. Yet here, many thousands of years ago, a dramatic event took place. The scene is deeply moving.

We talk with the police on sentry duty. No, apart from the passholders there have been no tourists. Who would want to climb up here in this weather? They keep quiet about the results of their measurements the previous day. At a greater or lesser distance uniformed Italians are stamping through the snow, fiercely waving to each other from time to time. Evidently they, too, are carrying out measurements.

From below, from the direction of the Similaunhütte, a small human crocodile approaches. At first one can only make out dots, but soon they dissolve into separate figures. The police officer climbs down to meet them. An Innsbruck professor with his students; of course he is allowed access. Their faces are wreathed in wisps of steam – breath instantly condenses in the cold air. He and I have not seen one another for months. The period without lectures, euphemistically described as half-term holidays, is nearing its end. Everyone has been busy with research and publication projects. Lippert's face is tanned reddish brown after several days in the eastern Alps. Normally meticulously turned out, he is hardly recognizable under his grey stubbly beard. I envy the tall man for the way he tackles the rocky steps with a confident tread. In dealing with other people he tends to be a little stiff, and he leans his head sideways when he talks. He offers his cooperation in the Iceman project. We soon reach an agreement; he is an experienced mountaineer and will perform the second investigation at the Hauslabjoch. Together we study the topography of the site. There is little time to be lost before the winter snows start. The pilot starts his rotor, then he lifts off and flies another spiral over the Hauslabjoch. The people below grow smaller. We leave the saddle behind us.

At the institute I find myself back in the expected hubbub. A few dozen interviews have to be given, and all the urgent work has to be done as well. My colleagues, who have not slept nearly enough all week, are beginning to show signs of fatigue. Somebody rustles up two bottles of champagne. Then, at 5 p.m. we firmly close the doors and set off for a brief weekend. Except that this does not make a lot of difference: the reporters switch their attention to our homes. The telephone rings day and night: Tokyo, Berlin, Sydney, London, Buenos Aires, Vienna, New York, Cape Town, Munich. ... On Sunday the television cameras roll in our garden, with tomato plants as background. Up at the Hauslabjoch the sentries make the following entry in their service diary:

29.09.1991. Quiet Sunday with föhn. Weather did not permit any visit of the location by outsiders. Officials taking turns in two-man patrol to protect the site. No Italian official present.

11 Monday, 30 September 1991

The Iceman becomes an international affair. At the crack of dawn, on Monday, 30 September 1991, Henn, Platzer and I fly to Vienna. The commission at the ministry is staffed by heavyweights: specialist professors from the universities in Vienna and Innsbruck, top civil servants from the Ministry of Science and Research and the Foreign Affairs Ministry – the latter because of the still unclarified frontier question. The President of the Federal Ancient Monuments Office and a representative of the Federal Criminal Investigation Department are also present. On instruction from the Vice-Chancellor, Platzer chairs the meeting. Dr Eike Winkler of the Vienna Institute of Human Biology, one of our principal critics, has invited himself and is bowed out again.

At 10.30 a.m. the chairman opens the meeting. From the outset everything points to a prolonged discussion; everyone knows that there is an urgent need for prompt action and rapid decision-making. Surprisingly, the atmosphere is informal, unusual for this kind of complex conference. Each item on the agenda is to be discussed on a two-track basis, one in the event of the site being on Austrian territory, one in the event that it is in Italy.

First to ask to speak is the representative of the Ancient Monuments Office, who asks whether the find is, in fact, in Austria legally. It is pointed out that the Italian police, having been notified of the find, had agreed that their Austrian counterparts should undertake its recovery, as, in the opinion of the Italians, the location of the find was in Austria. As a policeman is an official of the state, whose statements are considered binding, the find is being lawfully kept in Austria.

In any case it should be placed under Ancient Monument protection without delay. Someone else asks whether the treasure trove legislation might apply, seeing that the finders, the Simons from Nuremberg, have already filed a claim to ownership through an Innsbruck attorney. In that case they would own one-half of the find. But is a corpse treasure? This will have to be clarified by lawyers and appropriate procedures taken.

Two officials are sent out to get in touch with the Federal Office for Calibration and Survey, whose geodesists are to survey the location afresh as quickly as possible. They are also to contact the Italians and, if possible, set up a bilateral survey commission for the purpose.

Discussions are also to be held with representatives from South Tyrol. There is a chance that, on the strength of the autonomous status of South Tyrol and even if the location of the discovery were on its territory, agreement can be reached for the find to be studied at the provincial university in Innsbruck which serves both western Austria and South Tyrol.

When the restoration is discussed there is some reluctance about sending the objects abroad merely because appropriate equipment does not exist at home. In the end reason is stronger than patriotism. Unanimously the commission recommends that the finds be handed over for provisional safekeeping, maintenance and preservation to the Roman-Germanic Central Museum in Mainz.

Article 93 of the Austrian University Organization Law provides that for the undertaking of scientific research in a specified field for an indefinite period, or for specific research projects, *ad hoc* research institutes may be set up within the framework of a university for the duration of the work in question. The commission members are also unanimous in their belief that the volume of work generated by the Iceman project cannot be handled alongside the normal tasks of an institute engaged in administration, teaching and research. It is decided to recommend to the Vice-Chancellor, as the administrative head of the Ministry of Science and Research, that such a research institute be established. This would create a specific centre able to deal with such archaeological finds in Austria, not least because further finds could be expected during the next few years. Someone mentions the 'Men in the Salt'.

The fact is that the Iceman is not the first prehistoric corpse to have been found in Austria. As early as the sixteenth, seventeenth and eighteenth centuries the bodies of prehistoric miners who had met with accidents were found in the salt mines of Hallein and Hallstatt. The Salzburg Chronicle reported from Hallein:

Anno 1573 on the 13th of the winter month a terrible Comet-Star appeared and on the 26th of that month in Salzburg Dürnberg, six hundred and thirty shoes [210 metres] deep in the whole mountain, a man of nine ells [160 cm] long, with flesh, bone, hair, beard and clothes entirely undecayed, if somewhat flattened, his flesh all smoked, yellow and hard as a dried cod, was cut out, and for several weeks lay by the church for all and sundry to see: But finally began to rot and was buried. He must have been entombed in the mountain before the memory of man, grown into it, and preserved from rotting by the salt all that time, while shoes, clothes and wooden picks were also found.

In Hallstatt the thirteenth weekly report for 1734 said:

Three weeks ago the shuttering of the Kilb shaft in the Kaiser Josephberg collapsed, but did not completely wreck the outflow, so that last week and this week the brine could be run off, but when after these weeks the overseers as usual had inspected this shuttering because of the draining costs and the down-shaft, it was found that the down-shaft nearly three rods thick mostly consists of dead rock, which also exceeds the drainage costs, and in that down-shaft horizon there was seen a dead man who presumably, judging by his appearance, was buried more than four hundred years ago and massively grown into the rock, yet one still sees of his smock several stains, as well as *s. v.* shoes on his feet, and this caused a very bad odour in the gallery, which can be smelled even before that down-shaft.

Added to the sober report of the chronicler is an ecclesiastical note:

Praised be Jesus Christ for ever and ever, also Maria. In this year 1734 on the second of April the said dead body was brought from the salt mountain here to Hallstatt (buried the following day), whereof after an oath an evil spirit of a truly possessed female person told me, the undersigned, of herself and four others, dismissed with the grace of God, he that names and acknowledges Astototh, that such a body had been buried in a rockfall one hundred and fifty years ago with fifteen persons, as a travelling pot merchant named Andrä Liezinyer of forty-five years of age, married to Barbara Pieröckhin, who in consequence was widowed with three children, the first was named Johann, the second Mathias and the daughter Maria, all of them good Catholics. Pater Mathias Capuchin S.t. Missionary.

The history of the discovery of 'men in the salt' has a great deal in common with that of our glacier man – except that, given the state of scientific knowledge at the time, it is excusable that the great age of the finds was not immediately recognized. Moreover, the Church's methods of criminal investigation for establishing the victims' identity may seem to us rather questionable today. That the Devil was involved was assumed from the outset which is why, in the description of the dead man's footwear, the incantation formula *s. v.* (*salva venia* or 'with permission'), was used, as the name of the Devil with all that appertained to him was regarded as unutterable. After all, there might have been a cloven hoof inside the shoe.

The evidently very ancient finds of miners who met their death in the salt mines of the Saltzkammergut naturally excited the imagination of contemporaries. They found a literary reflection in Ludwig Ganghofer's novel, *The Man in the Salt*.

Modern archaeological research is intensely preoccupied with these finds of prehistoric men. Documents from ancient mining archives made it possible to locate with considerable accuracy the galleries in which the bodies were then found. There is no doubt that the corpses date back to the Early Iron Age, the so-called Hallstatt period, and were approximately 2,500 years old. Regular studies have been conducted for a number of years in the mines, and numerous remains have been recovered in the 'Old Man', as the old workings are called by the present-day miners. But archaeologists' hopes of discovering another dead miner dating from the Hallstatt period and preserved in the salt have not so far been fulfilled.

However, as such a discovery is a distinct possibility, the proposal that a special research institute be established for the scientific investigation of complex archaeological and anthropological finds meets with the commission's unreserved support. The first task of that institute would be to coordinate the investigation of the Iceman.

Next it is decided that further archaeological investigation be conducted at the location of the find. This will be directed by our colleague Lippert in view of his familiarity with the story so far and his experience as a mountaineer.

Then Henn makes an important suggestion with regard to our scientific investigations. In forensic medicine it is customary, in difficult cases, to call on two independent experts. The commission agrees that two scientists should be co-opted from every specialized field and that, if possible, they should apply different methods. This would ensure cross-verification of individual results – an aspect of exceptional importance in view of the significance of the Hauslabjoch find.

Because of the urgency and the situation's ever-changing profile, the commission decides to be permanently ready to convene. The next meeting will be announced shortly. At 2 p.m. Platzer declares the first meeting closed. The decisions taken by the commission represent the guidelines for the Iceman project.

There is a little time left before our plane takes off from Vienna. Platzer, Henn and I stroll through the Old City. The anatomist, being an old Viennese, acts as our guide. We pass St Stephen's Cathedral, in whose catacombs thousands of corpses are piled up. They say that in earlier times, whenever the weather was hot, the stench of the bodies decaying in the chambers below the church was often intolerable to the living. Because there was insufficient room within the city walls to bury the dead in ordinary cemeteries, the coffins with their contents were dragged into the crypts, the completion of the process of decay was awaited, and the bones and skulls then collected and piled up like briquettes in a heating plant, in order to make room for new coffins. We stroll through the narrow little streets with their shops. Henn scans the windows for icons, and I for old pottery. But we find nothing of interest. Besides, we lack the necessary concentration. Our conversation keeps returning to the project. Then the shuttle bus takes us to the airport.

Meanwhile the police up on the Hauslabjoch are patrolling and safeguarding the site of the find. Even before the official survey the Italian frontier staff have come to the definite conclusion that the location is, after all, on South Tyrolean territory. The Austrian police have therefore to show their service books, and their personal data are recorded and passed on to the commissariat in Bolzano. The Austrian police are instructed to leave Italian state territory. But as it has grown dark in the meantime, they are graciously permitted to spend the night at the Similaun refuge, but as civilians and off-duty. Later it would emerge that there was even talk of arresting the Austrian officials.

12 The New Survey of the Site

On the following day, Tuesday, 1 October 1991, the entire find is declared protected as an ancient monument. The order is issued without preliminary investigation because the risks inherent in any delay justify immediate action. The assumption of risk stems both from from the unclarified legal position with regard to the Austro-Italian frontier and from the fact that the items of equipment must be taken to a suitable restoration workshop without delay, in order to prevent their disintegration. Official permission to take the objects to Mainz is issued at the same time. Everyone now waits tensely for the result of the new site survey.

The day before, Manfred Neubauer, an engineer with the Federal Office for Calibration and Survey in Vienna, had been entrusted with the job even while the Iceman commission was still in session. He obtained the necessary papers, especially copies of valid frontier documents, and immediately left for Innsbruck. Tuesday is devoted to preparations for the operation, in cooperation with our local colleagues. Meanwhile a radio message is received: the representatives of the Italian delegation from the Military Geography Institute in Florence cannot arrive until Wednesday. The operation is therefore postponed by a day.

At 10.20 a.m. on 2 October 1991 the seven-member Austrian working party assembles in Vent, complete with their helicopter. Nothing even remotely connected with the Iceman takes place without representatives from the media. Journalists, photo reporters, cameramen of the ÖRF and the Austrian Press Agency surround the survey party. It is rather windy. And it is raining. A reconnaissance flight establishes that at the moment a helicopter landing at the Hauslabjoch is virtually impossible because of high winds and snow. A telephone call is received from the Similaun refuge on the South Tyrol side to say that the Italian officials are already on their way to the site to meet the Austrians. Towards 1 p.m. the weather improves slightly, and the helicopter can take off. Several flights are needed to take the people and their surveying equipment to the site. The helicopter chartered by the journalists helps with the transport. In sunshine, an icy wind and a temperature of minus 8° Centigrade the surveyors immediately get down to work. The depth of the new snow cover is 70 centimetres.

Up in the high mountains the frontier stones between Austria and Italy consist of flat, square concrete slabs cemented to rocky ground and numbered in sequence. At the Hauslabjoch the course of the frontier is shown on the documents as a straight line linking the frontier markers b-35 and b-36. Meanwhile Italian police arrive. They inform us that the Italian survey experts are on their way from Florence. The exact location where the body was found is confirmed by Vice-Brigadier Dal Ben on behalf of the Italians and by Customs Inspector Egon Fanger on behalf of the Austrians. A red-painted cross indicates the spot. Soon the surveyors effortlessly locate two crucial frontier stones, planted there seventy years earlier and now under nearly a metre of snow.

But first the officials of the Federal Office begin a geodesic verification of the course of the frontier. As a basis the frontier marker b-35 at an altitude of 3,283 metres above standard zero is chosen. From here they determine the directions and distances to the neighbouring markers b-34 and b-36 with the aid of a theodolite and a range-finder. Comparison of the new measurements and the data in the frontier documents shows remarkable agreement. The deviations for the angle formed by b-34 and b-36, as measured at b-35, amount to only eight centesimal seconds and for the distance from b-34 to b-35 to a mere 13 millimetres. For the line between b-35 and b-36, along which lies the site of the find, the difference at a distance of 254.43 metres amounts to only 27 millimetres. Nothing now stands in the way of the crucial determination of the location of the find. The surveyors warm their numb fingers in their trouser pockets.

Just then a message comes in by radio that, because of föhn on the South Tyrolean side of the main Alpine ridge, the Italian surveyors cannot fly up to the Hauslabjoch. Instead, they will try to drive to Innsbruck over the Brenner, and thence to Vent. With the agreement of the Italian police, the Austrians continue their measurements alone. The line between markers b-35 and b-36 is taken as base-line against which the location of the find is triangulated. Neubauer reads the scale indicator through the eyepiece of the range-finder. The intersection of the red cross on the stone on which the corpse had lain is exactly 92.56 metres from the state frontier, on Italian territory. At 5 p.m. the job is done, and the helicopter takes the surveyors back to Vent. At 6 p.m. Radio Tyrol broadcasts the first report on the result of the survey.

The representatives of the Italian survey commission, headed by Colonel Alberto Carcchio, are already waiting at the Hotel Post. The two parties sit down together. The Austrians submit their data to the Italians. They are jointly examined and mutually approved.

On the strength of these results the Institute for Pre- and Protohistory of the University of Innsbruck, which had filed an application for just this contingency, is instantly authorized by Provincial Governor Dr Luis Durnwalder, by Provincial Conservation Officer Dr Helmut Stampfer, and by *Commissario del Governo* Dr Mario Urzi to begin a further archaeological investigation. The frontier authorities are instructed 'to support' our excavation team 'in the execution of its task, bearing in mind existing regulations regarding entry upon Italian state territory and the possible landing of helicopters'.

13 The First Archaeological Examination of the Site

The tension of the past few days is gone. On Thursday, 3 October 1991, Lippert and I leave at the same time – he for the Hauslabjoch, I for Mainz. The restorers have packed up all the objects and prepared them for transportation. Not the corpse, of course; that remains in Innsbruck. The export papers are all ready. Fax machines prove a blessing, for what used to take days now takes only a few minutes. The objects in their containers are placed in the boot. Roswitha Goedecker-Ciolek drives the Kombi, while Markus Egg and I sit in the back. We choose the minor frontier crossing at Scharnitz, across the Karwendel, in the hope that it will be less busy. We are still worried that we might have to unpack everything at the frontier, but things go smoothly. The customs officers have been notified by telephone of our impending arrival. Naturally a few people collect at the barrier to catch a glimpse of the objects from the dawn of time. We show them the axe, the flint scraper and a few other less delicate items, and the customs men are happy. In the afternoon we arrive at the Roman-Germanic Central Museum.

I have kept in touch with the Mainz museum ever since my first major dig at the Magdalenenberg near Villingen in the Black

Forest in the early 1970s. The Hallstatt-period giant burial mound produced many rare and precious objects from well over a hundred burials. Nearly all the graves lay underneath tons of stone which had hopelessly crushed the delicate bronze ornaments and the elegant iron weapons. Yet, as if by magic, the Mainz restoration artists turned the fragments into magnificent gems. Even the splintered bracers of jet were restored to pristine beauty under their expert hands.

Virtually every kind of significant prehistoric artefact has passed through their hands; there is scarcely a single major European (or non-European) find which they have not restored, moulded, or examined. At the moment they are working on the bizarre and immeasurably rich grave furnishings of the Peruvian ruler of Sipán.

I am relieved that the Hauslabjoch finds have survived the somewhat risky journey to Mainz intact. As we carry them in, the prepared refrigeration tubs are already filling; it is late in the evening before the steel doors of the strongroom are closed. I fly back to Innsbruck because the official opening of the glazing works we excavated and restored in the tiny village of Abfaltersbach is scheduled for the following day and a host of VIPs is expected. Mayor Franz Aichner is worried that something might go wrong. He has an idea of what my diary looks like. Julia Tschugg comes out of the office:

'Sydney and Abfaltersbach are on the line. Which will you take?'

'Abfaltersbach.'

Lippert, as chief of the first archaeological examination at the Hauslabjoch, begins work on 3 October 1991. Time is running out: early winter storms may break at any moment. His team includes the archaeology students Norbert Leitinger and Gerald Grabherr, as well as the glaciologists Schneider and Markl. In addition, a survey team headed by Dr Gert Augustin and Dr Albert Grimm are taking part. The Similaunhütte serves as forward headquarters. Strictly speaking, the high-mountain season is now over, but Markus Pirpamer keeps the lodge open for the archaeologists.

The depth of the new snow cover at the Hauslabjoch is 60 centimetres. Initial examinations are confined to the spot where the corpse and the quiver were found, as well as the area beside and on the rocky ledge. First the fresh snow is shovelled clear. Then detailed photographs are taken and a thorough survey of the area of the find and its immediate surroundings is made. For that purpose the gully in the rock is triangulated into the local mountain landscape in accordance with the national grid and the entire area, as well as the various points where objects were found, is recorded terrestrially-tachymetrically. On this basis a vertical stratification plan can be prepared at a scale of 1:50.

The stripping work is done with the aid of a steam blower and a hair dryer. In the course of it the grass mat seen by Messner is found again on the boulder where the corpse was discovered lying on its stomach, so the exact position of the body can now be defined. In front of the northwest point of the boulder, along with fragments of a birch-bark container, some of its contents are also found – Norway maple leaves, tufts of grass and charcoal particles. On and close to the slightly raised rock shelf south-southwest of the spot where the body was discovered more scraps of fur and leather, cords of twisted grasses and wood splinters appear. At the base of this rock shelf, alongside a piece of long-haired pelt, lie two fragments of Alpine ibex bone. But in the area where the quiver was found nothing else is discovered. On 5 October 1991 work has to be suspended because of the weather.

The frontier issue is now gaining momentum, especially in the media, and assuming the proportions of a veritable frontier war. No official statements from the provincial governors are forthcoming, but the authorities are relaxed. The verdict of the bilateral surveying commission is unequivocal. On 8 October 1991, at 8.30 a.m. Dr Luis Durnwalder of the Autonomous Province of Bolzano/South Tyrol and Dr Alois Partl of the Austrian Province of Tyrol meet in the library of the Anatomical Institute of Innsbruck University. The case has the potential to be explosive, so press and camera crews are excluded for the time being. Discussions are behind closed doors. Partl and Durnwalder sit at the long table, facing each other; the entourage of the two politicians and the members of the Iceman Commission present in Innsbruck take the remaining chairs.

While to an outsider the atmosphere of the Anatomical Institute always seems somewhat macabre, the medics probably have long ceased to notice the all-pervasive smell of preserving agents. Under the window stand boxes of bleached human bones. A few skulls have been sawn open and their tops fitted with small brass hinges, like a little door to the brainbox.

As always, the tone of the conversation is cordial. Most people here know one another. Platzer, as the host, welcomes us and reports briefly on the state of conservation of the body which lies in its refrigeration box two floors below us, inviting those present to view it at the end of the meeting. From the corner a stuffed chamois buck, frozen in mid-movement, stares at us. His hind-quarters are held up on a cardboard imitation of a slab of rock, and his cheerfully raised tail towers higher than his horned head. But this does not really transport one to the mountain region of our glacier man.

Durnwalder stands and somewhat fussily straightens his jacket: 'As has now been established, the man was found on South Tyrol territory. I therefore assert our country's claim to ownership. Anything else is negotiable.' Sovereign must be answered by sovereign. Partl pushes back his chair and rises: 'There is no doubt whatsoever that the find belongs to our South Tyrolean friends. I believe, however, that investigations and scientific evaluations should be carried out at our joint provincial university, in other words here in Innsbruck.' With these few words a clear decision has been reached. The reporters can now be admitted. Durnwalder visibly radiates satisfaction: 'Well, as you can see, we're not quarrelling. There are no uncertainties at all. And one thing is incontrovertible: the find belongs to us, to South Tyrol. Eventually it will be up to us to decide where it will be exhibited. But this depends essentially on whether it is, in fact, fit to be exhibited or, indeed, suitable for exhibition.' Partl adds: 'And today we have agreed that the entire results of the scientific research will be published jointly by Tyrol and South Tyrol. In this way we are together rendering a service to mankind and science around the world.' In conclusion Durnwalder adopts a faintly pastoral tone and proceeds to the act of christening: 'The exact designation of the glacier find is, from today onwards, *Homo tirolensis* from the Hauslabjoch.'

14 Naming the Body

From the very beginning, and increasingly over the following weeks and months, the process of naming the discovery on the glacier developed a dynamic of its own. Names ranged from emotional outbursts to a mandatory scientific label, though even this did not go entirely unchallenged. The one thing that was undisputed from the outset is the fact that the actual discovery point does not bear a name of its own on topographical maps. The confusion of tongues was therefore only to be expected. The Innsbruck German-language scholar Dr Lorelies Ortner has studied in detail the business of finding a name for a hitherto unnamed find. She examined naming practice and naming results on the basis of 682 press and magazine articles spread among 595 German and 87 foreign newspaper items. By way of contrast, she also referred to 10 articles from scholarly publications.

For the scientific naming of archaeological finds location is of prime importance, i.e. the smallest administrative unit on whose territory the find was made. In the case of the Iceman this is undoubtedly the South Tyrolean municipality of Schnals (Senales). In the event of the name of the municipality being changed at a later date, either through amalgamation of municipalities or frontier changes, an appropriate renaming takes place.

However, as several archaeological locations may be present within the territory of one municipality, or may occur in the future, a further designation is always added to ensure an unambiguous identification. For this, only names entered on official maps are eligible. If, as in our case, the location of the find bears no name of its own, the nearest geographical designation is to be chosen. For the Iceman this is unquestionably the Hauslabjoch. As a rule a name remains unchanged, as appellations are normally safe from official interference, though sometimes they are of only short duration, e.g. Adolf-Hitler-Strasse or Karl-Marx-Platz.

As a single location may contain archaeological finds from different periods, the designation contains either the unit of time or the cultural sphere to which the find belongs, or both. Such designations may change in the course of research as was the case with the Iceman: he was initially assigned to the Early Bronze Age, but was later recognized as Neolithic.

For a precise designation the find category is added, although this is handled more loosely. It should certainly be short and concise, as for instance isolated find, settlement, treasure, necropolis and suchlike. As the particular category of our find had not so far been described in the literature, a new designation had to be found. We settled on 'glacier corpse'.

For naming in the humanities, just as in the natural sciences, the priority principle applies. In the first scholarly report, which appeared in the specialized periodical *Archäologie Österreichs*, the name Hauslabjoch alone is used.

The valid scientific designation of the find is therefore: *Late Neolithic glacier corpse from the Hauslabjoch, Municipality Schnals (Senales), Autonomous Province Bolzano/South Tyrol, Italy.*

The Hauslabjoch is situated at an altitude of 3,283 metres above sea level, embedded between high mountains in the midst of the Ötztal Alps. It takes its name from Franz Ritter von Hauslab, a colourful figure. The son of a family ennobled by the Empress Maria Theresa, he was born in Vienna on 1 February 1798. Even as an ensign in the 1815 campaign in France he so distinguished himself that, at the age of eighteen, he was assigned to the staff of the quartermaster-general. His first task was the mapping of the Ötztal glaciers, the upper Lech valley, the Bregenzer Ache and the Breitach. In 1817 he cartographically recorded the glacier region of the Ötztal Alps. In his maps, which he carefully shaded with watercolour, he was the first Austrian to introduce the French practice of drawing the mountains in horizontal strata. The system of stratification by colour, developed by him – on land: the higher the darker, on naval charts: the deeper the darker – was soon adopted throughout the world.

At the age of twenty-one he came to the Genie-Akademie in Vienna as professor for 'site draughtsmanship and terrain studies'; he taught there for thirteen years. Subsequently, he worked at the Austrian Legation in Constantinople, tutored members of the imperial house and accompanied the Sultan of Turkey on his European tour. As a military attaché he observed manoeuvres in Russia and in Turkey. At the time of the Siege of Vienna in 1848 he was already a major-general. Soon afterwards he helped to decide the battles of Szöreg and Temesvár in favour of the imperial army. Promoted to the rank of lieutenant-general, he took part in the war against France and Piedmont.

His scientific work is characterized by an incredible versatility. Although his mapping activities in the Alps, at sea (for the Royal and Imperial Navy), in the Balkans and elsewhere were his main interest, he dealt with historical, geographical and folklore subjects – such as the 'History of Vienna', 'On the Soil Formation of Mexico', or 'On Military Uniforms from the Beginning of the

Sixteenth Century' – with the same thoroughness. He was a member of numerous important academies of science and societies at home and abroad. Nor should his humanitarian work go unrecorded. Although, strictly speaking, an enemy general, he helped 3,000 rebellious Hungarians, who had fled to Serbia, return to their country. Greatly respected and decorated with the highest honours, Hauslab died at the age of eighty-five in his native Vienna, on 11 February 1883. In memory of this important and deserving figure, a saddle and a mountain were named after him in 1889 on the revised Austrian 1:75,000 map – the Hauslabjoch and the Hauslabkogel. Thus the Iceman acquires a name that is certainly not unworthy, at least in academic circles.

But other names, too, were proposed for the scientific designation of the Iceman. It was pointed out that the location of the find was, after all, 330 metres from the Hauslabjoch. Much nearer, at a distance of about 80 metres, was a small crossing of the ridge, known popularly as the Tisenjoch. That name, however, is not found on any official map. The Tisenjoch is situated on the main ridge of the Alps on the watershed between the Danube and the Po. It is an ancient crossing from the Val di Senales to the Ötztal. Even in the present century it was used as an alternative for driving sheep up to the Alpine pastures whenever unfavourable snow conditions made the steep and difficult southern slope of the neighbouring crossings too dangerous. Normally the Niederjoch was used. The name of the Tisenjoch, it was said, came from Tisenhof, a farmstead situated at the mouth of the Tisental, the Tisen valley, where it joins the Val di Senales. A further argument advanced in favour of the name Tisenjoch was that, according to local name researchers, 'Tisen' derived from a pre-Roman language and thus belonged to a prehistoric linguistic stratum. The name Tisenjoch, it was eventually argued, would possess symbolic significance in that it underlined the prehistoric link between the valleys and the regions north and south of the main ridge of the Alps.

The whole business is further complicated by the fact that South Tyrol is bilingual. Following its annexation after the First World War, all German names were given additional Italian ones. The Hauslabjoch was rechristened Passo di Tisa. If therefore an Italian speaker looks for that name on the map, he will also find the German name Hauslabjoch, but not the spot which is popularly known as the Tisenjoch. It seems that giving a name is often more complicated than giving birth!

Our Italian colleagues are, very sensibly, keeping out of this dispute by preferring the designation Val di Senales (in German, the Schnalstal), which, however, from a geographical point of view, is even more incorrect. Be that as it may. Certainly earlier researchers found it much easier to find a suitable name. Translated, Neanderthal man is simply 'New Man', even though he is a very ancient man.

Outside the world of the experts the search for an acceptable name produced even more colourful offerings. Ortner was able to collect more than five hundred different ones. Some enjoyed particular support, such as *Similaunmann*, relating to the geographical concept of a *Similaungletscher*, a Similaun glacier. Except that this glacier is a journalistic invention. There is no glacier of that name. There is only the Similaun, a mountain peak 3,607 metres above sea level, but some 3.5 kilometres from where the body was found. This peak gave its name to the Similaunhütte, the lodge or refuge which served as a starting point for all visitors attempting the ascent or the flight to the Hauslabjoch. Thus, this handy, easily translatable name soon got into the media and became a popular and irrepressible - albeit totally unsuitable – name for our Iceman.

The French went their own way in choosing a name for him. In France the label *Hibernatus* has come to stay. For aficionados of French cinema this name will hold no mystery. It is the title of the film in which Louis de Funès finds his wife's grandfather frozen into the ice of the North Pole: thawed out, the old man comes to life again and speaks.

Interestingly enough, the German press has suggested that in America the Iceman was jocularly known as 'Frozen Fritz'. It's true that Americans colloquially, or derisively, often refer to Germans as 'Fritz' – so the homophone link with *pommes frites* is probably no accident. But this claim must not be taken as gospel until the term 'Frozen Fritz' is actually proved to have been used in the American media.

Only one nickname has found world-wide acceptance: 'Ötzi'. Used without the article and written with a capital letter even abroad, it has become a 'given name' ready for inclusion in dictionaries.

Until now there has only been one instance in archaeology when a prehistoric human find was given a familiar name. When the roughly three-million-year-old skeleton of a young female Australopithecus was dug up in Ethiopia in 1976, it was christened 'Lucy' because someone played a tape of the Beatles' song 'Lucy In The Sky With Diamonds' in camp the night the fossils were found.

According to the Viennese reporter Karl Wendl, he invented the name 'Ötzi' by conflating 'Ötztal' and 'yeti'. He wanted to get away from gruesome terms such as body, corpse or mummy and introduce something a bit more attractive. 'This desiccated, horrible corpse must be made more positive, more charming if it's going to be a good story,' he thought. Wendl passed on his new name to the Vienna *Arbeiter-Zeitung*, and in the issue of Thursday, 26 September 1991, a week after the discovery, the newly coined name 'Ötzi' first appeared. From there 'Ötzi' began his triumphal progress across the globe.

15 Dating the Find

The Hauslabjoch find grew older at every stage of the initial scientific examinations. Naturally enough, this led to irritation and speculation among the public. The Simons, because of the ski-clip lying nearby, guessed at an age of ten to twenty years.Then Capsoni's name was mentioned: he lost his life in 1941. Police Officer Koler, thinking the axe looked a little antiquated, suggested the nineteenth century. Messner's estimate was five hundred to three thousand years, which came much nearer to the truth. But his press agent did not go beyond five hundred. The ethnographer Haid expressly referred to the Hallstatt period, i.e. the ninth to fifth century BC. I myself dated the find at first sight at four thousand years or more, with the result that for a while it was assigned to the Bronze Age. Then the first, still unadjusted, data leaked out from the Paris laboratory, and on 4 December 1991 *Libération* announced: 'L'homme des glaces ... un âge de quatre mille deux cents trente ans.' We hadn't heard anything about that yet. To my great astonishment an Italian correspondent in Rome gave me those figures over the phone. Eventually, the adjusted data came through: 5,300 to 5,200 years old.

With every scientific technique some teething troubles have to be overcome. Today the carbon-14 method is a reliable tool for dating prehistoric finds. A prerequisite, of course, is that the remains contain organic matter, i.e. originally live matter, such as bones, charcoal, parts of plants, leather, etc. Objects from inorganic materials containing no atmospheric carbon, such as stone, metal or pottery, cannot be dated this way. The finds from the Hauslabjoch were therefore ideal subjects for the carbon-14 method.

Radiocarbon or carbon-14 dating was developed between 1946 and 1947 by the American physicist Willard F. Libby and his co-workers. It is based on the premise that all living matter contains the element carbon. There are several variants, or isotopes, of the carbon atom, one of which is radioactive. It is known as C^{14}, or carbon-14. This radioactive carbon isotope decays uniformly. However, the radiation it emits is so slight that it does not harm the organism. Libby calculated that a given quantity of carbon-14 isotopes decay to exactly one-half of the original value over 5,568 years. Every human being, every animal and every plant during its lifetime absorbs carbon-14 isotopes contained in atmospheric carbon dioxide. At its death, this absorption process stops. The carbon-14 contained in the organism at the time of death decays without being replaced. If, therefore, one has a crumb of charcoal, a bone splinter or a scrap of leather from a prehistoric find, one determines how much undecayed carbon-14 there is left in the sample. As the initial quantity is known it is possible to calculate how many years must have elapsed for the radioactive carbon-14 isotopes to have decayed to the amount remaining in the sample.

Admittedly, for various reasons a number of errors at first crept into this method, which was why it was not initially very accurate. Over the past few decades, however, work has gone on to perfect the technique and to eliminate, as far as possible, all sources of error. Above all, strenuous efforts were made to buttress the carbon-14 results with dates arrived at by other procedures. In this respect, dendrochronology, which operates on the principle that every tree growing in a moderate climatic region each year forms an annual ring, has proved especially useful. In years of low precipitation the rings are thinner, and in rainy summers thicker, than in normal years. The result is a weather-conditioned historically unique sequence of varying ring widths. If one lets, say, slices from a newly felled tree, from recent roof timbers, from medieval houses or from Roman bridges, overlap at the right spot, this mutual overlapping yields a standard chronology reaching far into the past. This annual-ring calendar by now has a continuous sequence of some 10,000 years. With this annual-ring graph the carbon-14 dates can easily be calibrated.

Why, then, were the pieces of wood from the Hauslabjoch, the bow, the axe-helve and the backpack, not dated by means of this very accurate – sometimes to the very year – method? The answer is that in order to fit a wood sample into the annual-ring calendar at the right spot it must show at least fifty annual rings. The object with the most rings is the axe-helve, but even that has only twenty-seven. The dendrochronological method is therefore not suitable as a sole means of dating the Hauslabjoch find, but was used for calibration of the carbon-14 dates.

Whereas Libby needed substantial samples for his carbon-14 measurements, modern equipment requires only 5 to 6 milligrams. This is a crucial step forward: many archaeological finds – and that applies in particular to our Iceman – are so precious that only minute samples may be taken. For the progress of the project it was, of course, necessary to achieve the most accurate dating possible. Although my first estimate was not now disputed, scientific methods clearly carry greater credibility. In view of their own experiences, the forensic experts insisted on taking samples of the corpse itself, as well as of the equipment. They wanted to be absolutely certain that the two belonged together.

From the damaged region at the body's left hip small particles of bone and tissue fibres, already loosened by the pneumatic chisel, were removed. As for the equipment, we chose a few grass-blade fragments, which probably came from the shoulder-cape. The anatomical samples were sent to the Research Laboratory for Archaeology and the History of Art in Oxford and to the Institute for Medium Energy Physics of the Eidgenössische Technische Hochschule in Zürich. The botanical material went to the Svedberg Laboratoriet of Uppsala University and to the Centre des Faibles Radioactives in Paris. The samples tested by four great laboratories in Britain, Switzerland, Sweden and France yielded a surprisingly consistent dating, in the main ranging from 3365 to 3041 BC. Only the dates from the Swedish laboratory are slightly more recent, at 3053 to 2931 BC. Altogether there were six separate datings; the average calculated shows that the Hauslabjoch find dates back to approximately 3,300 to 3,200 years BC.

16 The Second Archaeological Examination of the Site

The winter of 1991–2 saw some of the heaviest snowfalls in Tyrol for years. At the Hauslabjoch approximately 700 centimetres of snow fell. Fortunately, there was some protracted fair weather in the summer of 1992, so that about 500 centimetres had melted again by the time the second examination at the site of the find, this time a large-scale operation, began. As the splendid summer weather continued during out work, roughly another 100 centimetres melted away before the onset of the first winter snowfalls. Even so, a residual snow cover of approximately 100 centimetres remained in the gully. In the summer of 1992 the Iceman would have had no hope of emerging from the glacier at all. Nor can a glacier low level as had existed at the time of discovery in September 1991 be expected for some time to come. One must, therefore, accept the well-nigh unbelievable fact that over the past 5,000 years the chance of finding the Iceman existed for only six days. Just how glacier formation will progress in the future remains to be seen.

So it is not surprising that, even though thousands of mountaineers cross the glaciers during the summer months, the search for further bodies in the ice has remained unsuccessful. Only a modest find was made: at the Stubai glacier a cat was released from the ice which the journalists christened 'Stubsi'. They told us over the phone, 'It looks just like Ötzi,' meaning it was mummified. Assistant Professor Leitner from our institute and Sabina Kneussl, a prehistorian who works at the 'Iceman Institute', set off at once with their freeze-box.

The cat looks as if it had frozen in its sleep. The skin is leathery and totally hairless, only the deeply rooted whiskers survive. The body is extensively dehydrated, so that the bones are clearly visible under the skin. Leitner and Kneussl carefully packed the body and placed it in their box. Then they drove back to Innsbruck as fast as possible. On the way they ran into a radar speed trap: 'We are from the Ötzi Institute and we have a glacier cat on board,' they explained, and were allowed to race on.

Now that the frontier question was clarified, the second examination of the site became the responsibility of the South Tyrol Ancient Monuments Office in Bolzano. Four people were in charge. For the South Tyrol office they were the archaeologists Lorenzo Dal Ri and Hans Nothdurfter; the University of Trento sent Professor Bernardino Bagolini of the Institute for the History and Culture of Europe; and our own university delegated Lippert, as in the previous year. Work began on 20 July 1992 with the manual removal of the remaining snow in the gully; this was about 200 centimetres deep. Because we wanted to obtain uncontaminated ice and sediment samples, we deliberately refrained from using diesel-driven machinery in order not to pollute the site needlessly with particles from the exhaust fumes. Altogether over 600 tonnes of snow had to be removed. At an altitude of 3,210 metres, this operation was probably the greatest physical achievement connected with the Iceman project. For more than three weeks four men tirelessly shovelled and carted snow.

On 10 August 1992 the archaeological work proper began. What glacier ice remained was partially removed with ice-picks. Between boulders and in crevices the ice could only be removed with steam-blowers and hair dryers. De-icing, especially in the non-draining lower parts of the gully, was particularly difficult. Because of the perfect and hot mountain weather considerable quantities of meltwater from the snowfield adjoining the gully in the north kept filling it again. Although some of the water was collected in plastic pipes and carried away from the excavation area, it also became necessary to cut a channel into the ledge bounding the gully on the southeast, to ensure that the base water drained away too.

In the course of this operation a number of minor finds were filtered out, such as grasses, moss, various leaves, charcoal particles, hair and parts of insects. Examination of the sediment in the gully also revealed, alongside further remains, anatomical traces such as pieces of skin, muscle fibres, blood vessels, human hair and a fingernail. These human remains are partially attributable to the injury to the Iceman's hip done by the pneumatic chisel the previous year. Among the major finds mention must be made of the end of the bow, snapped off during the official recovery in

1991 and recorded photographically at the time; its discovery now made it possible to determine the exact location of this find 5 metres south-southwest of the corpse.

At the southeastern base of the boulder against which the body had rested a fur object was found; for the time being it was thought to be a cap because it was found under the man's head.

Following the completion of the second examination on 25 August 1992 the site at the Hauslabjoch was considered exhausted from the archaeological point of view. It was returned to nature and to the tourists.

A trying aspect of Iceman research is the lecturing. Enquiries arrive every day. Local associations, Rotary and Lions Clubs, adult education institutions, medical congresses, museums, universities and industrial firms from all over the world seek information on the latest research from an authoritative source. As yet we have hardly any pictures. Elisabeth Zissernig keeps pestering the people who were at the site before the official recovery of the body. Time and again I have to go to Mainz to discuss the progress of restoration work. The Iceman Commission is in permanent session. In order to ensure reliable information, and above all to correct the often absurd reports in the media, we decide, as far as our other obligations allow, not to turn down any invitations to lecture. We have to split up: Egg, Leitner, Lippert, Nothdurfter and I set out on lecture tours.

The weekly chart and the diary fill up. On Monday Iceman Commission and teaching, plus a visit from our South Tyrolean colleagues at the Bolzano Ancient Monuments Office, four press and two television interviews. On Tuesday a lecture in the Great Hall of Vienna University. On Wednesday four hours of teaching, in the evening a committee meeting. On Thursday a lecture in the Great Hall of Berlin University. On Friday, also at Berlin University, a colloquium about the Iceman. On Saturday I work at my institute. Sunday free.

Then an unpleasant illness intervenes and I am hospitalized. At my own risk I discharge myself prematurely. During the following hectic weeks I experience the worst days of the year as the after-effects of my operation still give me a lot of trouble.

On 19 September 1991 Erika and Helmut Simon discovered the Iceman, when they spotted his head and shoulders emerging from the residual glacier in a rocky gully at the Hauslabjoch. With commendable presence of mind Helmut Simon took this picture. It was the last frame of his film.

The humidity band on the Iceman's skin, approximately 6 cm wide, immediately above the surface of the ice, marks the level of ice melted that day between sunrise and 1.30 pm. Glacier ablation on this scale is extremely rare and occurs only in specific weather conditions.

Panorama of the Ötztal Alps. The red circle marks the spot where the Iceman was found, very near the main ridge of the Alps. The picture illustrates better than words can do the great difficulties faced by those recovering the Iceman and re-examining the site.

The Iceman after the first recovery attempt by police officer Anton Koler and refuge keeper Markus Pirpamer on 20 September 1991. The compressed air in the pneumatic chisel lasted only half an hour. In that time the body was freed down to its hip. The head shows an injury, leading to the original suspicion of third-party criminal responsibility.

The discoverers: Erika and Helmut Simon on the summit of the Similaun (3599 m) on 18 September 1991. On their descent they found themselves among crevasses, which forced them to spend the night at the nearby Similaun refuge. This unscheduled delay resulted in the discovery of the Iceman the following day.

Left: Alois Pirpamer, head of mountain rescue in the Upper Ötztal. Together with Franz Gurschler he completely freed the Iceman on 22 September 1991 in preparation for removal on the following day.

Right: Markus Pirpamer, son of Alois Pirpamer and keeper of the Similaun refuge. It was he who notified the authorities about the discovery of the body. When he and Blaz Kulis first visited the site in the afternoon of 19 September 1991, they found the items on the rock shelf – the Iceman's bow, pannier and axe.

The forensic expert Professor Rainer Henn (left) on his arrival at the site on 23 September 1991. Fresh snow had fallen during the night, and the low temperatures meant that the body was again frozen fast. The police officer/air rescue worker Roman Lukasser removes the snow. To the right is the ÖRF camera crew.

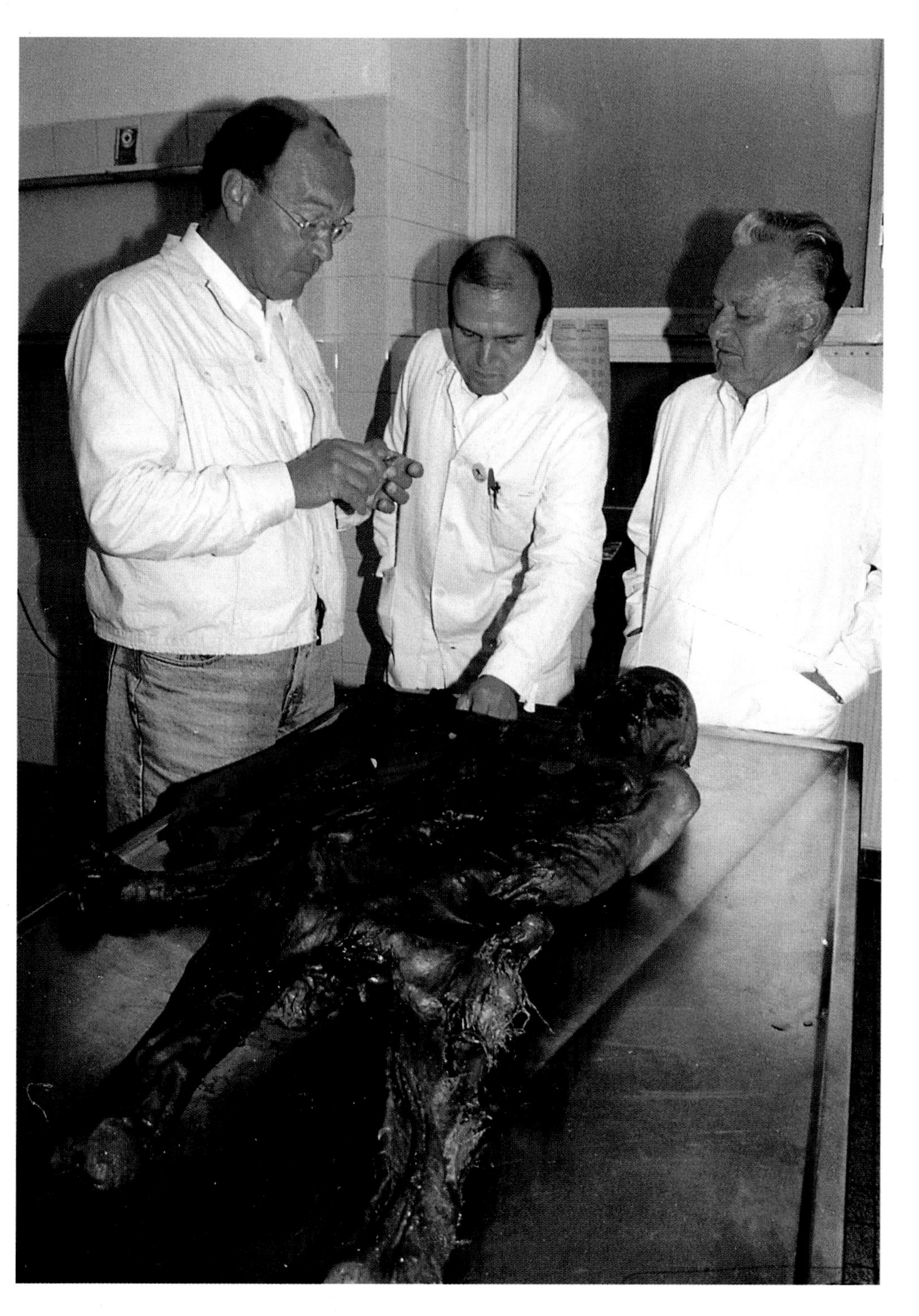

The preliminary archaeological assessment on 24 September 1991, in the dissecting room of the Institute of Forensic Medicine at Innsbruck University: Professors Rainer Henn (right) and Konrad Spindler (left). The dating suggested – at least 4,000 years old – met with universal disbelief.

Reinhold Messner (right) and Hans Kammerlander (left) at the site on 21 September 1991. The Iceman's leggings can be faintly made out through the meltwater. In the background, top right, the bow leans diagonally against the rock, its lower end firmly held in the ice. To the right of the oblong stone, beneath the ski-pole held by Messner, lies the crushed birch-bark container with its contents scattered about. Kammerlander supports himself with his right hand on a section of the pannier.

Kurt Fritz raises the Iceman's head, and his face is seen for the first time.

II
The Iceman's Equipment

1 The Archaeological Situation at the Hauslabjoch

Observations made during the days following the discovery, during the official recovery operation and the examinations of the site agree that the Iceman's equipment was distributed among three spots in the find area. First of all there is the spot where the body itself was found. The dead man was resting on a rounded piece of rock, his head towards the southwest and his feet towards the northeast, his body extended. The left foot was lying below the right. The back was facing upwards. The right arm was extended from the body at an angle of approximately 60 degrees, reaching deep into the ice. The right hand was either trapped or frozen under a small oblique slab of stone. The left arm was resting unnaturally, at a right angle to the axis of the body under the shoulders, pointing northwest. The head was lying half a metre higher than the feet. This inclined position resulted in the head, neck and shoulders emerging from the ice, with the back of the head the highest point of the body. In the suture region of the parietal bone and the occipital bone there was damage to the soft tissue which had been sustained after death.

The man was resting in the ice fully clothed. We assume that his headgear fell off as he stretched out, dying, in the gully, as it was recovered some 70 centimetres below the head. The clothing was not completely preserved, especially in the area of the back, because, in the state in which it was found, the leather (fur which had lost its hairs) was very soft and tore at the slightest strain.

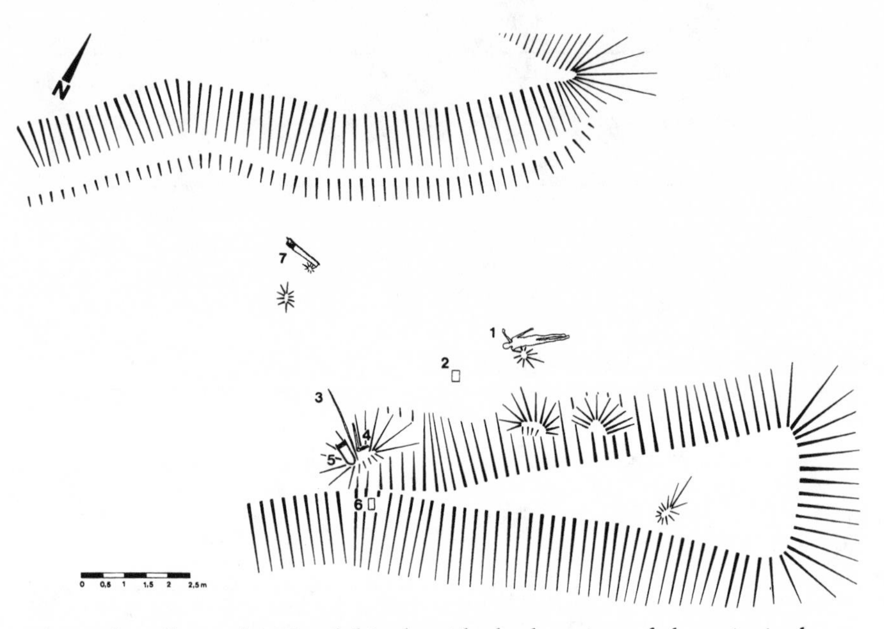

The rock gully at the Hauslabjoch with the location of the principal finds: 1: Iceman, 2: first birch-bark container, 3: bow-stave, 4: axe, 5: frame of the back pannier, 6: second birch-bark container, 7: quiver

During the body's emergence from the ice the wind blew one piece after another of the outer clothes off the body's back, so that the man seemed naked to the first observers. The dying man carried with him some items of his equipment. It appears that, up to the last moment, he held his birch-bark ember container in his hand, dropping it as he fell. It came to rest barely 2 metres south-southwest of his head.

Some 4 to 5 metres south-southwest of the head of the corpse there was a second group of finds, this time on the raised rocky shelf which bounds the site on the valley side. Here the axe, parts of the wooden frame of a backpack, remains of fur, cords, fragments of a second birch-bark container, and a few other small items were found. The remains of fur probably don't represent another piece of clothing. More likely they belonged to a pelt bag fixed to the frame of the backpack.

The quiver was recovered at a third location about equidistant from the other two, so that the three spots represent the corners of a triangle with sides 4 or 5 metres long.

The Iceman's Equipment

Before we attempt to reconstruct the dramatic events of the Iceman's final days and hours in the lonely mountain region of the Hauslabjoch, the individual pieces of equipment need to be described in detail, for they tell us a great deal about our man's life and death. Unusually, in this case the archaeologist finds himself in the position of describing a collection of finds straight from life. As a rule we deal only with dead bodies excavated in their graves, over whom the burial rites of their respective cultures have been performed, almost a kind of burial-ground archaeology. The uniqueness of the Iceman lies in the fact that he died far from his native village and remained lost for ever. Enveloped in snow, he and his equipment were preserved by the ice for thousands of years.

Under normal circumstances he would certainly have had an honourable burial. Perhaps modern archaeologists would even have found his grave. However, in mineral soils the organic substance of a body is quickly consumed. Even for modern burial sites one tries to choose a soil that is poor in chalk and dry, a soil in which bodies soon decompose and coffins quickly rot away. If our Iceman had died in such a place, all that would have survived would have been the blade of the axe, the flint artefacts, and the stone bead. At best, in a chalky soil, the skeleton, the bone awl, and the six items made from stag antlers might have been dug up. Under such exceedingly favourable circumstances the burial of our glacier man would have represented one of the richest grave finds from the Late Neolithic. The metal axe alone would have caused a minor sensation among prehistorians. But that would have been without a proper burial. If his heirs had buried him, they would certainly not have placed his entire property, all the items he had with him at the time of his death, into the grave with him – probably just the stone bead or some flint tools. His burial would have been merely one of the anonymous series of Late Neolithic burials and as such would not have attracted any particular attention.

2 The Bow-stave

The bow-stave is the largest item of equipment belonging to the Iceman. With one end resting on the base of the rocky gully, it rose at an angle of some 40 degrees diagonally towards the rocky shelf, where the upper end was more or less caught between two vertical stone slabs. At the time of discovery the lower end was still stuck in the ice to a length of about half a metre. It was clear that the bow-stave had not simply been dropped, but had been leaned against the rock with some care. Without a doubt it remained unmoved, by man or ice, from the time it had been put down before the Iceman's death until its recovery. Markus Pirpamer and Blaz Kulis first saw it when, shortly after the Simons reported their discovery, they went up to the Hauslabjoch. Four colour photographs confirm the state of the find at the time of its discovery – those taken by police officer Koler on the following day and another, one of two taken by Kurt Fritz, showing Messner and Kammerlander crouching behind the body. On the last the bow-stave is recognizable in the top right corner, its slanting position clearly visible. Finally, there is the photograph taken during the recovery by Henn, when Wiegele broke off the bow-stave. The broken-off end was found in the course of the subsequent examination in the late summer of 1992, when it was accurately measured, photographed and drawn. The position of the bow is, therefore, extremely well established.

Visitors generally described the wood sticking out of the ice as a stick or a rod. The fact that it had been worked was universally appreciated. As it proved impossible, despite often rather vigorous attempts, to pull it out of the ice, it was thought that it might be a lance whose barb prevented its extraction. Messner alone was reminded of bows he had seen in New Guinea. During the very first inspection of the equipment, while it was still in the forensic department, the fragment was conclusively identified as a bow-stave, even though it differed in appearance from the usual shape of prehistoric bows. After long discussion we eventually agreed that this was an unfinished item. Nothing of the kind had been found before. Bows from that period are known, but they are all finished weapons. Such bows, like modern ones, have a thicker middle section which the archer grips with his hand. The ends are usually flattened at the back of the bow, at right angles to the line of flight, to allow the bow to be properly bent. Provision is made at the horns for fastening the string or for hooking in a bowstring loop. None of these characteristics is present in the Hauslabjoch bow. We examined the ends of the bow meticulously under a reflected light microscope, but found no traces of bowstring impressions. The item thus provides a unique insight into the manufacture of prehistoric bows.

The wood chosen by the Iceman was yew (*Taxus baccata*). This low evergreen tree provides the most suitable wood: indeed, all prehistoric and historical bows are fashioned from yew. As late as the sixteenth and seventeenth centuries major quantities of yew bows were exported from Tyrol to England for the English army – which is why yew has become a rarity in the Tyrol and is now protected. While the yew grows virtually everywhere in Europe, in the mountains its habitat does not extend beyond an altitude of 1,400 metres. Its broad needles contain a virulent poison, which protects the tree against many herbivores.

Yew wood is both tough and elastic. Even under great stress it hardly ever splinters, and in contrast to other conifers it is entirely free from resin. By careful trimming the Iceman gradually fashioned the slender trunk into the desired shape. The belly of the bow, the side facing the archer, was evenly rounded. The shoulders taper from the middle towards the ends, so that the item always has a horseshoe-shaped cross-section. At a length of 182 centimetres the bow-stave is clearly more than man-sized. With only a little more work it could have become an effective weapon. But the Iceman was not granted the time to finish it.

3 The Axe

The axe was discovered by Markus Pirpamer and Blaz Kulis, along with the bow. But the very next day it disappeared into the air-raid shelter of the Sölden police and reappeared at the Forensic Medicine Institute only after the official recovery operation had been completed. Its original position was not photographed because Koler turned it a little to its side for his two snapshots, so it would show up better against the lighter-coloured rock. Originally, it was said to have lain in the cleft of the rock, which was otherwise filled with remnants of fur and cords, where the upper end of the bow-stave, too, was resting. The end of the shaft was pointing downwards, the axe-head upwards, and the cutting edge towards the east. No one who had the chance to view it before the archaeological experts knew what to make of it. Ideas on its age ranged from the nineteenth century (Koler) down to the Iron Age or the Hallstatt period (Messner and Haid respectively), although Messner knew the axe only from descriptions. Besides, because of the rust-coloured water patina, the blade was originally thought to be of iron. Some people correctly identified it as an axe, while others thought it was an unusual ice-pick. A mechanical (as opposed to chemical) examination of the material, conducted at the Forensic Medicine Institute, suggested copper.

For the haft the Iceman again chose the superb wood of the yew. He selected a longish piece of trunk from which a strong branch stuck out almost at right angles. Then he trimmed the trunk into the haft and the branch into the shafting for the blade. The natural connection of trunk and branch gave the axe handle maximum durability. This method of shafting has a long tradition in prehistory, extending from the Neolithic well into the Iron Age. Only during the last few centuries BC did the Celts invent the shaft hole for their iron axes, hatchets and hammers – or perhaps they adopted the idea from their more highly civilized Greek and Roman neighbours to the south. The new technique made it possible to abandon the trunk-and-branch method of shafting.

Quite a few axe handles have been found at prehistoric wetland settlements, the so-called lacustrian or pile-dwellings. But hardly ever has one been found together with its blade. There is one other peculiarity that distinguishes the Hauslabjoch axe from the mass of comparable finds: it possesses the only known yew-wood haft. As a rule the equally suitable ash was used, and, less often, oak and beech.

The spherically worked head of the haft is separated by a stepped indentation from the slightly but elegantly curved handle. Its total length is 60.8 centimetres. At the rear end the haft has an almost circular cross-section, changing towards the head to an upright oval; the durability of the shaft is thus perfectly ensured by the way the wood is worked. To the casual observer these fine details would scarcely be noticeable; only exact measurement testifies to the very remarkable skill of the carver. At the rear the cross-section of the shaft is 2.8 centimetres by 3.1 centimetres; at the oval-shaped front it is 2.4 centimetres by 3.4 centimetres, virtually anticipating the shafting ratios of modern axes and hatchets.

The shafting, growing out of the haft head, is shaped like a fork to hold the blade. This is, of course, concealed under the shaft binding. A longitudinal slit, 6.8 centimetres long, envelops the axe-head with an accuracy of fractions of a millimetre. When the blade is inserted into the slit only its cutting edge projects, 2.6 centimetres out. This means that nearly three-quarters of the length of the blade is held in the shafting fork, ensuring a firm grip. The cement between wood and metal is a thin layer of birch tar, the all-purpose glue of prehistory.

The shaft binding consists of narrow strips of leather or hide. For the modern scientist it is exceedingly difficult to decide whether tanned leather or raw animal hide was used. Raw hide is certainly more suitable, for if such strips are firmly wound when wet, on drying they will contract. Moreover, the colloids contained in untanned hide ensure additional gluing. Thus the blade would be very firmly held. There are three or four strips, slit at their ends to form loops and inserted one into the other. The free, slightly tapered end of a strip was then wound a few more times round the haft behind the head-step, so preventing the whole thing from slipping.

The rather small blade, only 9.3 centimetres long, has a narrow trapezoid outline. The neck is straight and shows a flat 0.4-to-1.8 centimetre ridge. This is necessary because with a sharp neck-ridge the blade, while in use, would be driven into the shafting fork and would split it. The edge is very slightly curved, with small points at its tips. A newly fashioned axe-head does not possess the necessary sharpness. As in the honing of a scythe, the cutting edge is thinned by careful hammering, so that as little material as possible is lost. During this process the cutting edge extends a little. The edge curves and the corners form points. In the case of the Hauslabjoch axe, the traces of hammering were smoothed out by careful grinding. After that the edge was presumably stropped. When blunt, the axe was reground in the same way. Our blade reveals very slight, barely perceptible dents; it was shortly due to be reground. The narrow sides of the blade have slightly raised flanges to prevent the blade from slipping longitudinally when the axe is used. These edges were presumably raised into flanges by hammering. In this process the narrow sides were faceted three times. Here, too, possible hammer traces have been smoothed by subsequent working. The blade thus shows a bi-concave cross-section which very accurately matches the cross-section of the empty space in the shafting slit of the wooden haft.

The blank of the blade was cast. This means that a certain quantity of metal was melted in a palm-sized thick-walled ceramic melting pot by heating it with bellows to at least 1100° Centigrade; the molten copper was then poured into a mould which stood upright with the cutting edge facing downwards and the neck upwards. It is not clear what kind of mould was used here, but as the blade was cast in an upright position we might think, for instance, of a double mould of stone, usually sandstone, or of well-dried unkilned or kilned clay. These can be used several times. Such clay or pottery moulds can be made from an original or from a wooden model. If a wax model is used the mould has to be smashed after casting; the process is known as the lost-wax method, or as casting *à cire perdue*. Surface analysis of the axe-head by the X-ray fluorescent method at the Roman-Germanic Central Museum's laboratory showed that the metal is 99.7 per cent copper, 0.22 per cent arsenic and 0.09 per cent silver. Judging by the trace elements arsenic and silver the metal may well have come from one of the copper deposits in the Alps. To put it simply, the blade of the Hauslabjoch axe was most probably made of local copper.

Pure or fairly pure copper, however, is not very easily cast. Often the mould is not entirely filled. Thus, because of oxygen absorbed during melting, porous castings occur: when the metal shrinks during cooling, small cavities are caused, known as shrinkholes. A particular danger area is the funnel of the mould, as the metal last poured in is most affected by the loss of volume during cooling. At this very spot, i.e. at the neck of the Hauslabjoch axe-head, a rather large shrinkhole has formed. Externally it appears in the shape of depressions and in the X-ray it can be seen as a lentil-sized cavity. Fine cracks extend from it into the metal body. The coppersmith evidently tried to tidy up this flaw with a very fine pointed graver. Hammering would not have helped, as pure oxygenated copper is rather brittle and cracks easily. This is what happened during the honing of the cutting edge, where the X-ray likewise shows a fine crack. Seeing that copper hardens under hammering, which is of course desirable for optimal use of the tool, but on the other hand runs the risks of cracking, the workman must strike an exact balance to ensure his product is successful and does not break. This requires experience and a delicate touch. The man who fashioned the Hauslabjoch axe certainly had expert knowledge of the peculiarities of his material and, with the means available to him, produced an excellent tool.

Even the preliminary analytical data indicate that the Hauslabjoch axe was not made of deep-mined copper. Instead, secondary copper minerals were used, such as green malachite or blue azurite. These form thin crusts on the surface of many copper deposits and can easily be scraped off and collected. All that would be needed for our axe would be a malachite or azurite crust of about 1 square metre – azurite occurring much more rarely in nature.

When the axe was found one strip of the leather or hide binding was torn, with the result that it began to unwind. This damage was made good during restoration in Mainz. The Hauslabjoch axe is the only prehistoric axe preserved complete with haft, blade and binding.

4 The Back Pannier

Along with the axe and the bow-stave, but lying higher, on the uneven surface of the rock ledge, Markus Pirpamer and Blaz Kulis discovered further pieces of wood, which they picked up, inspected and replaced. Thus when Police Officer Koler photographed the pieces of wood the following day they too were probably no longer in their original positions, even though the two emphasized that they had put them back exactly where they lay. The botanists identified them as two short boards of larch (*Larix decidua*) and a hazel rod (*Corylus avellana*) 1.98 metres long broken into four pieces. These fragments are a medium-length piece from the thicker end, a short and a longer middle piece, and a long piece from the thin end. Both ends are rounded by trimming and all broken surfaces fit together, so the hazel rod is complete. If the four fragments are laid end to end it can be seen that the hazel rod was bent into the shape of U.

In Koler's photographs the two end pieces lie at an acute angle to each other, with the two boards obliquely between them. The short middle piece has not entirely parted from the long thinner end. The longer middle piece cannot be seen, possibly because it was outside the frame of the picture. One of the photos of Messner and Kammerlander also shows part of the hazel rod. Kammerlander is supporting himself on the long section of the thin end as though on a crutch. As a handle he uses the short middle piece, which is by then alarmingly snapped but not yet entirely broken off. Later the two pieces must have definitely parted company, because they were handed in separately. The following day, Sunday, Alois Pirpamer and Franz Gurschler collected the bits and pieces lying scattered about and took them to the hotel in Vent, from where they were brought to the Forensic Medicine Institute by the undertaker Klocker on Monday afternoon. This batch contained the two larchwood boards and three parts of the hazel rod. The fourth piece, which was then still missing, the one we have called the shorter middle piece, was seen *in situ* by the glaciologists on Wednesday, but not recovered and brought down by Sölder and Lochbihler until Friday.

The narrower of the two small boards is 40.5 centimetres long, 4.6 centimetres wide and 0.6 centimetres thick. The other is 38.3 centimetres long, 6.1 centimetres wide and between 0.9 and 1.2 centimetres thick. The growth rings run parallel to the cross-section of the boards. Only the edges, not the surfaces, exhibit traces of working. This means that the two small boards were not cut out of solid wood, but from split chunks such as might result when a larch is struck by a gale. Then the trunk, as it were, twists out of its annual rings, and from the root-stock, which remains firmly in the ground, long flat chunks and splinters stick up, usually pointed at the top. If one is lucky enough to find an overturned tree like that, the necessary small boards can easily be cut from it. Both boards are doubly notched at each end. However, of the eight notches, only one is completely preserved. In the other seven the notched side has broken off at the head-ends of the boards, so that these ends now look like tongues. If the boards are fitted into the notches cut into the ends of the hazel rod, a U-shaped structure is obtained, closed at the bottom by the two parallel boards.

The jointing between boards and U-shaped rod was undoubtedly by means of the numerous cords found by and on the ledge, and especially in the crevice; mostly they are torn into fairly short pieces. 'Any amount of string,' was Messner's comment on the find. Proof is provided by the impressions of the cords on the hazel rod. We interpret the reconstructed form as the frame of a back pannier. Comparable panniers were being used by the population of the Alps well into the present century for carrying loads; in the Tyrol they were known as *Kraxen*. The light metal frames of modern backpacks are designed in a very similar way.

All the remnants of cord consist of grass. This may not sound very strong, but don't forget that during the period of shortages after the Second World War the string sold for wrapping up parcels was twisted from paper. Until quite recently, rural ropemakers, too, commonly used grass in the manufacture of string, cord and rope. The largest cord fragment from the Hauslabjoch is 87 centimetres long, but it is torn off at both ends. Some pieces show simple knots. One of the ends, immediately behind a knot, is fanned out as a tassel. The thickness of the cords is at most 6 to 7, occasionally 8 millimetres. They are always plied in S-twist from two twisted strands. Clearly, a uniform, well-tested and highly effective process was used in manufacturing these cords, one that remained unchanged over thousands of years right into the most recent past.

In the crevice in the rock and on the ledge numerous scraps of fur were found, most of them collected by Alois Pirpamer and Franz Gurschler. Unfortunately, at the Forensic Medicine Institute these scraps were put together with the remains of the clothing discovered during Henn's recovery operation. This created a considerable problem in the restoration of the remnants of fur, and in the attempt to fit the pieces together. It is now certain that what was found next to the body were mainly pieces of clothing, whereas the scraps of fur found along with the axe, the bow and the frame of the pannier must be part of some other object, probably a fur sack tied to the pannier's frame.

The strikingly bad condition of the back pannier can be explained by the fact that the location at which it was found, on the ledge, is higher by about 60 centimetres than that of the body. These objects began to emerge from the ice earlier than the corpse, and were subject, on the exposed shelf, to more intense solar irradiation and to stronger wind action than the finds at the bottom of the gully.

5 The Ibex Bones

In the course of the first archaeological examination at the Hauslabjoch at the beginning of October 1991, Lippert discovered two small bone splinters at the very base of the rock shelf. They were slightly below the spot where the axe had lain, approximately at the level of the ice surface at the time. The anatomical and zoological investigations were carried out by Professor Angela von den Driesch and Dr Joris Peters of the Institute for Palaeo-Anatomy, Domestication Research and the History of Veterinary Medicine at Munich University. They found that these pieces, only 3.1 and 4.5 centimetres long, were the lateral apophyses (outgrowths) of a fourth and a fifth cervical vertebra respectively. They have the same bluish vivianite efflorescences observed on the corpse as it thawed out in the dissecting room. Much more difficult was the zoological classification. Luckily, the Munich institute possesses one of the biggest comparative collections of vertebrate skeletons in the world, more than 14,000 of them. It was possible to assign the two splinters to a male Alpine ibex (*Capra ibex*). Von den Driesch explained: 'Confirmation of the identification became possible only after examination of several skeletons of male Alpine ibexes. Only with the fifth compared individual was a complete match of shapes found.'

The neck muscles of the ibex are massively developed (they have to carry the huge horns, which often weigh as much as 15 kilograms), and so provide excellent meat. Just as nowadays, the flesh of the animal was eaten not only fresh, but also preserved. In addition to pickling in brine, smoking and air-drying were customary methods of meat preservation. Cold smoking in particular is thought to have been widespread: the meat is cut into long strips, along the muscle fibres, and hung across the rafters, where the cooled smoke of the hearth fire turns them into dried meat. Once the meat is hard and dry it can be stored almost indefinitely; it is highly nutritious and because of its light weight it is ideal for long expeditions. Of course, it requires sound teeth and some strength for chewing, and the teeth get worn down by it. Frequent consumption of dried meat is a plausible explanation for the high degree of abrasion found on the Iceman's teeth.

On the other hand, the two pieces found are only splinters of bones. The meat surrounding them must therefore have been gnawed off. If our assumption is correct, that they represent the remains of a food reserve, then this must mean that the man, having reached the gully by the Hauslabjoch, at least lived long enough to consume his last piece of meat and to spit out the gnawed bone fragments.

The ibex was once common in all mountain landscapes of Eurasia and North Africa, but was all but eradicated by merciless overhunting. Certain parts of its body played a major role in folk medicine and in superstition. Its magnificent horns, up to a metre long, are a proud hunter's trophy and its meat, if properly marinaded, is delicious. Fully grown males reach a weight of up to 120 kilograms. However, only a small colony in the Gran Paradiso range of the Graian Alps saved the ibex from extinction. There now exist several successful repopulation programmes, including some in the great nature parks of the Alps.

Hunting ibex is regarded as exceptionally dangerous. It was originally hunted with bow and arrow, then from the twelfth century AD with the crossbow, and since the invention of gunpowder, from the end of the fifteenth century, with firearms. Unless this is a hunter's tall story, the *Jagdschaft*, a thrusting lance up to 7 metres long, was also used in the Middle Ages and in early modern times. Such hunting lances are nowadays on show only at Tratzberg Castle in the Tyrol. The few accounts to have come down to us describe ibex-hunting as involving strings of beaters, often over several kilometres, who would corral the game until it had no way of escape. Suitable rocks were known as 'cornering walls'. Then the hunter would have to approach to within rifle range of the animal, or within thrusting range if he was using a hunting lance. Sometimes he would approach it by abseiling. Given the ibex's almost proverbial climbing skill, this required supreme mountaineering skill. And there was another danger for the hunter. Occasionally the ibex attempts to save itself by a reckless leap. It will try to pass between its pursuer and the rockface, and in so doing push him off the rock to his death.

Even the ancient Romans used ibex preparations in medicine. Marcellus, minister to the emperors Theodosius I and II, wrote down the following recipe in his book *De medicamentis* at the beginning of the fifth century AD:

In the seventeenth moon collect some ibex droppings. Take as much as you can grip with one fist, provided the number of turds is an odd one. Place them in a mortar and add twenty-five finely crushed peppercorns, then a measure of wine, about two-eighths of best honey and two sextaria of best very old wine. After crushing the ibex droppings mix everything well and place it in a glass vessel. However, to ensure that the effect is successful, you must do this in the seventeenth moon. When applying the medication, start on a Thursday. Prescribe it throughout seven uninterrupted days, moreover in such a way that the patient drinks it in a chair facing east. This potion, as written down, applied and observed, even if the patient is in all limbs and his coccyx sick, contracted, paralysed and hopelessly ill, ensures that he walks on the seventh day.

In the Middle Ages and in more modern times, too, various parts of the ibex's body enjoyed great popularity, not only among quacks but also among learned physicians. With the increasing rarity of the animal prices rose to astronomical levels. Individual horns were mounted in gold and used as beakers. The old innkeeper in Zell in the Zillertal in Tyrol was the proud owner of a splendid pair of ibex horns, for which he was once offered a complete equipage of carriage and horses in exchange; he refused. All the ibex's organs were regarded as having miraculous curative powers – flesh, skin, hair, tail and goatee, blood, lungs, heart and liver, bladder, trachea and, last but not least, droppings. In the seventeenth century a special department was set up in the episcopal court pharmacy in Salzburg reserved exclusively for ibex medicine. Ibex preparations were believed to cure nearly all complaints, for instance 'maternal gravidity and other incidents of womenfolk', sciatica and inflammation of the joints, bladder stones and gout, heartburn and sore throat. It was sudorific and diuretic, and effective even against poisoning and witchcraft. The most desirable parts, needless to say, were the horns themselves, the eyeballs, the heart, and the bezoar stone from its stomach, which was credited with aphrodisiac qualities. The ancient mystique surrounding the ibex is also reflected in the fact that, under its other name of capricorn, it has given its name to the tenth sign of the zodiac and to the tropic which the sun reaches at the winter solstice. Such an unbroken chain of symbolism must have its roots in the mists of prehistory. Surely it was no accident that the Iceman on his dangerous mission carried with him some ibex meat.

During the second examination of the site in August 1992 a

piece of ibex horn about 5 centimetres long was discovered in the area where the body was found. It has been worked at both ends and looks like a small box. Its function is not quite clear. Presumably it was some kind of amulet or had therapeutic significance.

6 The Second Birch-bark Container

The first birch-bark container (*Betula sp.*) was actually seen by the Simons as it lay, close to the corpse's head. Then the remains of a second birch-bark container were discovered by Gerlinde Haid when she visited the site with the Messner party, lying on the rock shelf a little to the south of the spot where the parts of the pannier's frame were found. One may assume that the second container was tied to the pannier frame during transportation or was actually part of the contents of the fur sack. Gerlinde Haid took the fragments of birch bark with her to Innsbruck and handed them over to us on 25 November 1991. The bottom section of this container had been carried by the wind from its original location on the rock shelf over some 20 metres into the nearest ice-free gully, where it was found by Markus Pirpamer on the morning of the Messner visit. He also pocketed his find and in the afternoon showed it to the visitors at his lodge. The ethnographer Hans Haid immediately photographed it twice. The item, along with two bunches of fur, was then kept in the hut larder. On the following Wednesday, during a live transmission from Studio Tirol – I had just returned from the aborted helicopter flight to the Hauslabjoch – Hans Haid showed me one of the pictures. I told the two Alpine archaeologists Sölder and Lochbihler to collect the piece from the Similaunhütte, along with the scraps of fur which Pirpamer had collected, and on Saturday 28 August they handed it over to us at the institute.

On delivery all the fragments of the second birch-bark container proved to be totally desiccated, significantly warped and extremely brittle. This is largely due to their position on the ledge, where the objects which emerged from the ice much earlier were exposed for longer to the harmful effects of night frost, alternate soaking and intensive ultraviolet radiation. In the Mainz conservation workshops the fragments were successfully rendered supple again by steam treatment, so they could be bent back into their original shape.

The inner surface of the birch-bark container from the rock shelf retains its natural colour, a yellowish-white. It is worth emphasizing this because the inside of the container found near the corpse looks quite different. Its shape suggests a cylindrical box. The bottom is an oval 15 centimetres by 18 centimetres. Unfortunately, major sections of the sides are missing. Its height was at least 20 centimetres. The body of the container was formed by a rectangular strip of bark, with a series of holes on its shorter sides. One of the long sides (the lower part of the container) also had holes punched into it at short intervals. The strip of bark was bent to form a tube, the edges were allowed to overlap slightly and were sewn together. This was then sewn to the base piece, which also has holes along its edge. Unfortunately, none of the material used in the sewing of this container has survived. Equally, there is no sign of a lid.

The archaeologist is disappointed to find that the Iceman did not carry any pottery vessel with him, because it is customary to classify prehistoric cultures and communities according to their ceramic legacy. Had our man carried pottery with him, it would have been easy to determine, within a very short time, which corner of the Alps or the pre-Alps had been his home. As it is, his origin had to be arrived at in much more complicated ways, and even so can only be delimited rather vaguely. Yet it's obvious that a supple birch-bark container is much more useful in difficult terrain than a fragile clay vessel. At the moment nothing can be hazarded about the contents of the container from the rock ledge, as its inner surface has yet to be examined for micro-traces.

7 The First Birch-bark Container

In the preceding chapters I have described the objects found at and on the rock shelf on the basis of our present state of knowledge. The next few sections will deal with the articles the Iceman was carrying on his person – the implements, tools, etc., from the spot where the corpse was found. At this point I must emphasize the provisional nature of many of the statements made here. Time and again during our research we have been compelled to adjust our views – and it will be quite a few more years before the work is completed. Many findings will have to be narrowed down and made more precise, others, perhaps, will have to be revised. The uniqueness of the discovery makes unprecedented demands on the scientists involved. Often the research methods themselves still have to be developed, and special instruments and apparatus built. As to how long it will be before we know everything there is to know about the Iceman, that is best answered by reference to the famous burial ground of Hallstatt in the Salzkammergut: excavated a hundred and fifty years ago, to this day it is still the object of intensive research.

When the Simons discovered the body, they also noticed a birch-bark container lying barely 2 metres to the south-southwest of the corpse's head on the glacier ice. Helmut Simon picked it up and, he says, put it back exactly where he found it. When Markus Pirpamer and Blaz Kulis walked up to the Hauslabjoch in the afternoon they too saw the container, but didn't touch it. It remained untouched until the following day, when police officer Koler photographed the site before attempting to free the body with the pneumatic chisel. This is the only picture which shows the container still undamaged. It was certainly intact when Koler left the gully: before he boarded the helicopter he had inspected and then carefully replaced it. He was the last person to see it undamaged. From the descriptions of those who had seen it an approximate picture emerges. It was something like a flattened tube of birch bark, about 20 centimetres long and 10 to 15 centimetres in diameter. No one clearly remembers having seen a lid or a base, but they did recall a zigzag seam along the side. The contents were said to have been wet grass or hay, some of which fell out when the container was picked up.

No one knows who stepped on and crushed the birch-bark container the following day. The Saturday morning visitors admitted that they had paid no attention to the container. When Messner's party reached the gully in the afternoon and started taking pictures, it was already wrecked. Fragments of birch bark and tufts of grass were scattered about the spot where the container had lain, flattened underfoot. Nor can anyone now say for sure exactly who collected the remains – Markus Pirpamer with the refuse bag on Sunday or Henn's recovery crew with the body bag on Monday. In any case, they were not salvaged completely, because, on the Friday morning of the following week, Sölder and Lochbihler saw several more birch-bark fragments trapped in the ice in the area where the body was found, before snow covered everything again. Some more fragments were found during the two examinations of the site in October 1991 and August 1992, as were bunches of grass and leaves which had formed its contents. One leaf had been blown southwards, roughly to where the bow-stave projected from the ice.

From the scant fragments handed over to the restorers in Mainz the original shape of the birch-bark vessel can no longer be satisfactorily reconstructed. But it is likely to have been of much the same size and shape as its counterpart from the rock shelf. At least the material used to sew it with has survived; this shows that the sheets of bark were held together by thin strips of bast. But whereas the inside of the second container seems to have kept its natural colouring, the inside of the container found with the body is blackish. This observation is matched by the fact that flakes of charcoal clung to the leaves which were inside it. Clearly, the first birch-bark container must have been a vessel for carrying live embers.

Analysis of the Hauslabjoch finds of vegetable material was undertaken by a research team headed by Professor Sigmar Bortenschlager. The Botanical Institute of Innsbruck University is situated at some distance from the other university buildings, at the foot of the Nordkette range, beyond the Inn, in the middle of the idyllic Botanical Gardens. Away from the hectic rush of the city, Bortenschlager and his colleagues, including Dr Werner Schoch from the Quaternary Timber Institute in Adliswil, Switzerland, examined the samples. The remains of the leaves from the ember-carrier are all from the Norway maple (*Acer platanoides*). Their condition was so good that chlorophyll could still be extracted with alcohol. In other words, they were still green. The fact that only the flat parts of the leaves are present, while the tough stalks are lacking, indicates that they were picked fresh from the tree. Embedded in these maple leaves were needles of pine (*Picea abies*) and juniper (*Juniperus sp.*). Traces of grain also adhered to the leaves, fragments of glume and husk of einkorn (*Triticum monococcum*) and wheat (*Triticum sp.*). This proves that, while still alive, the Iceman must at least occasionally have been at a place where grain was threshed, when some threshing remains must have found their way into the birch-bark container. We shall return later to this remarkable observation.

The botanists managed to collect over 3 grams of charcoal flakes from the leaves, the size of the particles varying from 0.5 to 5 millimetres. Six different types of wood were identified - including elm (*Ulmus sp.*), which must have originated at lower altitudes. Other identifiable remains belonged to species from the sub-Alpine zone or the treeline region. The remains of reticulate willow (*Salix reticulata*), which the annual-ring radius shows to have been thin-stemmed, come from the region above the treeline. The heterogeneous composition of the charcoal spectrum indicates that the man lit campfires at very different altitudes, ranging from the low mountains to the Alpine level.

8 The Dagger with Scabbard

The small dagger was salvaged by Henn on the morning of the body's official recovery. As he picked it up the scabbard came loose and dropped back into the water. As the ÖRF camera was filming the scene, the position of the object in relation to the body can be fairly accurately determined. Most probably the man was carrying the dagger in its scabbard in the region of his right hip. It is likely to have been fastened to his belt, i.e. at stomach level, only a few centimetres above the right iliac ridge. During the recovery operation the dagger was damaged by at least one ice-pick blow which struck it obliquely at its centre. As a result both the front part of the handle and the tang of the blade are damaged.

The object is surprisingly small. We call it a dagger because the blade has two cutting edges; if it had only one we would describe it as a knife. Overall it now has a length of 12.8 centimetres. The tip of the blade is broken off – probably something like 8 to 9 millimetres are missing. Nor is it clear whether the point was lost during the Iceman's lifetime or whether the damage was caused during recovery. The overall length of the blade, including the tang inserted into the wooden handle, was 6.4 centimetres when it was found. Had the blade been found without the handle, its small size would have caused it to be classified as an arrowhead. No one would have thought something so small could be the blade of a dagger. The material chosen, as usual in the Neolithic, is flint, which in this case has a grey, white and light-grey mottled colouring. The shape was worked from a rough-hewn crude blade by vigorous pressing with a retoucheur. This only takes a few minutes. Australian aborigines require no more time to turn the base of a beer bottle into an elegant point. On both surfaces the retouches cover the whole blade. Near the haft there is a shallow notch on each cutting edge; into these the mounting windings are fitted. Then the narrowed, very short tang ends in a shoulder. The base of the blade was probably rounded, but this cannot be conclusively established from the X-ray pictures because the end was splintered by the blow from the ice-pick.

The handle is made of ash (*Fraxinus excelsior*), an excellent wood used then as much as now for the manufacture of tool shafts, handles and hafts. Its length is 8.9 centimetres. In cross-section the handle is a rounded rectangle with dimensions of 2.1 down to 1.1 and 0.8 centimetres. The handle, too, shows some splintering and a transverse crack from the blow of the ice-pick. A notch was worked into the area of the haft: the tang of the blade was driven into it so hard that the long grain of the handle was slightly split. A thin thread of animal sinew was firmly and tightly wound around it. The strength of a thread twisted from sinew can stand comparison with nylon: very deliberately the manufacturer of the dagger chose, for this critical point, the strongest material available. Otherwise the handle is whittled with no particular care; its user was concerned less with beauty than with stability and suitability for its purpose. All the way round the knob of the handle runs a groove, which holds a grass cord twisted from two strands wound round a few times and tied in a simple knot. Now only 6 centimetres long with a torn-off end, this string presumably served to fasten the dagger, in much the way that we might nowadays carry a pocket-knife secured by a little chain. The central line of the blade diverges by a few degrees from the axis of the handle – presumably also the result of the ice-pick blow. Interestingly, there is no evidence of the usual adhesive between wood and flint.

The scabbard too was badly damaged, presumably by the ice-pick as well. It is 12 centimetres long and covers the dagger almost entirely; only the stubby knob shows. It is plaited from bast, most probably lime-tree bast, and it is no exaggeration to describe it as a minor masterpiece of plaiting technique. Did the Iceman with his coarse hands make it himself? One would be more inclined to ascribe such delicate work to gentler hands.

First a small mat was made; this was then folded in half and the long sides plaited together with a grass thread. The mouth of the scabbard is reinforced by double plaiting. Then, at regular intervals of 1.4 centimetres, there are seven wefts of strong thread. At its lower end the scabbard is firmly drawn together with a grass thread to form an oblong case which is pointed at the bottom and ends in a small lump. Repeatedly threaded through the longitudinal seam is a leather thong, 8 millimetres wide and torn off at both ends. This presumably attached the delicate scabbard to the belt. No comparable object has ever been found before.

9 The Retoucheur

The small implement in the shape of a stubby pencil of the kind used by carpenters gave us a lot of trouble before we thought we had solved the mystery. To start with, no similar – let alone identical – object was then known either from prehistoric or from historical or ethnographic finds. Prehistoric research is an empirical discipline. It involves, among other things, comparison and experience – more accurately the personal experience acquired by a scholar over the course of his life. Thus, if an object is discovered whose significance or function has not yet been described in the specialized literature, other ways have to be found to solve the mystery. One can, for instance, scan the inventory of recent primitive peoples living at a similar stage of civilization as the excavated culture, and if the search is successful, one can ask them what purpose their object serves. Often a glance at historical cultures is useful too, because many articles developed in prehistoric times continued in use until the beginning of the industrial age. The colossal changes brought about by technological progress mean that many articles once in daily use, which would have been familiar even a few generations ago, are now totally forgotten. When none of this produces results, experimental archaeology occasionally helps. A new implement is made from the same materials and one then experiments with it in the hope of discovering what it might be most useful for. Interpretation of our implement was further complicated by the fact that at first we followed a false scent. At its sharp point there is a small lump of some rather hard material, which initially led us to believe that it was for striking sparks. We were led towards this interpretation by the fact that among the objects found loose at the Hauslabjoch there are two items looking rather like tree fungus, which at first we thought were tinder. Scientific examination, however, proved us wrong. The man's real fire-lighting equipment was, as we shall see presently, discovered in an entirely different place.

The exact position of the implement was not recorded, either on film or by photograph, during Henn's recovery operation on Monday. Henn and Lukasser had lifted the corpse from its icy bed. The resulting trough then kept filling with meltwater, which the men tried to divert through a drainage channel. During that time Wiegele kept probing the gully with his ice-pick and salvaged various remains of clothing and equipment, which, more or less unexamined, he dropped on a little heap next to the ice trough. These finds were photographed three times by Gehwolf and also filmed by ÖRF, and the pictures show the pencil-shaped implement as well as the dagger lying on the snow, on the edge of the heap. The implement must therefore have been found shortly after the dagger. Together with the other objects it arrived at the forensic institute in the body bag and was initially labelled 'spindle'.

The tool has an overall length of 11.9 centimetres. The small lump in front projects by 4 millimetres. The wooden shaft therefore has a length of 11.5 centimetres; its diameter is 2.6 centimetres. It consists of a piece of lime-tree (*Tilia sp.*) branch, stripped of its bark. At its rear end the shaft is whittled straight but not very smoothly. There is a crudely cut groove, reminiscent of that on the knob of the dagger. It seems reasonable to assume that a cord to prevent it being lost was fitted into it and firmly knotted. It seems likely that this item, like the dagger, was fastened to the Iceman's belt with cords.

At its front end the implement is whittled into a point. The fine cutting facets can be fairly well made out. All in all, and as with the dagger, emphasis during manufacture was not so much on beauty as on efficiency. The wooden shaft is slightly damaged by three blows from an ice-pick, and slightly splintered especially at its rear end.

Much to our surprise, the X-ray showed that the projecting lump was in fact a long spike reaching almost exactly to the middle of the wooden shaft. It was evidently drawn vigorously into the medullary canal of the lime branch. The spike shows a length of 5.1 centimetres and a diameter of 5 millimetres. As it resisted careful attempts to pull it out, it proved rather difficult to determine what it is made from. Eventually the experienced animal anatomist von den Driesch succeeded in identifying it as a splinter of a stag's antler, which had been carefully rounded and smoothed. The fact that the rounded front end was hardened by firing proves that the implement was intended for heavy-duty use, and explains why it is not the bone colour of normal stags' antlers, but is charred to a blackish hue.

Conversations with flint experts confirmed our suspicion that the implement was most probably used for making, or sharpening, flint tools. The prehistorian Jürgen Weiner of the Rhineland Ancient Monuments Office explained to us the use of such a retoucheur: 'Manufacture of stone implements depends on acquaintance with the specific splitting behaviour of flint. Over a period of thousands of years the Stone Age stone-worker perfected that acquaintance, and by the Neolithic he had at his disposal a complete palette of effective stone-working techniques and methods. These enabled him to produce almost any shape of stone implement from flint. We distinguish seven basic techniques of stone-working – percussion flaking, pecking, pressure flaking, indirect percussion flaking, sawing, drilling and grinding.' As the Iceman was carrying a selection of flint implements, it is quite easy to reconstruct what tools and aids he needed for their manufacture and maintenance. The blank – one distinguishes between lump and slab flint – is first divided into blades and chippings. The preferred material for this is river roundstone of tough rock. Both percussion flaking and indirect percussion flaking – which requires a hardwood, bone or antler rod, as well as a wooden mallet – are used for this process. The fine work, on knife and dagger edges or on scraper and graver tips, is then done with a retoucheur. Different shapes of retoucher can be used, but the one from the Hauslabjoch seems particularly well suited. With its hardened working head the shape of the desired tool can be produced by pressing off minute shell-shaped flakes; a cutting edge or point can be sharpened in the same way. On the strength of his own archaeological experiments Weiner emphatically confirmed that the mysterious implement from the Hauslabjoch can be nothing other than a retouching tool. Once the working head was worn down, its effectiveness could be restored by paring off the point of the wooden shaft, like sharpening a pencil.

10 The Belt-pouch and its Contents

The Iceman's equipment also included a body belt with a pouch. Comparable belt pouches or bags are still often used by hikers, climbers, touring skiers and cyclists. In the Tyrol they are called *Banane* (bananas) or *Wimmerl* (blisters); English skiers call them bum-bags. In the Middle Ages, as in modern times, belts with a money pocket fitted to the inside were also worn, to protect against loss or pickpockets. Unfortunately, the Iceman's leather belt has not survived intact: it is torn off at both sides of the sewn-on pouch. The small pouch with its strap was discovered when Wiegele probed for remains with his ice-pick. By the time the small heap was first photographed the pouch was already there. Up at the Hauslabjoch the collection of finds was not inspected in any detail. Down in Innsbruck, at the Forensic Medicine Institute, the pouch's contents were seen, through a tear in the front, to include a flint implement and a blackish mass. Teissl took a photograph of the item when it arrived, in which the flint implement peeps out of the tear by about 1 centimetre – conclusive proof that it was part of the pouch's contents. Shortly afterwards it was extracted and separately preserved.

The belt fragment is 50 centimetres long between its torn-off ends. As with the rest of the Iceman's fur and hide items, identification of the material was difficult. Professor Willy Groenman-van Waateringe of the Institute for Prehistoric and Early Historical Archaeology of the University of Amsterdam, which worked on the fur and hide finds, identified it as calf leather. The narrow strips used as sewing material are of the same leather. As the aurochs, the ancestor of our domesticated cattle, became extinct in 1627, it is no longer possible to decide whether the leather came from a wild or domesticated animal.

The belt is pieced together from two lengths, 28 and 25 centimetres respectively; they are overlapped by 3.5 centimetres and sewn together. Sewn to them in turn, by means of narrow calf leather strips 4 to 8 millimetres in width, is a third leather band,

20 centimetres long and 6 centimetres wide. Together with the belt it forms the pouch. The flesh sides of the leather strips always face towards the body. The upper seam of the little pouch is left unstitched at the centre, creating an opening about 7 to 8 centimetres wide, both ends of which are reinforced by decorative seams. One leather strip hangs free for some 15 centimetres, with its end cut off at an angle. By pulling it alternately through the loops of the decorative seams the little pouch may be closed. When found, however, it was open. On the narrow sides of the pouch the body belt continues in both directions up to the points where it was torn off. The shortest part is still 5 centimetres long and at its end shows clear gathers, as if a string or a thong had been wrapped around it. The longer portion is 25 centimetres long and also shows gathers 5 centimetres from the pouch, where a grass cord is wound around it twice. Both ends of the cord hang free by roughly 8 centimetres and are then torn off. While one of the ends gets neatly thinner, the other has a uniform thickness of 2 millimetres, just like the cord that is wrapped around the knob of the flint dagger. One can therefore easily imagine that, to the right and left of the little pouch, the dagger and the retouching tool were fastened to the belt with cords. The longer part of what is left of the belt is decorated or reinforced with two parallel tacking seams. There the leather has a small tear, which has been mended with a narrow strip of leather. Altogether the belt must have been subject to some stress, because behind the little pouch there are two longish tears, which have been repaired with fine, twisted sinew thread.

Two further fragments, also of calf leather, almost definitely belong to the belt fragment with the little pouch. One of them is 75.5 centimetres long and 1.5 to 4.3 centimetres wide. The other is 57 centimetres long and from 2.4 to 4.1 centimetres wide. Both become steadily narrower from one end to the other, and are torn off at the tapered end. As for the shorter fragment, it seems that only a very small piece is missing at the narrow end, probably no more than 1 or 2 centimetres. Groenman-van Waateringe has no doubt that all three fragments belong to the one body belt, even though the torn ends do not fit together. In consequence, the belt would have had a length of at least 182.5 centimetres, which means it would have been wrapped around the man's hips twice. In such a reconstruction the pouch lies almost exactly in the

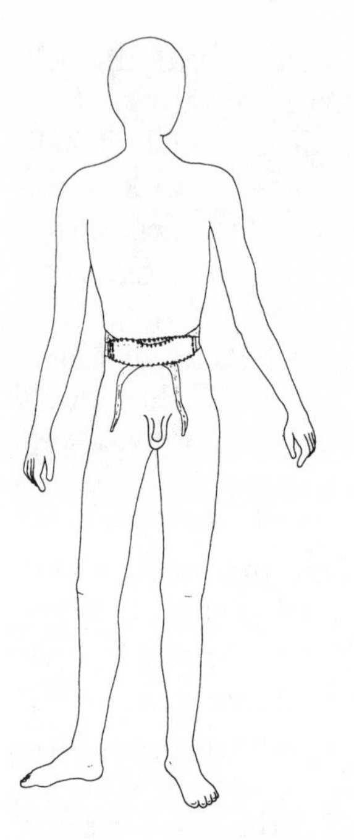

How the belt-pouch was worn. Made of calf-leather, it was probably just under two metres long. The small pouch was placed against the front of the body, the two ends of the belt were then passed back and brought forward again over the hips and knotted together behind the pouch. The loose ends were left dangling. The pouch has an opening at the upper edge: this could be closed with a narrow strip of leather. When the body arrived at the Forensic Medicine Institute in Innsbruck a large chunk of ice was still stuck to the loin region. The two loose ends of the body belt had to be thawed out of it. This made it possible to reconstruct the manner in which the belt was worn.

middle. When the Iceman put on his belt he would place the pouch on his stomach, take the open ends round to his back, bring them forward again and then knot them in front. The position of the finds suggests that the dagger hung down on the man's right and the retouching tool perhaps to the left of the pouch. The significance of the single hole in the shorter end-piece, as well as of the arc-shaped row of five holes on the longer fragment, is still to be determined.

There were five objects in the little pouch; from the X-rays and computer tomographs they proved to be three flint implements, a bone awl and a piece of tinder.

The first flint implement to be taken out of the pouch at the forensic institute is described as a blade scraper. Mottled grey to dark-grey, it was made from a very strong blade, probably triangular in cross-section. It bears deep retouches all round. Its length is 6.7 centimetres, its maximum width 1.35 centimetres and its thickness at least 0.75 centimetres. It is a tool with many uses including cutting, carving, planing, smoothing and scraping. One edge shows traces of so-called 'sickle gloss', a fine silica deposit often found on flint instruments used as sickles for cutting grain stalks. We may assume that the Iceman used it for cutting the grass of which some of his equipment consists.

The second implement is rather unusual among the known shapes of flint artefacts and must be assigned to the category of drilling tools. This drill, too, has presumably been worked from a blade of triangular cross-section with the retouching tool, presumably from the same raw material as the scraper, that is a whitish to grey-speckled flint. It is 4.9 centimetres long, has a maximum width of 1.3 centimetres and is 0.8 centimetres thick. One working end is slightly broadened, the other is long, narrow and pointed. We can take it that the latter was good for drilling holes. If one considers the possessions of the Iceman, this tool could well have served to drill the stone bead or the holes in the quiver stiffening.

The third and smallest flint implement is a thin blade broken off, possibly intentionally, at the top. On the front of the lower end (its ventral surface) we could identify a trace of the percussion bulb, i.e. the spot where the blade was pulled away from the rest of the stone. With a delicately measured blow or pressure the bulb was almost entirely removed. Along the edges very fine retouchings can be seen. The raw material for this item was of a similar, perhaps even the same, substance as that of the other two implements, a flint mottled grey to whitish. The minute piece is 1.7 centimetres long, 1.2 centimetres wide and has a maximum thickness of 0.2 centimetres. With this blade very delicate carving and cutting tasks could be performed. It probably was especially suitable for preparing the arrows, cutting notches to the bowstring and making shafting slits for the arrowheads.

The fourth tool taken from the pouch consists of animal bone, but the species from which it came cannot be determined for certain. As the implement was cut out of the unidentifiable shaft of a metatarsal bone, and was considerably changed through intensive working, the species can only be narrowed down to goat, sheep, chamois or female ibex – both domestic and wild animals. As, however, the bone substance, the *compacta*, is harder in wild species, their bones were preferred in prehistoric times for making tools. In line with the natural form of the bone the awl has a curved cross-section. One end is rounded off in a spatula shape, the other is ground to a sharp point. Both ends have secondary fractures or deformities, which may, perhaps, have occurred during the recovery operation. This tool makes it possible to punch very fine holes, for instance in fur or leather. It could also serve as a tattooing needle. The length of the awl is 7.1 centimetres.

The 'black mass' accounted for most of the contents of the belt pouch. When removed it was found to consist of four irregularly shaped flat tough tuberous objects. They proved to be impregnated with water, yielding and soft under slight pressure. The upper surface was finely scarified in a cellular manner, while the underneath exhibited a shallow fluted grooving. Hence the original belief that this might be a cement, something like birch tar, could not be sustained. Birch tar, the widely used glue of prehistory, turns hard and brittle some time after application, as found, for instance, in the shafting of the Iceman's axe or in the cementing of his arrowheads and arrow-flights. The samples examined by Professor Friedrich Sauter of the Organic Chemistry Institute at the Technical University of Vienna had, when damp, a shiny iridescent surface reminiscent of coal, changing from brownish black via anthracite to black. It became significantly paler, in fact brownish, as it dried, and lost its sheen.

Examination under the scanning microscope revealed a tangle of small fibre parts which gave the impression of matted fibres, evidently of biogenic origin. It seems that the main portion of the black mass consists of the middle layer, the *trama*, of a tree fungus, the 'true tinder' fungus (*Fomes fomentarius*). In a few places relatively fine particles were also identified, which, with darkfield illumination under the stereomicroscope, clearly exhibited crystalline structures of brassy to gold-coloured brilliance. The chemical nature of these crystals was established by X-ray fluorescent spectroscopy which showed sulphur and iron to be the main constituents. This proved that the minute crystals were pyrites.

It seems reasonable to suppose that the Iceman carried a fire-lighter in his belt pouch. From the tinder-fungus growing as a parasite on trees a readily ignitable tinder used to be prepared in the old days by cutting the soft fungal mass into slices, hammering it vigorously, steeping it in nitre (potassium or sodium nitrate) and letting it dry. To make fire a lump of pyrites was struck against flint; the shower of sparks was caught by the tinder fungus and by careful blowing made to glow. In historical times pyrites was replaced by steel. Although there was no pyrites in the belt pouch, the striking traces in the form of minute pyrites crystals on the tinder show that the Iceman occasionally carried pyrites with him in his little pouch.

Studies of Neolithic fire-lighters from grave finds have shown that these may consist of as many as six items in varying numbers and combinations. A complete set of fire-lighting equipment consists of: a piece of pyrites; a flint core; one or two flint blades; a bone implement; a shell; and the tinder. Fire-lighters, or parts of them, have so far been found exclusively in the graves of males – a factor which tallies with the Hauslabjoch find. Moreover, each fire-lighter surely had a container, a bag, a pouch, or a case. In graves, however, these containers did not survive, presumably because they were made of organic material, such as fur, leather, tree bark, bast or textile, and so decayed in mineral soils. The belt pouch of the Iceman is the only known prehistoric example of a fire-lighter container. The fact that the fire-lighting utensils were originally kept in a bag or similar case can be deduced from the fact that in grave finds the items are invariably lying very close and tidily together. Not infrequently their containers also held other implements, tools or replacement materials which had nothing to do with fire-lighting.

Pyrites, incidentally, is not a very stable material. Under certain conditions it can disintegrate or oxidize, which complicates, and sometimes falsifies, the evidence. This applies even more so to tinder itself, which has usually totally disappeared.

The numbers and shapes of flints belonging to a prehistoric fire-lighting kit vary a great deal. Frequently we find a badly worn flint core. In addition, however, there are often one or two flint blades or flakes, but as a rule these bear no traces of fire-striking. Sometimes flint implements, such as arrowheads, drills, etc., are found instead of unworked flint. Implements made from the metatarsal bones of small ruminants, such as sheep, goats or deer, are also common. Shells of river molluscs are rather less frequent; so far they have been found in only four Neolithic fire-lighting kits. But here too the problem of preservation must be borne in mind.

Experimental archaeology has made remarkable strides in the reconstruction and assessment of the probable use of prehistoric fire-lighters. The Kiel prehistorian Norbert Nieszery began with a six-part fire-lighting kit, not counting the case. Although a variety of natural substances can be used as tinder, the best is still the true tinder fungus, although only the fruiting part (*trama*) between its outer crust and lower tubular layer is suitable. This must therefore be cut out of the fungus, which usually grows on diseased or dead beech or, more rarely, birch. Hence Neolithic fire-lighting kits include implements with cutting edges, such as flint blades or tools. The well dried tinder material is then ground until it acquires a fibrous, flaky, cotton-wool-like consistency. The resulting enlargement of its surface greatly enhances its glowing capacity. To protect it from damp it is advisable to collect the tinder in a little dry bowl – which is evidently what the mollusc shells were for.

However, it is certain that the Iceman did not carry any shells with him. So what would he have used instead? We are reminded here of the well-known cup stones which often feature in European landscapes. By cup stones we mean largish, slightly rounded boulders, whose upper surface exhibits several, sometimes artificially created, shallow hemispherical hollows. They are usually found on low hills providing a good view of their near surroundings. There can be no doubt that the shell-like depressions on these stones are mainly tinder bowls. The elevated position causes the dew to evaporate quickly under the sun's rays. The stony base prevents any unintended spread of the fire. In the mountains there are enough natural hollows in the rock suitable for the purpose – which is why artificially worked cup stones are not found there. A noticeably greater frequency, on the other hand, is observed in definite grazing areas, which means that the shepherds are probably responsible for hacking out the cup marks. The long grazing period would have provided them with enough leisure for that. In prehistoric settlements and villages, just as on adjacent arable farmland, there are, at least in original archaeological find locations, no cup stones, because there the fire was lit inside the dwellings.

Examination of the Iceman's fire-lighting equipment provided us with a welcome opportunity to assign a practical function to the all too frequent mystery attached to the cup stones.

Once the prepared tinder material is placed into the shells or bowls, it is ready for the spark. The flint core is repeatedly and forcefully struck against the lump of pyrites. (Pyrites lumps used like this for any length of time show a fairly deep groove.) Flint core and pyrites are missing from the travelling kit of the Iceman; conceivably, they could both have been lost through the tear in the front of the belt pouch. At least the crystals adhering to the tinder mass prove that a knob of pyrites originally formed part of the contents. The three flint implements found in the belt pouch were certainly not used for striking sparks, as they do not show the characteristic traces of wear, but the blade scraper might well have been used to scrape the tinder fungus. It would take several strikes for the tinder to start glowing. Even those with experience do not succeed at the first blow because heat must first be generated by friction. As a result, a thin grey-coloured layer of pyrites dust forms on the tinder material, which in turn impairs the glowing capacity of the fungus. The tinder may, therefore, have to be turned several times or loosened up. This is where the bone rods come in useful, such as the one found in the Iceman's belt pouch. Presumably he used the rounded spatulate end of the bone awl to stir the tinder.

As soon as a spark has caught in the tinder the glow can be magnified by careful blowing. With other easily glowing or flammable materials – Nieszery recommends washed and dried reed-mace wool, hammered willow bast, mosses, down feathers, thistle-down, bullrush or juniper pith, as well as hay and small twigs – it is possible, with an appropriate supply of oxygen, to make the glowing materials flare up. The material, now blazing like a torch, is finally brought to the prepared spot.

This certainly reliable but rather cumbersome prehistoric procedure for making fire – (a prerequisite is dry weather, unless there is a roof overhead) – explains why the Iceman had an ember-carrier with him. His fire-lighter represented an additional safety measure in case his embers were extinguished.

11 The Two Birch Fungi

Two spongy structures mounted on fur strips were also salvaged during Henn's recovery operation. They first appear on the photograph taken by Gehwolf of the heap of objects beside the gully. This makes it certain that the Iceman had been carrying these items on his body, even though it was impossible to determine precisely how. Nevertheless, the fact that they are threaded on narrow strips of fur suggests that they were fastened to some part of his clothing, his belt or his wrist. They were taken to Innsbruck in the body bag.

One of the samples has a broad wedge-like shape with diameters of 4.5 to 5 centimetres. It is covered with small carving marks, clearly indicating that it was cut from a larger piece. One side has been worked fairly smooth, and it has an oval hole in the middle. The surface is irregularly corrugated, almost wrinkled, and when saturated with water is a dirty white to grey with local small dark occlusions. Its consistency is something between cork-like and leathery. The object is mounted on a fur band, now 14 centimetres long and, on average, 9 millimetres wide and 1 millimetre thick, the hair of which, as in the other fur finds, has mostly fallen out. Minute tufts of blackish animal hair remain only at the point where the fur strip enters the object and at its lower knotted end. The upper end is torn off. The longer, free, end reveals five small holes at irregular intervals, through which a second, narrower, band 12 centimetres long and 5 millimetres wide has been threaded. Whether this was intended as an ornament or for reinforcement is not clear. A later tear on the main band suggests the latter, unless of course the tear is of more recent origin.

The other sample consists of the same material, but has been cut in a more spherical shape. Its diameters are 4.6 to 5.3 centimetres. This too has a hole in the middle and is threaded on a simple strip of fur, 22 centimetres long, with a knot in its lower end, and with its upper end torn off.

Scientific examination of these two objects is under the direction of Dr Reinhold Pöder of the Microbiology Institute of Innsbruck University. The mycologists did not find it too difficult to prove that both objects are parts of the fruit of a bracket or tree fungus. Far more of a problem was the identification of the species, as both samples represent sterile fruiting-body tissue. The macroscopic and microscopic characteristics normally used for determination were lost in the original working. Following the application of numerous morphological and anatomical selection procedures the samples were narrowed down to two species – the larch fungus (*Lariciformes officinalis*) and the birch fungus (*Piptoporus betulinus*).

To determine from which of these two fungi our objects came, a chromatographic method had to be developed, by which the Hauslabjoch samples could be compared with the collection at the institute. Only a comparison of the band patterns finally ensured definitive identification. The two objects from the Hauslabjoch

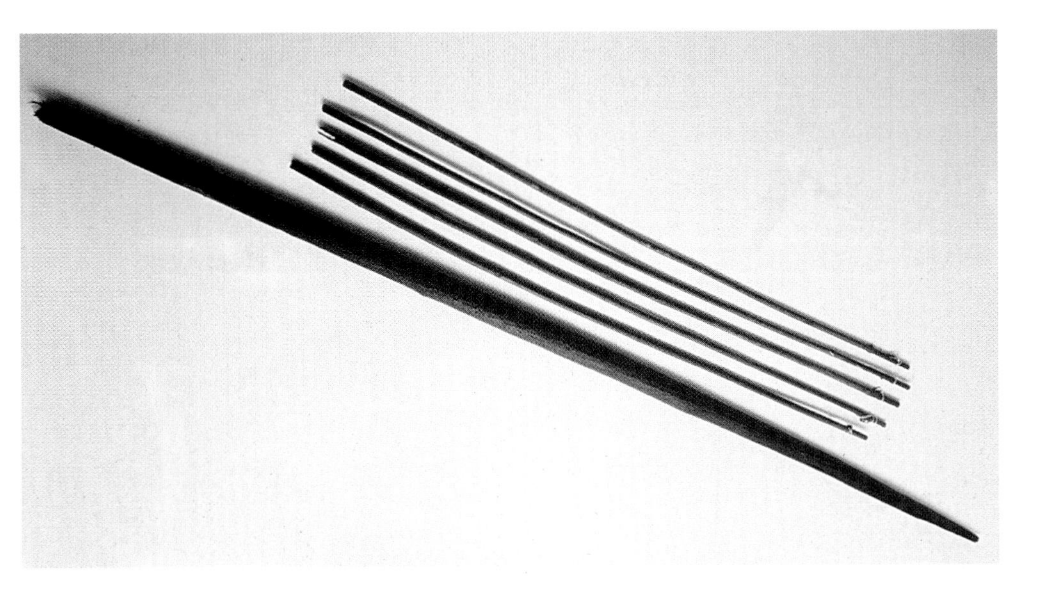

The bow-stave, broken during recovery on 23 September 1991, and five crude arrow-shafts taken from the quiver.

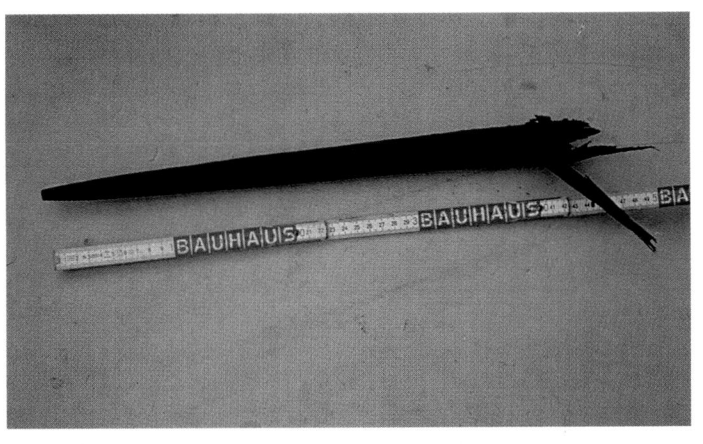

The broken-off end of the bow-stave, found when the site was re-examined in August 1992. The length of the complete bow is 182 cm. Its owner was 160.5 cm tall.

The close-up clearly shows the sharply faceted carving marks on the bow-stave. To work the tough yew wood the Iceman probably used the cutting edge of his copper axe. By comparison with other Neolithic bows, which are all carefully smoothed, the carving marks indicate that this bow-stave was unfinished.

The Iceman's axe, 60.8 cm long, consists of a wooden haft, a leather or skin binding and a copper blade. For the hafting fork, not visible under the lashing, the carver chose the fork of a branch. For this photograph the blade was pulled out slightly from its fixing, but its correct position can be seen on page 213.

Two small larch-wood boards. These were cut not from solid wood but from the splintered trunk of a tree blown down in a gale. Originally each end had two notches; most of them have broken so that the ends now seem to have tenons.

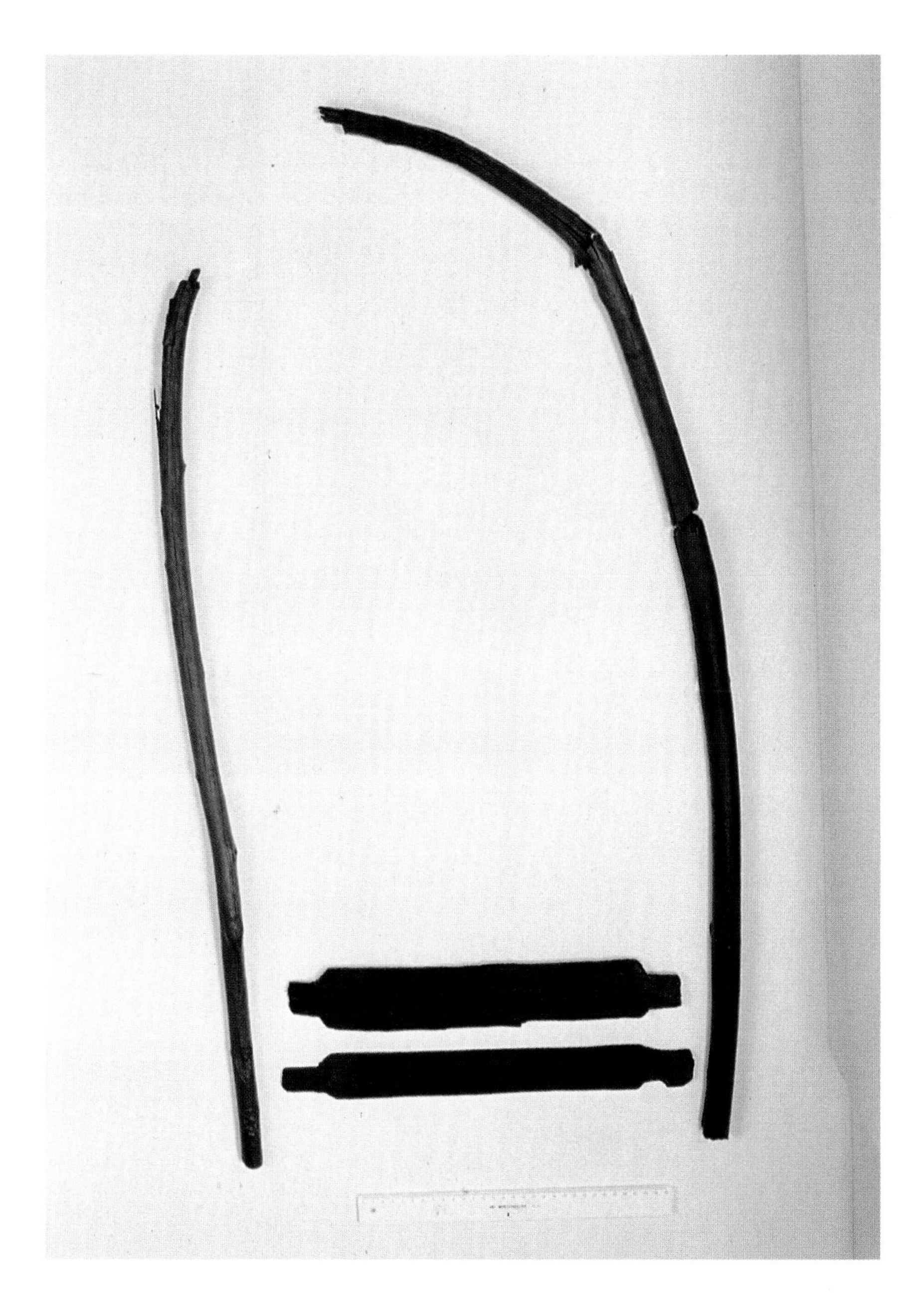

The frame of the pannier the Iceman wore on his back. It consists of a U-shaped hazel rod 198 cm long, each end of which has two notches, into which the double-notched small boards were fitted and tied with cord. The imprint of the cords still shows on the hazel rod. The entire rod has survived, though it has been broken into four parts.

Numerous cords were recovered from the rock shelf, along with the wooden parts of the pannier. These were used, among other things, for lashing together the sections of the pannier. All the cords are made from two strands of grass twisted together. The longest fragment measures 57 cm.

Above right: Fragments of the birch-bark container found on the rock shelf. They can be fitted together to form a cylinder with a flat bottom. The container was presumably tied to the pannier or carried in the fur sack itself.

Right: Freshly picked leaves of Norway maple. Only the leaf-faces are here; the stalks have been torn off. The leaves still contain chlorophyll (extracted in the laboratory with alcohol), proving that they were stripped fresh from the tree. They were in the birch-bark container where, with grass, they served as insulation when it was used for carrying embers.

An implement shaped like a pencil stub, interpreted as a retoucheur. A retoucheur is used for finishing flint implements, to give the object its final shape. It can also be used for sharpening points and cutting edges.

An X-ray of the retoucheur. Driven into the medullar canal of a lime-wood grip, the stag-antler spike with its rounded working end shows up dark against the surrounding, longitudinally-grained wood.

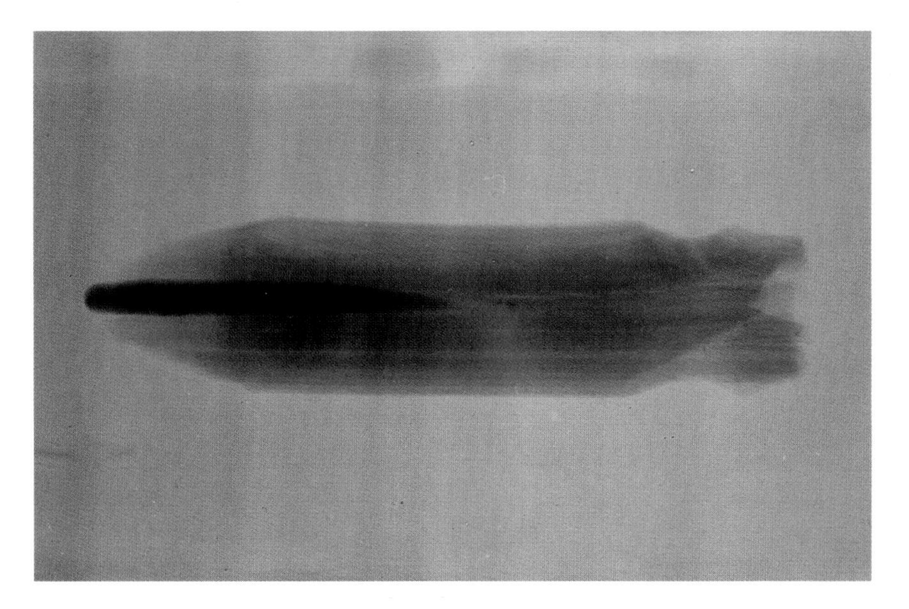

The small dagger with its scabbard. The blade consists of flint, worked with a retoucheur on both surfaces; the haft is of oak. An animal sinew thread – as strong as nylon thread – was used for the haft binding. A grass cord is wound around the pommel to prevent the dagger being lost. The finely knotted scabbard is plaited from strips of bast and sewn up at the side with a grass cord. A narrow leather strip enabled the scabbard to be attached to the belt.

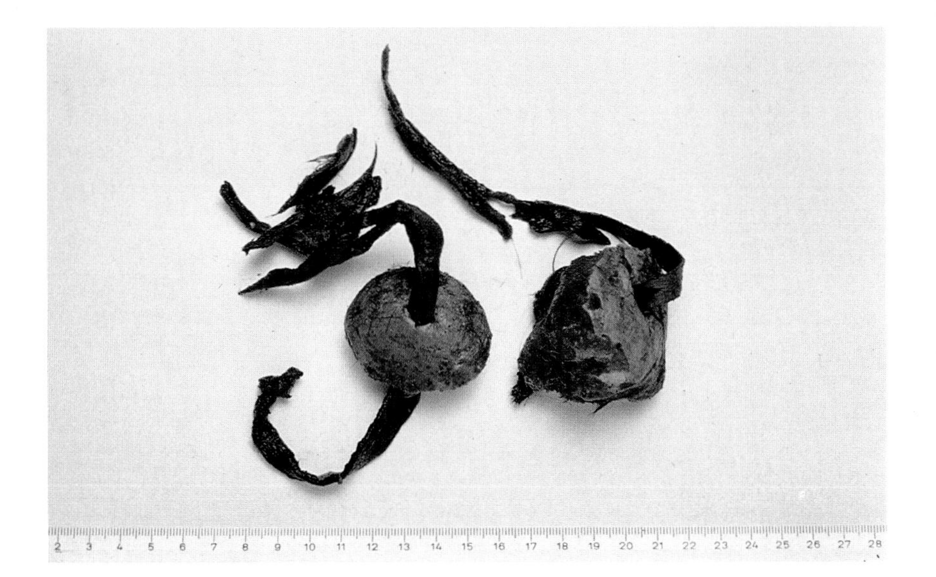

Two pieces of birch fungus, drilled through the middle and threaded onto fur strips. The Iceman presumably carried them on his left wrist, which still shows corresponding pressure marks. The birch fungus contains bacteriocidal agents and was probably the Iceman's travelling 'medicine kit'.

A tassel: the only item found at the site that could tentatively be classified as an ornament. The flat bead is of white marble. The narrow strips of fur twisted into ringlets were probably not decoration, but for repairs.

had been cut from the fruiting body of the birch fungus. This result came as a surprise to us, as we had assumed it to be tinder. The birch fungus, in fact, is most unsuitable for this purpose as its tissue is not highly flammable. The solution to the riddle had to be sought elsewhere.

It is well known that the birch fungus contains, among other things, an antibiotic substance, polyporic acid C. This is highly active against some types of mycobacteria, including the tuberculosis agent. As long ago as the first century AD the Greek physician Dioscurides recommended tree fungi as an astonishingly effective medication. Saint Hildegard of Bingen, who lived in the eleventh century, lists six different species of tree fungi which allegedly provide relief from a wide variety of diseases. Even at the beginning of our century some pharmacists still offered tree fungi as medical preparations. They served to staunch blood and, taken by the tablespoon, were considered helpful in bladder conditions and other complaints. Alexander Solzhenitsyn, the Russian Nobel Prize winner, in his shattering book *Cancer Ward*, describes how, in a desperate attempt to relieve their sufferings, the patients cling to every straw. He mentions the birch fungus. This was not a real fungus, it is explained, but a kind of canker growing on old birch trees. It is described as a tumour, an ugly growth, curved like a hunchback, black on the outside but light brown inside. Peasants would drink it as an infusion, dried, finely ground and then boiled in water. In this way, preventively as it were, they had been trying to cure themselves of cancer for centuries. The idea of a nature cure with anti-carcinogenic effect was encouraging to the cancer patients, who hoped that the black, misshapen growth on the white birch trunk could restore their health.

Writings about the birch fungus include a few apocryphal references to its allegedly hallucinogenic properties. But this has yet to be proved either medically or pharmaceutically. It need not, then, enter into our hypotheses about the significance of the fungi among the Iceman's belongings.

All folk medicine has its origins in prehistory. Over hundreds and thousands of years remedies were passed on from generation to generation. The modern pharmaceutical industry has frequently analysed the active constituents of traditional medicines and makes use of them to this day, where synthetic forms cannot be produced. Seen in this light, the Iceman with his modest but no doubt effective travelling medicine kit, is not all that remote from ourselves.

12 The Tassel with the Stone Bead

Along with the two birch fungi, a tassel was recovered which the Iceman presumably wore at the front of his body. This is the only item of his equipment that could possibly be classified as an adornment. At the top of the tassel is a bead of brilliant white marble, polished into a flat circular shape, 2.4 centimetres in diameter and 4 millimetres thick. The hole in its centre could have been made with the drill-like flint instrument from the Iceman's belt pouch. As now, any article worn as an ornament is also, or even principally, endowed with an imaginary significance, in which the choice of material plays a part. In antiquity, for instance, various precious and semi-precious stones were credited with specific magical or medicinal properties, or were worn as talismans against evil. The Iceman's stone bead can almost certainly be regarded in this light. This view is supported by the strikingly bright colour of the fine-crystalline marble, which may have lent the material an imaginary value.

Threaded through the central hole is a narrow strip of fur, which is repeatedly knotted and whose ends are once more tied together. Through this loop nine more fur tapes have been drawn. These little thongs are each twisted into a spiral. For them to retain their – certainly intended – tubular shape, they would have had to be twisted while wet and afterwards dried. They are rather loosely held in the loop and have therefore slipped about, so that the tassel now seems rather irregular. The overall length of the ribbons averages 40 centimetres, which means that the tassel itself dangled for some 20 centimetres below the bead. Unwound, the ribbons are 6 to 10 millimetres wide and a little over 1 millimetre thick. Twisted into tight tubes they have an average diameter of 3 to 5 millimetres.

Personally, I think it very likely that apart from its decorative function, the tassel had an entirely practical and down-to-earth purpose. In this way the Iceman carried with him a bundled supply of spare thongs. Looking at his possessions in general we find that narrow strips of leather or fur were used in various ways – on his shoes, his grass cloak, the scabbard of his dagger, his quiver and his axe. Given the heavy demands made on his equipment in difficult terrain, an adequate supply of repair materials seems almost indispensable.

13 The Sloe

When the two Alpine archaeologists Sölder and Lochbihler visited the Hauslabjoch in a gale, they recovered a sloe, the fruit of the blackthorn (*Prunus spinosa*), near where the corpse was found. Lying as it was on the surface of the ice, it had almost certainly been moved from its original position by the tumultuous events of the previous day or by the meltwater. With its short spiky branches the blackthorn often forms impenetrable hedges along the edges of forests and copses. It blooms early in the spring, its small, almost spherical fruit ripening in the autumn. Because of their high acid content and bitter substances sloes are not edible raw. 'They draw your shirt-tails in at the back,' is the popular saying in the Tyrol. After the first frost they are more or less fit for consumption, but still sour enough to make one grimace. Only a very thin layer of flesh forms over the pip. For that reason they are not a satisfying food, but rather a garnish, highly valued since ancient times for their vitamin and mineral content. Even though prehistoric man was unacquainted with these concepts, he was surely familiar with the health-giving properties of trace-substance carriers. It is not therefore surprising that, given favourable conservation conditions, sloe pips are found regularly at the cultural levels of prehistoric and ancient settlements. In the rare instances of organic material having survived in graves, we sometimes find vessels containing sloes. Admittedly, in contrast to our find in the glacier ice, the flesh of the fruit in those cases has always decayed. Among other examples, an oakwood coffin excavated at Oberflacht in Württemberg at the famous Allemanic necropolis of the sixth and seventh centuries AD was found to contain a lathe-turned ash-wood bowl containing about a hundred and fifty sloe pips.

14 The Net

During the first re-examination in the gully in October 1991 and just west of the boulder against which the body had rested, Lippert found the remains of a net lying on the ground. Conditions at the site suggest that these pieces had been somewhat washed about by meltwater. Nevertheless it seems highly probable that the Iceman also carried the net on his person. The scant remains no longer convey any idea of its original size. Like all the other cords from the Hauslabjoch except one, it consists of grass strings, plied from two strands in Z-twist. Their average thickness is only 1.8 millimetres. Even so the net is very wide-meshed, the space from knot to knot being roughly 5 centimetres. It is too wide-meshed to have been used for carrying things: any contents would slip through easily. It might well have been a fishing net. As, for reasons that we will come to later, it seems certain that the Iceman stayed in the valleys for long periods, he quite probably fished in the waters of the lower-lying settlement regions. But here, too, doubts arise. Much of his catch would have escaped through the mesh and only fish weighing more than a kilogram would have been caught. Fragments of nets found in prehistoric coastal settlements and described as fishing nets by archaeologists usually have a much closer mesh.

During my prolonged excavations in Portugal I mostly employed as labourers local men who ran small farmsteads. They lived with their families in modest huts without electricity or running water. A moped was to them the epitome of modern comfort. Otherwise their lifestyle hardly differed from that of earlier centuries. In their lunch break they made a small fire and over it cooked a scant meal. They would pour water into a tin pot, seasoned with the coarse salt of the lagoons. A potato or two, a carrot, and sometimes some green vegetables were sliced into it. Then they reached into their pockets and produced songbirds, mostly sparrows. They wrung their necks, plucked them and dropped the tiny creatures into the boiling soup. Once they brought a kingfisher, which, to their total incomprehension, we paid them to free. I did some detective work and found that nearly every one of our helpers had stretched out a net behind his house, hidden among the shrubs, out of which each morning he would collect the wriggling birds.

The remains of our net from the Hauslabjoch exactly matched those nets. Mesh width, thickness of cord and probably also overall size agreed quite surprisingly. This empirically acquired explanation may well be the most likely one for the function of the Iceman's net.

15 The Quiver and its Contents

Earlier I described in detail the discovery, salvage and removal of the quiver by a team of glaciologists under Patzelt on Wednesday, 25 September 1991. What is important is that the quiver, lying about 5 metres from the corpse, did not begin to emerge from the ice until the preceding day. When the glaciologists arrived at the spot it had not yet been seen, let alone touched, by anyone. The fine layer of dust (*Cryoconite*) which remains after the melting of glacier ice and which covered the exposed part of the quiver was completely intact. When the object was found, some three quarters of its length were still trapped in the ice. Only the quiver's open end, with the shaft ends of the arrows, was visible. While it was being freed, they discovered that the quiver was resting with its upper part on a slightly inclined slab of stone, so that it and its contents had been very slightly bent by the weight of the glacier. There can be no doubt that the quiver with its contents, frozen as it was to the stone, had lain unmoved at this spot since it was placed there over 5,000 years ago, until the very moment of its recovery by Patzelt and his team.

The quiver consists of a longish rectangular fur sack, narrowing a little towards the bottom. As with the other fur objects, the outward-facing hair has largely disappeared, evidently as a result of slight decay during the mummification process, and the material now looks like leather. Only in a few folds and in the area of the seam have scant tufts of hair survived, especially where they were caught in the sewing. According to Groenman-van Waateringe's examination the colour of the hair is light brown. Presumably the hide of a stag-like animal such as red deer or roe deer was chosen – less probably elk, though in prehistoric times it was still native to the meadowlands of the Tyrolean river valleys.

The main part of the quiver consists of a piece of fur cut, first of all, into a narrow trapezoid shape. This was folded over and sewn up along the sides with fine leather bands. For some reason this seam faces outwards, while the seam at the lower, short end of the quiver is turned inwards. Along the side seam the quiver was stiffened with a hazel rod, 92.2 centimetres long and with a diameter of 1.4 centimetres, which had been carefully stripped of its bark and smoothed. A longitudinal groove with a V-shaped cross-section was cut into the rod; the groove stops short 4 centimetres from the lower end. The base of this V-shaped groove has twenty perforations, at regular intervals of 4.7 centimetres. It is interesting that the Iceman carried with him all the tools necessary to make this stiffening – the blade scraper for cutting off the ends and for removing the bark, and the drill-like flint tool, its broadened cutting end for routing out the groove and its point for making the perforations. The long seam of the quiver was then placed into the groove in the hazel rod and fastened, through the holes, with fine leather bands. The upper end of the rod is level with the quiver-mouth – which is itself of complex construction, as we shall see. The lower end projects by 4.2 centimetres, which is why the groove was not continued.

When the finds were delivered to us it was immediately obvious that the stiffening strut had broken in three – a long lower piece, a short middle piece and an even shorter upper piece. All three fragments fit together at their fracture points: nothing has been lost. At this point things begin to get interesting because our next observations are crucial in clarifying the Iceman's fate, what eventually led to his death. The middle piece was *not* found with the quiver, which was recovered by the archaeologists with meticulous care. Instead it was found at the spot where the corpse lay; drifting in the meltwater, it was seen by Messner's party and initially thought to be a flute – a whole four days before the glaciologists discovered the quiver. Messner's group left the piece where it was. On the following Sunday it was collected by Alois Pirpamer and stuffed into a refuse bag together with the other remains lying scattered about. On Monday the bag, by then in the undertaker's coffin, arrived at the forensic institute. As the part of the quiver to which the snapped middle piece should have been attached was still firmly lodged in the ice when Patzelt's glaciological team arrived on Wednesday, it follows that when the quiver was laid down over 5,000 years ago this middle piece was already missing. The man was carrying it separately on his person, presumably intending to repair the quiver when a suitable opportunity arose.

This means that when the quiver was taken to the Hauslabjoch it was already severely damaged. When it was found, the short upper segment of the stiffening strut was twisted by 180 degrees, so that its rounded end was pointing down and the fractured end up. Having lost its reinforcement, the quiver must have stretched a little during transport. This was discovered during the restoration work in Mainz, when the middle piece was fitted into place. Despite the fitting fracture-faces, the gap between the upper and lower segment could not be completely closed; a space of roughly 2 centimetres remained.

Quiver-mouth, closing cap and carrying device were evidently fashioned in a complicated manner, though because of further damage it is not possible to judge quite how. On the front of the quiver, at the top, the fur has been cut away to a level 17 centimetres below the top of the hazel rod, leaving only a narrow portion still attached to the rod. On the rear the fur extends beyond the trapezoidal shape, to continue as a strip the full width of the quiver, 7.5 centimetres above the quiver-mouth. There it is torn off diagonally, from the edge of the cut upwards towards the seam-side. The missing segment, which cannot be identified among the other finds, must have been the carrying device of the quiver, perhaps a shoulder strap. It is not clear at which point of the quiver the other end of the carrying strap was fastened. Perhaps the projecting lower end of the stiffening strut has something to do with it. Far more probably, however, the end was fastened to the broken middle section.

On to the strip of fur left attached to the rod in front a flap has been sewn with a narrow leather strip. Its upper edge, level with the narrow upper side of the strip of fur attached to the rod, forms the quiver-mouth. The flap is roughly semicircular in shape and is 17 centimetres high. For reinforcement and decoration it has a seam of tacking stitches along its edge. Its surface is similarly enlivened by three diverging tacked seams. An irregular tacked seam, running obliquely upwards from the lower inner corner, was probably intended mainly for reinforcement.

The greater part of this flap has been torn off diagonally upwards from the lower corner. The severed piece, however, unlike the carrying device, has been preserved for us. When it was discovered, the quiver was lying on the rock front down, and the part where the flap should have been was partly frozen to the ground and partly enclosed by ice. When the quiver was delivered to us, the flap was already torn off, but without question it was not in the area where the quiver was found. The missing piece – fitting perfectly – was found among the items which reached the Forensic Medicine Institute in the coffin on Monday. Unfortunately, the fragment cannot be identified anywhere in the verbal accounts or photographs recording the site before and during Henn's recovery operation, so it is impossible to decide whether it was found with the body or with the group of articles lying at and on the rock shelf. Be that as it may, it was clearly not with the quiver and must, like the middle piece of the stiffening strut, have been separated from it even before the quiver was set down 5 metres away from the other two find locations. In this instance, too, a repair was clearly intended but was never made.

Though no closing cap has come down to us, there is an indication that such a cap existed, fitted a short way above the quiver-mouth on the carrying strap running up from the back of the quiver. On the front of the quiver, 7 centimetres below the cut-away edge, there are, close together, two small parallel vertical incisions through which a narrow leather strip has been drawn horizontally and knotted to form two ribbons. It is torn off at both ends and is now only 11 centimetres long. We believe that with it the quiver's cap, pulled down forward, was closed. However, there is the possibility that the quiver had no closing cap at all and that some other object, such as the tassel with the stone bead, was fastened to it – but we do not consider that very likely.

If, therefore, one wished to get at the contents of the quiver, one first had to open what we presume to have been a cap and hinge it upwards. Next one had to turn the flap sideways, and only then could one reach the fletched shaft-ends of the arrows. This double closure protected the delicate arrow-ends. More importantly, when pulled out they would not have brushed against the inside of the quiver, which could have easily damaged the fletching.

The findings of the glaciological examination during the recovery, the exact reconstruction of the circumstances of its discovery and the very detailed examination of the quiver all prove that, while in use, it came to the Hauslabjoch in a badly damaged condition and was put down 5 metres from the boulder on which the corpse was found. Its carrying strap and cap had been torn off earlier and did not find their way into the gully. The flap, likewise torn off before the man arrived at the Hauslabjoch, and the broken middle piece of the reinforcing strut, to which the carrying strap may have been attached, were carried by the man separately. Of the middle piece we know that he had it on his person. (Where the flap was found is not known – whether with the corpse or on the rock shelf.) At any rate, the man intended to perform a repair; otherwise he would hardly have carried the severed fragments with him. His sudden death put a stop to his plans.

The remarkable observations made on the quiver itself were confirmed and complemented by the archaeological analysis of its contents. We already knew from X-rays and computer tomography pictures that, in addition to the fourteen arrow-shafts visible in the quiver-mouth, it contained a number of other objects. Its opening was therefore awaited with great suspense. This took place, after extensive and conscientious preparations, on 19 December 1991 in the workshops of the Roman-Germanic Central Museum in Mainz, in the presence of numerous scholars involved in the Iceman project. It held some astonishing surprises.

The quiver contained two arrows ready to be shot, twelve partly finished arrow-shafts, a coiled string, four bundled stag-antler fragments, an antler-point and two similarly bundled animal sinews. Like everything else associated with the Iceman, the quiver and its contents are unique. Definitive evaluation and interpretation will take some considerable time. In consequence, the account which follows, like much that has been said earlier, is somewhat provisional. Some interpretations will be rendered more precise in the future, and some opinions given here may have to be revised.

The shorter of the two finished arrows has an overall length of 85 centimetres. It consists of a wooden shaft, a flint arrowhead and the fletching. A long shoot of the wayfaring tree (*Viburnum lantana*) was chosen for the shaft. The long, straight and very tough shoots of hard sapwood which are characteristic of this shrub made it, along with the cornel tree (*Cornus sp.*), which has similar growth characteristics, the favourite source of arrow-shafts in prehistory. Only at the beginning of the Iron Age, when improved tools became available, did shafts split from heartwood come into use. The wooden shaft of the arrow was cut straight at both ends, stripped of its bark and carefully smoothed all round. In the Neolithic there existed fist-sized kits made of two oblong pieces of sandstone, laid flat on top of each other, which could be described as arrow-shaft smoothers. Into the two flat surfaces semicircular grooves were ground, which, placed upon each other, formed a tube. If an arrow-shaft was evenly drawn through this opening, first one way and then the other, an excellent rounding of the wood was achieved. Admittedly, there was no such tool among the Iceman's possessions. He must have achieved the same effect in a different way. The main body of the arrow-shaft is 0.9 centimetres in diameter, although the forward section of the shaft, about 9.5 centimetres in length, is slightly thicker, at 1.1 centimetres, to ensure that the arrow is nose-heavy during flight, which greatly improves stability. The forward end is deeply notched and slightly split to allow the tang of the arrowhead to be driven into it. The head, 3.9 centimetres long and 1.8 centimetres wide, consists of whitish-grey flint. Its leaf shape was formed by retouching on both faces. The cutting edges are slightly asymmetric. The pointed tang is recessed. Half of the flint point and the forward section of the shaft are thickly cemented with birch tar over a length of 4.3 centimetres. Chemical examination of the shafting adhesive (also directed by Professor Sauter of Vienna) has established that the pitch needed for making the tar was obtained by carbonizing either the wood or the bark of a birch tree (*Betula sp.*). Wood pitch has a viscous to oily consistency and is obtained by heating while atmospheric oxygen is excluded, i.e. by dry distillation, for instance in a charcoal pile. If this substance is subjected to prolonged heating in an open vessel, an asphalt-like tar forms which hardens during cooling. With prolonged storage it turns brittle and crumbly like sealing wax, which is why minor sections of the material used as cement in the Hauslabjoch finds have crumbled off. Finally, a thin thread, probably consisting of untwisted animal sinew, was wound around this part of the shaft; it is slightly embedded in the birch tar.

The tail end of the shaft, with the exception of the nock for the bowstring, tapers by less than 1 millimetre over a length of 13.1 centimetres. This shallow recession was filled in again with a thin, evenly-spread layer of birch tar. Into this the three-fold radial fletching, exactly parallel to the shaft and the same length as the layer of tar, was cemented. Suitably trimmed halves of feathers were used, with the narrow-trimmed shaft-strips forming the underside. The relative circumference of the feather segments suggests that only primaries from the wings and tails of biggish birds were used.

On the basis of pigmentation tests carried out by Professor Jochen Martens of the Zoological Institute of the University of Mainz the feathers could have come from any of the following bird species: black woodpecker, Alpine rook, Alpine crow, common raven, capercaillie, bald ibis, golden eagle, black vulture, Egyptian vulture and griffon vulture. (The German magazine *Stern* was a little hasty in speaking of eagle feathers in its issue of 9 July 1992.) All these birds were found in the Iceman's presumed habitat. Professor Ellen Thaler of the Alpine Zoo in Innsbruck has made available to us a complete set of comparative material; the feathers of the bald ibis, long extinct in the Alps, come from a small colony which has survived in the Atlas Mountains of North Africa. Under the direction of Dr Gabriele Wortmann of the German Wool Research Institute in Aachen a method is now being developed to narrow down even further the possible species to which the arrow feathers could have belonged.

Once it had been stuck on with birch tar, the three-part radial fletching was spirally wound round with a thread as fine as hair. This was twisted from two strands and has a thickness of no more than 0.15 millimetres. It probably consists of animal hair, possibly sheep's wool.

Into the tail end of the arrow the nock to receive the string was cut. This has a depth of 12 millimetres and a width of 4 millimetres. Such fine work could only have been done with a delicate tool. Among the Iceman's equipment the only possible candidate would be the razor-sharp flint blade which he kept in his belt-pouch.

The second finished arrow is markedly longer than the first; it measures 90.4 centimetres. The length of the fletching – otherwise worked in exactly the same way – is slightly greater, at 13.8 centimetres. The leaf-shaped arrowhead, also worked on both surfaces with the retoucheur, is 3.8 centimetres long, 1.6 centimetres wide and 0.5 centimetres thick and appears to be of the same, or at least a similar, variety of flint as that of the first arrow. It is also fastened in the same way, except that the birch tar does not touch the point but merely covers the wooden shaft. At 14 millimetres the nock for the insertion of the string is slightly deeper.

In one important characteristic, however, the arrows are different. The longer one is a so-called composite arrow, meaning that the wooden shaft is made of two sections fitted together and therefore has an intended breaking point. The short front part of the shaft, 10.5 centimetres long, was shaped into a point at its rear end and inserted into the hollow front end of the rear section, which is 72.5 centimetres long. The junction of the two parts is covered with birch tar over a length of at least 4 centimetres and bound with the same thread material as the fletching. The glueing and binding is clearly longer on the rear section, because it is there that, owing to the tenon of the foreshaft, the greatest risk of fracture existed. Both parts of the shaft are 1.1 centimetres thick. Unlike all the other arrows in the Iceman's quiver, the foreshaft is made not of viburnum wood but of the wood of the cornel tree.

The question arises whether this arrow was made from two broken arrows, which would be entirely conceivable, or whether it represents a genuine composite arrow. Composite arrows are common among primitive peoples. The elaborate method of manufacture, as described here in detail for both arrows, should make it clear what a valuable property a finished arrow was for the archer, and that even fragments, if they could be reassembled, had a certain value. North American Indians, on the other hand, use arrows with detachable foreshafts as weapons of war. If an enemy tries to pull the arrow out of his wound, he merely detaches the rear section, while the point and foreshaft remain inside. However, as our arrow consists of two kinds of wood, it seems more likely that it was put together from parts of two spent arrows than that it was constructed as a composite arrow from the outset.

The twelve partly worked arrow-shafts in the quiver were all made of shoots of the wayfaring tree. They were stripped of their bark, with small side-shoots being removed at the same time. The painstaking smoothing found on the finished arrows had not yet been carried out on the blank shafts; some slight thickenings at the leaf-nodes are still visible. The ends are cut off square, and the thicker ends, towards the root, have notches about 2 centimetres deep cut into them to receive the arrowheads. It is interesting that the Iceman did not complete one arrow at a time. Instead, he performed each separate operation, in a sort of serial production, on all the arrow-shafts. Presumably, the next step would have been to cut the twelve nocks for the bow-string. This would have been followed by the smoothing of all the shafts. Such a procedure would save him continually changing tools.

The length of the blank shafts, which corresponds to the intended length of the finished articles, fluctuates between 84.5 and 87.8 centimetres. For the benefit of readers interested in prehistoric numerical mysticism and prehistoric systems of measurement I am deliberately listing the individual lengths, even though this may seem a little ponderous: two at 84.5, one at 86.8, two at 87.2, two at 87.3, two at 87.4, one at 87.7 and one at 87.8, as well as a shorter one, only 69.2 centimetres long, with one end splintered, clearly as a result of a secondary fracture. The missing section was not in the quiver, so it must have been broken off previously. The sacred number of twelve speaks for itself, but strangely enough, the total number of arrow-shafts matches one of the most exciting archaeological finds of our time. The gold-adorned Celtic ruler from Hochdorf in Württemberg had fourteen arrows as grave furnishings.

When we opened the Iceman's quiver we found to our amazement that eleven of the twelve blank shafts were in good condition, while the two finished arrows were broken. In both cases the fracture was near the point. The composite arrow had been subjected to such force that even the flint arrowhead was broken. Its tang is still in the fore-end of the shaft. There are additional fractures, in one shaft repeatedly about the middle and in the other more towards the rear. It is surely no accident that, of all the arrows, the two ready ones were broken, while the blank ones were undamaged. If all the arrows had been in the quiver at the time of a presumed violent incident, the finished arrows would either have been partially protected by the blank shafts acting as a sort of splint, or else the blanks would have been damaged too. The only possible interpretation is that the two finished arrows had already been damaged before the blank shafts were placed in the quiver. I shall return to the conclusions to be drawn from this, as well as to the condition of the quiver itself, at the end of this section.

On the basis of archaeological statistics, flint in its different varieties was the usual material for arrowheads in the Neolithic. To what extent other materials were used we do not know. We do know that primitive hunter societies also used wooden arrowheads, especially for hunting birds. But given the conditions of site and conservation they are extremely rare among prehistoric technological samples, if indeed they can be identified as such at all. We also know that bone and antler played their part as raw materials for arrowheads. Elongated twin points of bone and antler have been found again and again, especially at Swiss pile-dwelling settlements; these have on them shafting remains which suggest arrows. Their effectiveness is demonstrated by an arrowhead from Lattringen on the Bieler See in Switzerland, found firmly lodged in the sacrum of a stag, having completely pierced the bone. What is unclear, however, is whether the use of bone and antler was a second-best solution in the absence of sufficient flint, or whether that material was actually preferred for specific hunting purposes. Anyway, the Iceman carried with him a bundle of prepared stag-antler fragments, from which he could have carved at least eight arrowheads of the kind found in the Swiss lake-dwellings. In spite of these qualifications we can work on the assumption that flint was the common and preferred material for arrowheads in the Neolithic period and that the Iceman did not have enough raw material at his disposal to arm all his arrows with flint heads.

The four antler fragments were strapped together to form a bundle 16.5 centimetres long. They were at the very bottom of the quiver, immediately next to the side seam, so that their ends touched the bottom. To take them out the whole of the quiver had first to be emptied. The long tines are all only roughly worked and seem unfinished. Their length varies: one is 15.1 centimetres long, another is 16.1. The other two are markedly shorter, but their length cannot be established as one end of each is hidden under the binding. Their oval cross-sections measure fairly uniformly 5 to 6 millimetres. In each case one end is trimmed down to a fairly sharp point, while the other end is irregularly cut at an angle or rounded. The binding is of bast, most probably from a lime tree, wound in broad strips around the bundle, partly edge to edge and partly overlapping. These fragments are probably the raw material, or replacement material, for some implements among the many that could be made using stag antlers.

The X-ray first identified, among the contents of the quiver, an elongated, slightly curved antler point. This implement, 20.7 centimetres long, had been carved out of the integument of a gently curved antler branch. The fore-end has a well-smoothed sharp point, while the other end is rather crudely worked, with an almost square cross-section, measuring 1.2 by 0.9 centimetres, over a length of 8 centimetres. The end is cut in the shape of a flat cone and then broken off, so that its surface shows a rough elevation at its centre. As with the other implements, we may suppose numerous uses. The curved pin might be used, for instance, for knotting or repairing fishing or bird-trapping nets of the kind the Iceman carried with him. It would also be suitable for splicing ropes and cords. Very similar tools were used in the past for skinning large animals, such as stags, cattle and pigs. In short, a multipurpose tool.

The two sinews in the quiver were also bundled with bast, most probably from a lime tree. Judging by their size and appearance they are probably the long Achilles tendons of a cow or a stag, freshly cut from the animal. Both were folded at their thick ends. The tapering thin ends were slightly twisted and turned over, so that the bundle measures 17.2 centimetres long. Unrolled, the sinews have a length of 23.8 centimetres each. If after drying the tendons are frayed into fibres and then twisted, a very durable thread is obtained. Such fine threads have been used, for instance, for sewing most of the Iceman's fur clothing. He also used sinew threads for the shafting of his dagger and arrowheads. Perhaps he intended to fashion his bowstring from the two pieces, unless the string also found in the quiver was intended for that purpose.

The string is wound into an oblong irregular tangle measuring 14.1 centimetres. It was twisted from two strands. According to the botanists it is made of tree bast, unlike all the other cords found at the Hauslabjoch, which consist of grass. The fact that the Iceman had carefully stowed it away in his quiver also indicates a special function. One end is knotted; there the diameter is 7 millimetres. Towards the other end the string is thinner, only 3.5 millimetres. Its unrolled length can be roughly estimated, at a probable 1.9 to 2.1 metres. Such a length would perfectly fit the 1.82-metre bow-stave. Nowadays it is difficult for us to judge whether animal sinew or bast makes a better bowstring; perhaps both materials are equally suitable. It should, however, be pointed out that the string with its relatively thick cross-section would scarcely, at least in its present state, fit into the 4-millimetre nocks at the end of the arrows. On the other hand, it could become thinner when stretched. As the Iceman was about to make himself a new bow, it would be quite natural for him to stow away his indispensable bowstring with particular care. Whatever the final interpretation of the tangle, its possible use as a bowstring should not be entirely ruled out.

The time has come to consider the strange discoveries made in the course of recovery, restoration and scientific examination of the Iceman's equipment and possessions.

The man carried an unfinished bow with him. It would be unreasonable to assume that this was going to be his first bow ever. We have to proceed from the assumption that he lost a bow he had previously owned, and therefore had to make a new one. So he found a suitable piece of wood from the trunk of a yew. While he was resting at his camp in the evenings he worked on it. Work had progressed quite a way when death caught up with him. He was an experienced maker of bows. He had stowed away in his quiver the string of his old bow – if indeed that is what it is – in order to use it again for his new bow.

The reinforcement strut of the quiver was broken, and our man carried the loose central piece about his person. He put the quiver down 5 metres away from the spot where he died, at the bottom of the gully. Missing from the quiver was the carrying strap and the cap, both of which he had lost earlier, before he reached the Hauslabjoch. No doubt he intended to complete and repair the quiver at a suitable opportunity.

What applies to the quiver also applies to its contents. It contained twelve blank shafts and only two finished arrows, which were already broken when the blanks were placed in the quiver. That, too, had happened before he sought refuge in the rocky gully. Before setting off into the mountains the man must have used up a considerable number of arrows. But while he succeeded in providing himself with wooden shafts for new arrows, he did not manage to acquire suitable, or at least adequate, material for making the arrowheads.

Taken together, this suggests that the Iceman had suffered a misfortune at some point, though probably not more than a few days or weeks before his death – a disaster involving considerable violence. It could have been a conflict with hostile humans or a fight with wild beasts. Even an accident not involving third parties would be a possibility, such as a fall from a rockface or into a crevasse. What seems very odd, however, is that he then took with him only some of his damaged property, while leaving behind other items, such as the carrying belt and the cap of his quiver. Evidently there was time only for a semi-orderly retreat. This, of course, would suggest that his presumed adversary did not force him into headlong flight, but nevertheless made him leave the scene in something of a hurry. These reflections narrow down his opponents to human beings.

Although he subsequently managed to acquire some replacement material – new wood for his bow and arrow-shafts – he seems nevertheless to have felt himself pursued. To venture into the mountains without full equipment was no less risky then than it is today. Any Neolithic mountaineer must have realized that, just as should every modern mountaineer. Even so, despite the most modern equipment, on average two hundred mountaineers suffer fatal accidents in the Alps every year.

Be that as it may: the Iceman, in view of his earlier misfortune, must have been in a state of physical and mental stress as he set out on his precipitate ascent to the main ridge of the Alps. Besides, the time he chose – or perhaps was forced to choose – for his climb was by no means ideal. Summer was approaching its end. A sudden early onset of winter can be fatal in the high mountains – as in fact proved the case.

Finally, his food supplies were not exactly optimal. All we can prove is a single measly sloe and probably a chunk of meat, perhaps pemmican. His second birch-bark vessel suggests a possible reserve of liquid.

There must have been some reason for him to risk the ascent to the main ridge with these massive handicaps. His clothing at least, from the grass cloak through the fur items to the grass-upholstered shoes, reveals the experienced mountaineer. There can be little doubt that the disaster he had suffered was one of the reasons why the Iceman ultimately lost his struggle against an all too often hostile nature.

III
The Iceman's Clothing

1 The Iceman's Clothes compared with other Prehistoric Textiles and Clothing

The clothing of the Iceman is unique in that the glacier ice has preserved it for us, at least in fragments. For the first time we have an insight into an almost complete set of Neolithic clothes. The most ancient complete sets of clothing hitherto found in Europe date back only as far as the Bronze Age. They were recovered from oak-coffin burials under grave-mounds mainly in Denmark, their conservation due to the tannic acid naturally found in tree coffins. These graves with their impressive pieces of Nordic Bronze Age clothing all belong to the second half of the second millennium BC. Until now they have provided our most important evidence of early European textile and costume history. In nearly all cases the textiles are woollen. Remains of leather or fur clothing are an exception. Clothes made of linen, whose existence may be presupposed in the Bronze Age, have been found only in traces. This vulnerable material is thought to have totally dissolved despite the action of tannic acid. In northern Europe a very distinctive costume was worn at that time, with a highly developed weaving technique testifying to a considerable tradition.

Prior to the discovery of the Iceman we had no direct evidence concerning Neolithic clothes, as only minute amounts of original material have survived. Without exception, these come from the settlement levels of the circum-Alpine pile-dwellings. The small scraps are invariably of linen, as animal materials such as leather, hair or wool did not survive in the damp environment of the lakeside settlements. The find situation is thus the exact opposite of that in the north, where sheep's wool textiles have come down to us, whereas plant textiles like linen have not. Occasionally, mixed wool and linen textiles are encountered, where, for instance, the woof is linen thread and the warp is wool. In the north, therefore, only the woollen warp would survive, while in the lake settlements of the Alpine region only the linen woof would survive. In the case of the Iceman, the superb conserving conditions of the glacier ice have made it possible for both types of material to survive.

Although the scant scraps of textile from the wetland settlements of Switzerland, southern Germany, Austria, northern Italy and Slovenia provide no indication of the style of the clothes, they nevertheless testify to the high level of the Neolithic weavers' skill in the settlement areas all round the Alpine arc, and confirm that woven materials were then customarily used to make clothes. To what extent articles of clothing made of leather, fur and wool were also worn was almost entirely unknown (the few exceptions will be touched upon below). The discovery of a complete set of clothing at the Hauslabjoch represents an invaluable contribution to the history of European dress.

Some qualifications, however, have to be made in view of the surprises presented by the find. First of all, it is astonishing that the Iceman's clothing contains no woven materials. This is certainly not due to any failure in preservation, since both vegetable and animal material have survived perfectly, a fact confirmed most impressively by the gossamer-thin threads, only 0.15 millimetres thick, which surround the fletching of the arrow-shafts and which were probably spun from woolly animal hair. Mention must also be made of the threads twisted from animal sinew, which were used as sewing and shafting material. All the other threads and cords consist of vegetable material, mainly of grasses and to a limited extent of bast.

In view of the knowledge available to us, it is tempting to say that the Iceman probably was not wearing the usual clothes of his time, but some special equipment intended for prolonged absences in the wilderness, all the way up to the high mountains. Again, it cannot be overemphasized that, in view of preservation conditions at the usual Neolithic sites in the southern Central European region and beyond, we know virtually nothing about the amount or the type of leather and fur clothing. We are here focusing on the fate of one male individual, and we must not make easy generalizations or push our interpretation in any preconceived direction. We must be very careful in applying our knowledge of living conditions at the time. The Iceman's clothing consists of a cap, his upper garment, a pair of leggings, a loincloth, a pair of shoes and a cloak. The body belt with its pouch is not part of the clothing proper and has already been discussed.

2 The Cap

The cap was found in the course of the second archaeological examination of the site on Wednesday, 19 August 1992. It lay, firmly encased in ice, at the bottom of the gully, right on its western side and at the base of the large boulder on which the corpse rested. As the man collapsed and came to lie on the stone, his headgear probably slipped off his head and was covered by falling snow, as he was himself. After its discovery and the surveying of its position it was carefully freed from the ice by means of a steam blower, with the meltwater being continually siphoned off. These operations were recorded in drawings, photographs and on film. After recovery the item, still lightly frozen, was packed in a sterile plastic bag. All the bags with finds were then placed in a watertight light metal box which had been kept buried in the snow. In this way it was possible to keep the temperature in the container more or less constant around 0° Centigrade. The box was transported as fast as possible to the South Tyrol Ancient Monuments Office in Bolzano, Italy, where appropriate cooling equipment had already been arranged.

Following the successful conclusion of negotiations between Bolzano and Innsbruck, the later finds are also to go to Mainz for restoration.

In the state in which it was found the cap was laterally squashed flat. Its height is roughly 25 centimetres, its shape that of a blunted cone. Like the rest of the fur items it is sewn together from individually cut pieces of fur. The outward facing hair remained attached, thanks to the careful recovery, but came off at the slightest touch. At the lower edge two leather straps are fitted, whose loose ends were knotted together. Presumably this was a chin-strap.

3 The Upper Garment

In view of its fragmentary state of preservation only a few deductions can be made about the cut of the upper garment. Its shape was probably similar to that of a cloak or a cape. The cap was certainly a separate item and was not attached to the upper garment as a hood.

The photographs taken of the body at the Hauslabjoch, and also those taken at the forensic institute before the body was washed, show a largish piece of fur, which, on the left, reaches from the shoulder along the neck to below the chin. If one assumes – and there is no reason not to – that the garment was in its original position with regard to the body, then we have a modest clue. It seems that the upper garment became narrower from the shoulders to the neck.

As with the body itself, so – with the exception of the cap – the hair of the fur garments has come out in places and is now present only as separate clumps. That is why the clothing has the appearance of leather. On the flesh side there are several traces of scraping, dating back to the cleaning of the hides.

The initial analyses conducted by Joachim Langer of the West German Tannery School in Reutlingen produced some signs of vegetable matter used during tanning of leather and furs. Naturally, vegetable tanning substances first spring to mind. These would have been extracted from, for instance, tree bark. In this form they were allowed to act directly on the raw hides, which produces durable furs and leather. In Roman times, and also in the Middle Ages and the modern period, it was customary to use vegetable substances for tanning alongside other agents. Excavations from those periods in a damp environment have not infrequently yielded leather articles, such as shoes, cases or quivers.

The extreme rarity of prehistoric finds of leather and fur suggests that effective tanning was not yet widespread. One has to assume that, in the Stone Age, the Bronze Age and the Iron Age, a simpler form of tanning was practised – primarily tanning by smoke. This is a relatively simple procedure. The raw hides are suspended in a way that will expose them to the action of cold smoke. Then the leather or fur is greased, dried and mechanically softened, for instance by chewing. Although smoke tanning produces serviceable materials, they do not have the durability of properly tanned leather.

Professor Groenman-van Waateringe has isolated the pollen which adhered to the fur clothing of the Iceman and made a surprising discovery. While some of the grains had clearly shrunk and been bleached, others seemed relatively fresh. This supported the view that the garment had been smoke-tanned; the pollen on the raw hides had evidently been changed by smoking. On the other hand, the pollen collected when the garment was worn was preserved in its normal state. As substances related to vegetable tanning agents, such as pyrogallic ester, are released in minute quantities during smoke tanning, this is the more likely explanation of the fluorescences found by the Tannery School.

The fur garments were sewn together from numerous small, mostly square, pieces. This is in line with the practice of modern furriers, who use this method to achieve a uniform nap, while at the same time ensuring a good fit. This method also makes it possible, by the use of differently coloured pieces of fur and by contrasting the direction of the nap, to achieve a patterned effect. The individual pieces are very neatly joined throughout by even oversewing. Here, the sewing material was predominantly fine threads twisted from animal sinew fibres. The garments had been worn for a long time: they show considerable signs of use, both on the flesh side and on the grain side. On the grain side a few bald patches have formed where the brittle hair has rubbed off. On the flesh side there are some clotted patches of the kind caused by human skin, fat and sweat. There are also a few mended seams, some of them awkwardly cobbled together with grass thread. Elsewhere there are some expert repairs, using fine threads from bast fibres twisted together with hairs. These observations lead one to assume that the Iceman was most probably not the manufacturer of the fur clothing, but that the simple repairs done with grass thread probably were his handiwork. This suggests a prolonged sojourn away from his native settlement.

Initial examination suggests that the skins used for the fur garments came predominantly from stag-type animals, presumably red or roe deer. Teats on two of the pieces of fur show that the skins of female animals were also processed. As these teats are very small, the skins presumably did not come from pregnant or nursing hinds.

It is not clear whether the fur garments were worn with the flesh side or with the grain outermost. Possibly the Iceman changed the sides according to the weather as the Eskimos do to this day: in winter they wear the grain side inside, in summer outside. Modern industry similarly offers reversible coats.

Further useful findings relating to the cut of the upper garment continue to come out of the restoration work. As the overall length amounts to at least 94 centimetres it must have reached from the Iceman's shoulders to his knees. There are no signs of sleeves having been set in, which suggests a cape- or cloak-type cut.

From a Danish bog, the Møgelmose, we have a poncho which, by pollen analysis, is also dated to the Neolithic. This consists of five large rectangular pieces of fur, the three middle ones sewn together so as to leave an opening for the head. The other two hang down over the shoulders, covering the arm slits at the sides. Judging by its size, this garment, if worn by an adult, would have reached roughly to mid-thigh. It has no opening at the front, so it had to be slipped over the head like a real poncho. When compared with the Iceman's upper garment, the important point is that the Danish find also has no sleeves.

However, the Iceman's garment seems to have a somewhat different cut from the Danish poncho. The front part is relatively well preserved because it was the most protected, as the body rested on it. There the hairs of the fur remain in place over considerable areas. The side-pieces are also quite respectable. Worst preserved is the back portion, which is badly tattered and entirely hairless. In spite of these provisos it is clear that the garment was open all the way down the front, from top to bottom. But the cut, far from being straight, is in fact very irregular. Along both edges there are loop-shaped or tongue-shaped projections. There are a few roughly circular incisions. The teats are towards the lower end, roughly at groin height, to the right and left of the opening. This very variable cut of the front opening of the garment is certain to be connected with its closing mechanism – but it is impossible to ascertain exactly what that was. Similarly, it is not clear whether, and if so how, his arms emerged through the sides of the garment, whether slits were provided for that purpose and whether, as in the Neolithic Møgelmose poncho, these were covered by fur pieces coming down from the shoulders. Finally, fitted sleeves cannot be entirely ruled out, but, if they existed, nothing is left of them. The illustration of the reconstructed garment therefore shows only one of several possible cuts.

The upper garment is cut off at the lower edge without any sign of reinforcing seam or hem – the same goes for the front opening. It is shorter towards the front and more or less rounded off towards the edges of the opening.

It is interesting that a patterned effect was contrived when the garment was made. The basic principle was that long strips of fur, from 8 to 16 centimetres wide, were sewn together. In the shoulder region these tended towards horizontal alignment: here dark-brown to black fur was used. In front, at the sides, and possibly also at the back the strips were aligned vertically, with dark and light strips alternating. The resulting striped pattern was not, however, very strictly observed – which may be connected with the numerous repair patches. It is evident that the alternate-colour strips are narrower in front than at the sides and back.

Immediately after it was made this garment must have been splendid to look at. The heavy demands of the Iceman's sojourn in difficult terrain caused a certain amount of wear and tear, though its owner was careful to keep it in good condition by regular repairs.

4 The Leggings

Eyewitness accounts from those who saw the corpse before its official recovery agree that the dead man's legs were covered by clothing. At the first recovery attempt the body was freed down to the hip and buttocks region. When Koler and Markus Pirpamer were working among the ice lumps in the meltwater, they did not notice any articles of clothing but Messner's party, who reached the Hauslabjoch the following day, saw more. The glacier ice had melted a little further, and in the morning the meltwater had been channelled away. Although the man's legs were still under water, they were no longer totally encased in ice. All the visitors to the site describe a leathery covering on each leg, roughly from the crotch down. A photograph confirms these observations, even though the position of the legs is rather blurred. Some of the visitors even claim to have made out seams. On the following Sunday, when the body had been almost completely freed from the ice, Alois Pirpamer only noticed a covering on the legs from the knees down. During Henn's recovery operation on Monday, 23 September 1991, even these remains disintegrated, so that the corpse was lifted from its icy grave wearing nothing except its right shoe.

From the scraps of clothing, handed in to us by various routes and described in the first part of this book, it proved possible to reconstruct two tubular shapes made of fairly large pieces of light brown fur, only occasionally cut irregularly. The Iceman was wearing not a pair of trousers but, rather like the North American Indians, two separate leggings.

The better preserved right legging is a long trapezoid-shaped piece of fur made up of several longitudinally arranged sections sewn together, widening towards the top. The upper end is in the shape of a flat arc. When worn, the legging tapered slightly downwards and would be like a long stocking without its foot. The less well preserved left legging matches its companion, except that its top is cut not in an arc but in a flat bell-shaped curve.

The length of each legging is about 65 centimetres. They therefore covered the thigh and calf, but without clinging closely; the fit was loose, allowing enough freedom of movement to bend the knee. The upper edge is reinforced by a narrow strip of fur, set in with a tacking stitch. At the apex of the curve, a band of fur barely 3 centimetres wide is sewn on vertically; further along this strap is divided lengthwise into two. The gathers on the body belt, to the right and left of the small pouch, show the spot where the legging was attached to the belt by this strap. Sewn on to the lower end of the legging is a fur tongue, which was pushed into the high neck of the shoe, roughly to the instep. When the shoe was laced up this made a firm connection, preventing the leggings from riding up, for instance during wading through deep snow.

How the leggings were worn. Despite their rather fragmentary condition, this reconstruction is fairly reliable. The leggings, tapering downwards, end in sewn-on oblong tongues which were pushed into the shoes. Sewn on at the top are double supporting straps with which the leggings were knotted to the belt. The hide was worn fur side out.

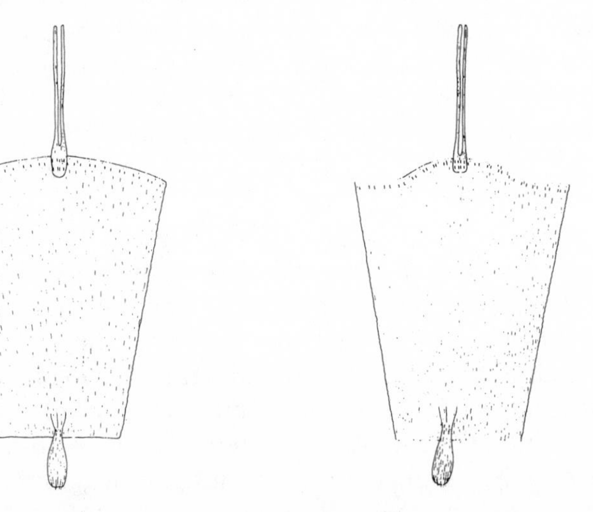

Reconstruction of the cutting pattern of the leggings. One has an arc-shaped upper end, while the other is bell-shaped. Both upper ends are reinforced by a simple tacking seam. Sewn to the centre top are the 'suspenders' consisting of slit straps. This is thought to be the most ancient leg clothing in the world. It proved so successful that the North American Indians, for example, preserved such leggings unchanged down to our own century.

5 The Loincloth

Of the Iceman's loincloth only the front flap survives, but it is in fairly good condition and more or less intact. As for the rest, we only have small scraps which cannot be exactly placed.

Nothing can be said about its position at the site of the find, as the fragments are not identifiable on the photographs taken before or during the recovery operation. The garment, therefore, can only be interpreted on its own internal evidence.

The loincloth differs from the upper garment and from the leggings in that it is made of leather, not fur. As its colour is very dark, the pieces belonging to it were easily separated from the other finds. The leather used is rather thin and soft. Careful examination showed that it contains no grain; only the flesh side is present. It is possible, therefore, that leather scraped thin, or split, was specially chosen for the loincloth. The species from which the leather came cannot be determined, as identification depends on the follicle pattern of the hair on the grain side. The surviving part is 50 centimetres long. Its width at the bottom is 33 centimetres, but towards the top it narrows to 23 centimetres. It is made up of long narrow strips of leather, joined together by oversewing using thread twisted from animal sinew fibres, the same as the thread used in the other items of clothing. The lower end is slightly rounded. The edges were left untidied.

The assumption that this item represents the front flap of a loincloth is based on several considerations. First, because of the position in which the body was found, the pieces of clothing worn on its front have all been preserved better than those on its back. Moreover, the scraps of leather which, by character and colour, belong to it are all folded in a longitudinal direction, as if they had been drawn through a narrow gap. This inevitably suggests the crotch. That the well-preserved part would hang down freely seems fairly certain, as it exhibits no other arrangements. It could, in fact, only have been slipped through from below between stomach and belt. In that case it would have reached down from the upper edge of the body belt to just above the knees. There are a lot of parallels in the field of ethnography for such an item of clothing. A loincloth passed through the crotch like a scarf protects the genitals, then it is gathered and brought upwards between the buttocks and slipped through the belt behind, as in front, and the

How the loincloth was worn over the belt and the leggings. Only the front bib of the loincloth is well preserved, the rest consists of fragments. Slipped under the belt, it hung freely to the knees. The loincloth consists of leather scraped thin, or split, and is almost black. The edges were not neatened. Its overall length was probably approximately 1.8 metres. In shape it resembles a long scarf tapering towards the middle.

excess length then falls free over it. This overhanging flap in front also provided additional protection for the vulnerable belt-pouch and, most importantly, kept the tinder dry.

6 The Shoes

The Iceman wore shoes on both feet. They were first observed by Messner and his party and reminded him of the footwear of the Lapps. The right shoe is better preserved, as it was still on the dead man's foot when the body was delivered to the institute. It was removed on Wednesday, 25 September 1991, by Markus Egg and Roswitha Goedecker-Ciolek for restoration. The much more fragmentary left shoe was easily identified among the objects in Klocker's coffin. Without doubt this, too, was dug out of the ice by Wiegele with his ice-pick on Wednesday, after the removal of the body, even though it cannot be identified in the records made during Henn's recovery operation. In the original position of the corpse, the right foot was resting above the left, and thus the intense solar radiation during those days helped to free it from the

How the cap, the upper garment and the shoes were worn. The fur cap, in the shape of a blunt cone, had a chin-strap. Of the nine items of clothing (excluding the belt but including the shoes) the upper garment is the worst preserved: only one third of it survives. We do not know whether it had sewn-on sleeves or, if so, what shape they were. The design suggested here, with lateral slits for the arms and rectangular pieces of fur hanging down over shoulders, is based on a poncho – also from the Neolithic period – found in a Danish bog. Alternating brown and black pieces of fur created a vertical pattern of stripes. The upper garment, originally no doubt rather splendid, shows numerous patchwork repairs.

ice. During the recovery attempt on the preceding Sunday the right foot, complete with shoe, was relatively easily freed. The left foot, trapped more deeply in the ice, could only be freed, according to Alois Pirpamer, after laborious hacking with the ice-pick. The two shoes are symmetrically similar.

Each shoe consists of an oval piece of leather, the edges turned up and bound with strong leather straps. Microscopic examination has shown that the material used was leather and not fur. The soles were presumably made of cowhide. Attached to the straps was a net knotted from grass cords; this covered the instep and the heel. This device was intended to hold in place the grass stuffed into the shoes for warmth. The cord-net also covered the loop hanging down from the leggings. Attached to the sole leather were the uppers, presumably of fur, which then continued up the leg roughly in the form of a boot. This was tied around the ankle with grass cords.

The oldest shoe found in western Europe before the discovery at the Hauslabjoch was unearthed as early as 1874. It comes from the Buiner bog in the Dutch province of Drente. On the basis of pollen analysis this shoe – it is impossible to say whether it is a right or a left one – has been dated back to the end of the Neolithic period. In absolute terms, this would be about 2500 BC. The pair of shoes from the Hauslabjoch is thus more than 500 years older. The shoe from the Dutch bog lacks a heel seam – a primitive characteristic – but this particular part has not been preserved in our shoe. The Dutch shoe, like all more recent ones, is a *Bundschuh*, a kind of peasant shoe whose sole and upper are made from a single, oval piece of leather. (Later a seam in the form of an inverted T was placed at the heel, ensuring a better fit.) This was tied around the foot with a leather strap, parts of which are preserved, which passed through slits about 2 centimetres long placed some 3 millimetres in from the edge. Even though nothing else of that shoe – such as an inner lining – remains, it is clearly constructed on a different principle from that of the Iceman's footwear, which consist of separate pieces of material sewn together.

7 The Grass Cloak

Over his fur clothes, and certainly also over the equipment he carried on him, the Iceman wore a cloak plaited from grass. Part of this was first discovered on Saturday, 21 September, when Messner lifted the head of the corpse. Beneath the body, which, in the region of its neck and chest, had frozen to the rock on which it was then still resting, those present noticed something plaited. This is described by Messner as 'cords or something plaited from string', by Kammerlander as 'remains of string, possibly some kind of pouch or provender sack which the dead man had hung round his neck', and by Fritz as a 'mat of plaited material, perhaps a basket'.

Fragments of this matted structure were hacked free by Wiegele with his ice-pick after the removal of the body during Henn's recovery operation on the following Monday, and dropped on the pile of collected items. Some of the pieces can be recognized in Gehwolf's three photographs. Together with the other finds of that day they came to Innsbruck in the body bag. A further, relatively large, piece of the cloak was discovered by Lippert in the course of the first examination of the Hauslabjoch site at the beginning of October 1991. Evidently missed during the earlier

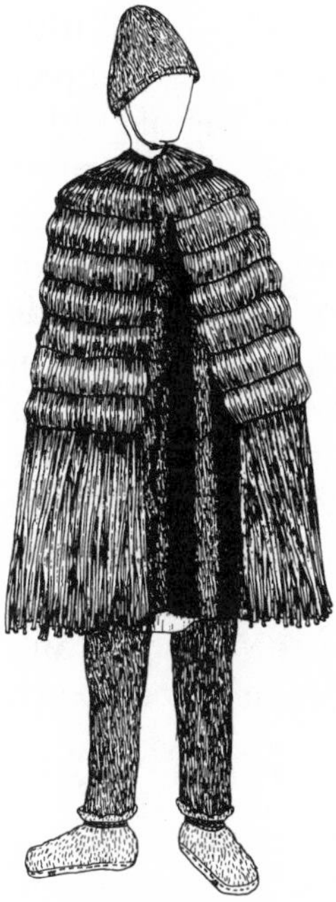

How the grass cloak was worn. The cloak was plaited from grasses over 1 metre in length and it is open in front. As it has only been preserved in a fragmentary state it is not clear whether it had lateral slits for the arms. The reconstruction proposed here represents the most likely cut – one that is supported by other finds. There is no doubt about the seven or eight horizontal cross-plaits in the upper part or the long fringe at the bottom.

salvage operations, a part of it had frozen to the southwestern, oblique to vertical, edge of the rock. It has been conclusively established that this was the same fragment as the one seen by Messner's party earlier on, when the body was lifted. With the help of this observation it was possible to reconstruct and pinpoint the position of the body with a good deal of accuracy. The records of this item show that the upper, plaited, edge of the cloak was lying round the dead man's shoulders and neck when he died, and from there it hung down more or less loosely. He had therefore put on his cloak before his death.

The restorers in Mainz succeeded in fitting together major sections of the cloak. Some parts, however, seem to be missing, especially from the back. Perhaps these were blown away by the wind when the corpse began to emerge from the ice and are never to be seen again.

The remains show fairly definitively that the original length would have been at least 90 centimetres. Laid over neck and shoulders it would have reached down, as the upper garment did, to the knee. It was made from very long grasses. The upper edge of the cloak, the one laid around the neck, is plaited in a simple twine arrangement. Knotted into it at regular intervals of 6 to 7 centimetres are several grass cords whose function is unclear. In front the cloak is open for the whole of its length. The two long edges are formed by grass cords, to which, threaded through horizontally, twine strings are knotted. From top to bottom the coat is articulated by twine woven in horizontally: it is only this that makes the garment a plaited article, as the vertically arranged grasses, which form the cloak proper, are not interwoven or twisted together. The horizontal twines are at intervals of 6 to 7 centimetres. Excluding the plaiting at the top edge, there are seven or eight individual horizontal twines, so that the plaited area of the cloak has a length of some 50 centimetres or a little more. Below that, the vertical grasses continue as a fringe for another 40 centimetres.

For a man who spends prolonged periods in the wilderness the construction of this cloak has many advantages. To start with, it is easily put on and taken off. While resting it can be used as a ground sheet or as a blanket at night. The loose plaiting, held together only by seven or eight horizontally inserted twines, makes the cloak very flexible and allows it to fit closely around the body. The open fringe at the bottom ensures that leg movement is not restricted. Whether the material was impregnated, for instance with fat, is still unclear. But even without such treatment grass and straw mats are highly water-repellent – one need only think of thatched roofs. This property is enhanced by the vertical arrangement of the grasses. The horizontal cross-strings are inserted at the largest possible intervals, to prevent water from piling up against them in heavy rain. Pastoral tribes have taken advantage of the heat- and moisture-insulating qualities of such straw or grass cloaks right down to the present century.

A travel account from 1729–30 describes such a scene in the Turin region of northern Italy:

> People of lowly standing, walking on foot, wore a coat of straw or thin reeds down to their calves, and firmly tied up around the neck, but nowhere else joined together, but hanging down free in its entire length, which caused the water that was falling upon it presently to drip off again.

One might almost think this learned traveller of the Rococo period had encountered the Iceman in his grass cloak. For a hunter, moreover, such a cloak would provide excellent camouflage.

IV
The Body

1 Preservation of Human Bodies by Artificial or Natural Mummification

Nature has arranged matters so that the biomass of every living creature is, after death, reintegrated into the natural cycle by way of a very variable food chain. Only man upset this sensible rule when he began to reflect on life after death. This resulted in highly differentiated burial patterns according to peoples and cultures. At one end of the scale is the total elimination of the dead body, whose extreme form is cremation. A half-way position is represented by the Towers of Silence in Persia (also used by Parsees in India), on whose uppermost platforms the dead are offered as food to the vultures. There, greater importance is attached to the survival of the soul, while its mortal frame is considered unclean.

At the other end of the scale, numerous religions believe in a corporeal survival of the dead in the world beyond. In some instances this leads to the attempt to preserve this corporeality after death. The best known example is the burial customs of the ancient Egyptians. But in many other parts of the world, too, an attempt is made to preserve the dead body by artificial means. Frequently, either deliberately or unwittingly, use is made of natural circumstances which impair or prevent the biological decay of the organs. If a dead body is thereby wholly or partially prevented from decomposing, we call it a mummy. There are also mummification processes which take place without intentional human intervention, for instance through extreme dryness, in ice, in salt, or in bogs. There is no doubt that the endeavour to let the dead be mummified caught on particularly among peoples where the natural conditions for this are present. What applies to the human body also applies to animal cadavers, which occasionally become the object of deliberate or accidental mummification.

In principle we can differentiate between three categories of mummies: (a) bodies in which the preservation of soft tissue is due solely to artificial measures, such as embalming; (b) bodies whose preservation is due solely to natural circumstances, such as bog or ice corpses; (c) bodies in which both artificial and natural effects favour the preservation of soft tissue, such as the mummies from the kurgans at Pazyryk in the Altai Mountains, and a considerable number of Egyptian mummies. As a rule, human and animal mummies are the result of burial. The few mummies resulting from sacrificial activity or judicial killing have to be viewed as rather unusual grave finds.

The practice of giving a dead person a dignified funeral is so firmly rooted in the customs of most societies, however diverse their backgrounds, as to be hardly without exception. To let an executed man decay on the gallows would nowadays be condemned as barbarically medieval. Even the body of a person who has been missing and presumed dead for a long time is normally recovered, often at great personal risk, in order to be buried according to custom. We might recall the salvage operations after the lost Franklin expedition around the middle of the nineteenth century, an operation which, for its part, claimed more than thirty victims.

A person who meets his death far from his family is, since he cannot be buried, subject in a much greater degree to the natural process of decay. After a few years, at most, nothing is left of him. Scavengers, micro-organisms and autolysis do their work. In such cases mummification is the exception. At the same time, prehistoric or historical bodies found away from burial grounds can supply far more information about the living conditions of past cultures than can those which have been properly buried. In all cultures burial is subject to a firmly established ritual, which has largely diverged from real life. For the archaeologist it is often difficult to judge to what extent grave furnishings and grave goods reflect the lifestyle of the people buried there.

If, on the other hand, the victim of an accident – someone who has not been given a burial – is recovered, with the remains of his clothes and equipment preserved, then we know what clothes and ornaments he wore in everyday life, and what implements he used. The shroud in a coffin rarely reflects the dead man's normal attire. Grave goods were chosen by his descendants according to cult rules and were often specially manufactured for the burial ritual. Needless to say, the prerequisite for a comprehensive evaluation of a discovered mummy is that it should be made available to science as speedily and as little changed as possible.

If we consider what conditions are necessary to provide optimal information from a discovered mummy, we have to concede that the likelihood of such a scenario is virtually nil. A person meets with accidental death. He has his clothes on and his equipment with him; his body cannot be recovered by his relatives; it is not attacked either by scavengers or by autolysis; together with his belongings it is subjected to conditions of mummification; it is rediscovered hundreds or thousands of years after his death; the find is handed over to scientists without delay.

If only a single link in that chain is missing, the discovery loses its value for watertight assessment. An example of such a less-than-optimal scenario is given by the bodies found in the salt mines. All the right conditions were present, except that discovery occurred too soon, before archaeological research existed. So far there has been just one instance of a prehistoric find which meets all the above conditions – that of the Hauslabjoch man.

2 Other Human Mummies

In addition to mummies from graves there is an exceedingly small number of human corpses which were mummified without any interment ritual. We have mentioned the men from the salt mines of Upper Austria; they were lost to science.

Comparable to them are two human mummies found in a South American copper mine. One was subsequently exhibited at fairgrounds as a 'copper man'. He was discovered in the Restauradora Mine near Chuquincamata in northern Chile when some prehistoric ore workings were put into operation again, and is now housed in the Smithsonian Institute in Washington DC. In this case preservation was due to the effect of bacteriocidal copper salts at extremely low air humidity. However, only the skin, which had become as stiff as a board, and the skeleton are left. All other soft-tissue organs have gone. The body shows no deformations. As an X-ray film taken in 1932 shows an undamaged skeleton, it is assumed that this miner was not killed by a rock fall but asphyxiated. His discovery provided an impressive insight into the living and working conditions of an Indian miner during the period AD 400–600. He wore his hair in plaits. His occiput was artificially flattened. Except for a loincloth of woollen rags he was totally naked. Around his ankles he had wound woollen threads tied up with fur. With him were found baskets for the removal of the ore while he kept his personal belongings in a carefully worked leather sack braced with wooden rods. For tools he had a heavy grooved mallet and a stone hammer. The miners who extracted the copper for the Iceman's axe must have used much the same equipment.

At this point we ought to mention the bodies found in bogs in northern Germany, southern Scandinavia, Holland and England, less often elsewhere in Europe, reports of which have been on record for a few centuries. Their number probably far exceeds one thousand. Yet only a few of these finds, hardly more than two dozen, have been scientifically investigated. Following the almost total suspension of turf-cutting since the 1950s, finds of human bodies in moorlands have become exceedingly rare. The preservation of soft tissues is due to their anaerobic position in a moist environment and to its effect on the reduction of bone substance – so that frequently only a kind of skin bag has been preserved with remains of the collapsed internal organs.

A large proportion of bog corpses can be dated from the centuries around the birth of Christ. According to ancient tradition, at least some of these finds are due to judicial actions. Some especially ignominious crimes, such as adultery or homosexuality, were punished among the Germanic tribes by drowning in the bogs. While some observations testify to previous execution by throttling or throat-cutting, a certain number of bog corpses may also be the result of irregular burials, for instance of suicides, or special burials. In all these cases grave goods or pieces of clothing are an exception. Often the bodies are naked or wrapped only in a blanket.

Some bog corpses, on the other hand, suggest an accident. A person who lost his way, or a fugitive, might sink in the treacherous bog grass and, without help, be unable to save himself. Such accidents happened all the time, from the Stone Age to the Second World War.

The body found in the Møgelmose bog in Denmark was discovered in the course of turf-cutting in 1857. It is thought to be that of a child aged between twelve and fourteen. According to the person who found it, the body was discovered standing upright in the bog. That is why it is assumed that the child, probably a girl, sank in the bog wearing her everyday clothes, here a fur poncho. There are no indications of foul play or deliberate drowning. Dated by pollen analysis to the Neolithic period, this is one of the most ancient bog corpses.

Discoveries of mummies suggesting accidents include, last but not least, the grave-wax corpses from the glaciers. With a few exceptions, such as the mercenary from the Theodul glacier or the shepherdess from the Porchabella glacier, these are all twentieth-century mountaineers or soldiers killed in action in the two world wars. Once their identity and cause of death have been established by forensic examination, and once a death certificate has been issued, they are, as a rule, handed over to their families for burial. Not infrequently glacier corpses are deformed and dismembered by the sheering action of the ice masses flowing downhill. As a result, the mortal remains of some victims are recovered piecemeal over many years. Thus on 3 September 1953 the remains of a human body, largely reduced to a skeleton, were discovered east of the St Pöltener Hütte at the base of a glacier called Prägratkees in east Tyrol, at an altitude of 2,700 metres above sea level. The soft tissue remaining was partially mummified and partially transformed into grave wax. Some items of equipment, such as a right boot, scraps of material, a wooden pipe and a pair of horn-rimmed spectacles made identification possible. He was a 37-year-old Englishman who had been missing since 3 September 1936. On 24 September 1990, thirty-seven years after the first find, further human bones, remains of clothing and parts of a left mountain boot were recovered in the same place. Forensic examination, jointly with police criminal investigations, proved that they belonged to the same individual. Glaciological observations at the site revealed that the body of the mountaineer, who had fallen into a crevasse, froze into the ice and was then dismembered by the forces within the ice. In view of the very low flow velocity of the Prägratkees glacier – about a metre each year – it is not thought that the glacier transported the parts of the body very far. The accident occurred at the beginning of a prolonged period of general glacier recession and it was in the course of this melting process that the first find was made in 1953. The rest remained concealed under a renewed advance by the glacier. As a result of further glacier recession from 1980 onwards the second find was eventually made in 1990 at much the same spot.

However, a glacier body is not always crushed by the sheering forces of the ice. Although the Iceman's excellent state of preservation must be regarded as a unique piece of good luck, there have been similar cases. In August 1923 a rope party consisting of two men and a woman were crossing the Madatsch glacier in the Pitztal in Tyrol. The woman crashed into a crevasse, the rope snapped, and her body could not be recovered. Twenty-nine years later, on 14 August 1952, her totally unblemished body, by then fully released from the ice, was found in the moraine debris at the base of the glacier. According to the findings of the local police, 'the corpse undoubtedly covered a distance from the site of the accident of presumably several hundred metres, inside the glacier'. Despite the brittleness of grave-wax corpses, the dead woman was brought virtually undamaged over difficult ground down to the mortuary chapel of Plangerosa.

The forensic report gives an impressive description of the appearance of a grave-wax body. Grave-wax formation is always linked to a humid environment with low oxygen content. The phenomenon, therefore, is found in bodies recovered both from ice and from water. Adipocere or grave-wax formation is very rarely found in bodies buried in the ground, and then only under very specific conditions. It is a chemical change in the body fat, starting with the subcutaneous fatty tissue, progressing deeper in the course of time, and sometimes attacking the body fat of the internal organs as well. After hydrolysis of the neutral fats the unsaturated fatty acids, such as oleic acid, are changed, through a shortening of the chains, into palmitic and stearic acid. Some of the fatty acids are also turned into calcium compounds. The remaining substance, however, is by no means a wax. Nor, strictly speaking, is it still a fat. The mass which, while maintaining the body's form, after desiccation surrounds the skeleton like armour can best be described as dead-body lipid; however, the terms 'grave wax', 'grave fat' and 'fat wax' have gained general currency over the more scientific 'adipocere' or 'lipocere'. This so-called grave wax seems to be relatively stable, but is occasionally, evidently by the sheering forces in the glacier, separated from the bones, so that eventually only the bone substance is left. This was probably the case with the 400-year-old mercenary from the Theodul glacier and his mules. Originally these were certainly grave-wax corpses. By the time they emerged from the ice, however, only the remains of bones were found.

In the case of the female corpse from the Madatsch glacier, on the other hand, the scalp, cheeks, lip area and chin formed a homogeneous mass of grave wax. The same formation was exhibited by the neck, the hemispherical breasts, the abdominal wall, the genitals and the extremities. Even the hands, down to the fingertips, had transformed into grave wax. On the back, too, the dorsal part of the skeleton was uniformly enclosed by a grave-wax armour. Autopsy revealed a greatly shrunken brain, which, as a light, relatively hard and dry mass, clearly showed the gyri.

When the thorax was opened one could see the very thin collapsed pulmonary lobes, discoloured blackish-grey. The pleural cavity, outside the foliate lung, was empty. Heart and liver had been converted into grave wax. According to the forensic report, it emitted an odour strikingly similar to that of dried milk powder. The intestinal loops and the mesentery were partially desiccated to look like parchment and partially transformed into grave wax.

Whereas in grave-wax corpses recovered from water the layers of the skin, the epidermis and dermis, usually rot away and the surface of the subcutaneous fatty tissue reveals a grape-like relief, the *Panniculus adiposus*, the skin of the dead woman from the Madatsch glacier was much better preserved.

Glacier corpses, however, do not invariably exhibit such a total transformation of the organs into grave wax. Occasionally dry mummification occurs. In the case of the male corpse from the Sulztalferner, mentioned in the Prologue, the soft portions of his head, torso and upper extremities, as well as his internal organs, had transformed themselves into grave wax. This transformation, however, ended at the forearms, which, along with the hands, seem dry-mummified. In the course of this process the soft tissue shrank, the skin became tighter and brown like leather. The prerequisite of mummification by dehydration is a position in a dry but airy atmosphere. According to forensic experts, this process can take from a few weeks to several months, whereas extensive grave-wax formation may take many years.

It is clear, therefore, that the mummification of the Iceman must have taken place under dry and well-ventilated conditions. Besides, the body must have been covered by a loose layer of snow from the start; otherwise it would have been attacked by fly grubs and scavengers.

In Central Europe corpses lying in the open are first subject to decay accompanied, especially in summer, by the feeding of insect larvae on the soft parts and by skin discoloration by green and black products in the blood pigment. Eventually the process is interrupted by desiccation, which is why numerous mummies also show remains of fly or beetle larvae, mites and chrysalis cocoons. Moreover, the body is attacked by carrion-eating animals: foxes, rats and mice gnaw at it. In the lowlands and in the mountains vultures, eagles and ravens, as well as smaller birds, pounce on the dead body.

The description of different mummification processes for humans supplies us with explanations for the preservation of the corpse from the Hauslabjoch glacier ice. Equally, the examples show that, even on a worldwide scale, this is a totally unique find. While parallels may be found for a number of individual aspects, the accumulation of such a host of amazing coincidences, due solely to natural causes, makes the find in the Ötztal Alps the archaeological event of the century.

3 The Conservation of the Iceman

Given the Iceman's prolonged stay in the glacier ice, the task facing us was to turn a natural conservation process into an artificial one in a single unbroken manoeuvre. Innsbruck University's Anatomical Institute has two refrigeration chambers in the basement; these were temporarily put at the disposal of the Iceman project. They are large enough to contain the mummy resting in a horizontal position on a trolley. The second chamber, though empty, duplicates the first, so that an immediate switch can be made in the event of a technical breakdown. Because the two existing chambers need to be available for the institute's normal activities, two new chambers were built in the summer of 1992, equipped with more powerful plant.

Each chamber is independently fitted with an efficient refrigeration plant and set to run at a constant temperature of minus 6° Centigrade. This is the annual mean temperature of the glacier at the Hauslabjoch, where it fluctuates between 0° Centigrade in summer and about minus 10° Centigrade in winter. Humidity in the chamber is kept at 75 per cent by means of ice that has to be brought in. The mummy is wrapped in a sterile operation sheet, on which a layer of crushed ice is spread. On top of that a stable plastic sheet is laid, creating something like a humidifying chamber with a uniform humidity of 96 to 98 per cent. As a heat buffer more packs of ice are placed on the plastic sheet, and another layer of plastic foil wrapped around. Although humidity within the glacier ice is 100 per cent, the few percentage points of the shortfall, which would be extremely difficult to remedy with the technical facilities available, can safely be ignored. The alternative would be freezing the body once more into a sterile block of ice. But then it would not be available for scientific research. Nor would it be possible to perform the necessary regular checks. The ice needed for conservation is produced, in sterile conditions and from distilled water, in the institute.

Each of the refrigeration chambers is equipped with six temperature sensors and two humidity sensors. One temperature sensor in each chamber is connected to an electronic device which triggers an optical and acoustic alarm in the event of a deviation. Two members of the institute are continually linked to the alarm device by pager, so that a possible alarm is passed on to them by telephone and radio. The electricity supply to the refrigeration chambers is backed up by batteries as a protection against power failure.

Two other electronically controlled sensors also monitor temperature and humidity. One of them lies directly against the body, the other is fitted to the chamber's inside wall. These transmit their readings to a recorder which stores the data every second and prints them out every hour round the clock.

When research work is in progress the body is moved to a sterile 'micro-flow box', into which sterile air is blown. This box resembles a large aquarium whose front pane can be raised on hinges. To increase the air humidity, crushed ice is introduced here as well, but this only achieves a concentration of about 60 per cent. For this reason and because of the rise in temperature any research stint must not exceed thirty minutes, including the unwrapping and rewrapping of the body. While they work the scientists wear sterile surgical outfits. After each stint the body needs a recovery phase of not less than forty-eight hours, to allow the temperature and humidity to stabilize at their original levels.

The purpose of this costly conservation, the first of its kind in the world, is to maintain the Iceman in a state which will enable future generations of researchers to perform studies which are beyond our capacity today.

4 Determination of Sex, Age at Death and Height

In the evening of 24 September 1991 the forensic team handed the Iceman's body over to the anatomists. At the time there was no way of telling how it would react to the planned conservation procedures, or whether, in fact, it could be preserved in this way. It was obvious that comprehensive data ought to be obtained as speedily as possible, and using non-invasive methods. Radiology and computer tomography equipment is available in Innsbruck. Under the direction of Professor Dieter zur Nedden from the Department of Radiology and Computer Tomography at the university conventional X-rays were produced, with the corpse lying on its back, of the head, thorax, abdomen, thoracic and lumbar spine, pelvis with hip joints, as well as knee and ankle joints. For the computer tomography examinations the entire body down to the ankles was scanned in transaxial sections. The 'slice' thickness in the torso region amounted to 8 millimetres, while for the skull and the knee and ankle joints it was 4 millimetres. In addition, the bones of the skull, together with the inner ear, were photographed in super-high resolution CT mode with 1 millimetre 'slice' thickness. Once these procedures were complete, the body could be entrusted to the artificial atmosphere of the refrigeration chamber.

The first questions facing the archaeologists after the discovery of prehistoric human remains are the sex, the age at death and the height of the deceased. The anthropological examinations of the Iceman are directed by Professor Horst Seidler of the Institute for Human Biology at the University of Vienna, Professor Wolfram Bernhard of the Anthropological Institute at the University of Mainz, and Professor Torstein Sjøvold of the Osteological Research Laboratory at the University of Stockholm.

Attribution of sex is based on the formation of primary and secondary sexual characteristics. A child's sex at birth is established according to its external genitals and then entered on the birth certificate. In doubtful cases chromosome investigation provides the answer. For subfossil or fossil human remains determination ranges, according to their state of preservation, from easy to impossible. With bodies whose external sexual characteristics are identifiable the diagnosis of sex causes no problems. This was the case for the Iceman, but with certain reservations. Distribution of body hair, which can reveal typical male or female characteristics, such as beard growth or pubic hair outline, was of no help as, except for four pubic hairs, all the head and body hair had fallen out.

The first guess at the sex of the body was made by the discoverers themselves. From the delicate appearance of the body Erika Simon concluded that it was that of a woman. That delicacy, of course, was primarily the result of desiccation during mummification. Cautious as ever, the forensic expert Unterdorfer recorded a second sex diagnosis on Monday, 23 September 1991: 'The external genitals foliated, as far as can be judged most probably male, desiccated.' Detailed examination of the external genital region by Platzer, the anatomist, revealed slight injuries, but enabled preserved foliated structures to be unequivocally identified as a scrotum and penis. Thus the question of sex was clarified from the anthropological standpoint, more particularly as all other characteristics relating to a diagnosis invariably indicated male sex. These include the relatively small pubic bone angle, the profile of the forehead with strongly developed superciliary ridges, the receding forehead, the rather angular eye-socket outline and the marked mastoid processes at the base of the skull.

The archaeological determination of sex can only base itself on equipment, on articles and grave goods found with the body. Weapons are regarded as typically male attributes. In our case the position is obvious. Bow, arrows, quiver and axe belong to the male sphere. Hence the anthropological sex diagnosis agrees with the archaeological one. There cannot be the slightest doubt that the Hauslabjoch corpse is that of a man.

The exact age at death is more difficult to determine, especially with adults. The Iceman no longer revealed any juvenile features because all growth processes were complete. Hence the man's age was certainly over twenty. The relatively high degree of tooth wear from chewing, very clearly visible through the parted lips at least in the front teeth, would normally indicate an age of between thirty-five and forty. However, specific living conditions and diet can cause considerable variations to that average. The degree to which the cranial sutures have grown together sometimes enables a person's age to be estimated. In point of fact, cranial sutures can be made out in the computer tomogram. But it is not yet clear whether these pictures show open sutures, i.e. ones not yet grown together, or whether the sutures are already closed and merely reveal paler structures from the inside of the skull. If, however, the findings relating to the seemingly open sections of the coronal and sagittal sutures were to be verified, then the Iceman's presumed age would have to be revised downward, to roughly twenty-five to thirty years. For the moment, then, we are considering an age range between twenty-five and forty. When he was evaluating the X-ray pictures, Professor zur Nedden observed certain wear-and-tear phenomena on the spine and on the knee and ankle joints, which would suggest the upper age range.

A more precise determination of age is expected from an examination of the cancellous (spongy) structure, the spongiosa, in the joints of the long bones; as a person ages this changes its bracing direction and becomes thinner and looser. The most up-to-date methods now operate with a standard deviation of five to seven years from the definitive age. The method combines age-conditioned changes in the hard bone substance, determination of mineral content and measurements of the marrow-space cross-section. However, it involves at least partially invasive action, which will not be performed until some future phase of the research. Until then we shall have to content ourselves with the statement that the Iceman lost his life at an age between twenty-five and forty, probably between thirty-five and forty.

The first measurement of the Iceman's height was taken immediately after his delivery to the forensic institute in Innsbruck. Measured *over the left leg* this gave a figure of 153 centimetres. Further measurements were made by Bernhard and Sjøvold. As the body was relatively extended, the distance from the top of the head to the soles of the feet could be measured directly – 159 centimetres. A figure 1 centimetre less is obtained if the body's main sections – leg, trunk, head and neck – are added up.

But even this second measurement was not reliable. As a result of dry mummification, the cartilage in the Iceman's joints and intervertebral discs has shrunk, so a certain, albeit slight, reduction of his height is almost inevitable. On the other hand, there is no reason for the bones to have shortened, despite 5,000 years in the glacier.

There already exist well-tried methods for establishing height based on the length of the long bones. The data obtained are multiplied by a certain factor, derived from measurements on living persons, in order to produce the original height. As the parameters were established on the basis of different modern populations, the results calculated by different tables diverge slightly. Because of the damage to the left thigh, it was possible to calculate almost exactly the length of the femur. Sjøvold's result was 41.4 centimetres, Bernhard's was 41.5 centimetres. Using this measurement, we could estimate the full height of the body according to the various formulas:

Formula devised by	Bernhard (41.5 cm)	Sjøvold (41.4 cm)
Olivier *et al.* 1978	158.7 cm	—
Pearson 1899	159.9 cm	—
Manouvrier 1893	160.5 cm	—
Trotter & Gleser 1952–8	160.9 cm	159.9 and 161.6 cm
Breitinger 1937	163.1 cm	162.4 cm
Lorcke *et al.* 1953	—	159.0 cm
Sjøvold 1990	—	158.8 cm

These ten results give a mean value of 160.5 centimetres. As we expected, this figure is greater than that directly measured on the corpse. At the time of his death, the height of the Iceman was 160 centimetres with a margin of error of 2 centimetres either way.

5 Head and Body Hair

As the Iceman seemed to be totally bald when he was discovered, rumours instantly sprang up that he had been shorn, he had shaved his head, his hair had fallen out due to disease, or he had suffered from genetic hairlessness. None of this is true. One of the most sensitive parts of the human body is the epidermis. After death, it starts to decay relatively quickly, even if dry mummification subsequently takes place. That is why even dry mummies, preserved by natural means, as a rule have no hair left. Of course, this is true of both human and animal cadavers. A good example is the dry-mummified cat, mentioned earlier, which was released from the ice of the Stubai glacier in 1992. It had lost all its hair, except for the very deeply rooted whiskers and eyebrows. Our forensic experts, who were first to examine the Iceman, were so familiar with this phenomenon that hair specialists were brought in for the subsequent examinations. The first question was whether the numerous tufts of hair handed in, which came in the main from the clothing and other pieces of equipment, also included human hair. These investigations are directed by Dr Manfred Wittig of the Federal Criminal Department in Wiesbaden and Dr Gabriele Wortmann of the German Wool Research Institute in Aachen. In addition to animal hair, sufficient amounts of human hair were, as expected, identified. Both head and body hair was present, and a few of the hairs are even thought to come from a beard.

One sample is actually a bunch composed of several hundred hairs. According to this, the hair was wavy and dark-brown to black in colour. The maximum length is 9 centimetres, but so far no complete hairs have been found. Most are broken; some have identifiable hair-roots and others have hair-ends. This leads us to conclude that the Iceman's hair was originally more than 9 centimetres long. The thickness of the hairs ranges from 50 to 90 micrometres.

Some of the hairs have a bulbous thickening in the region of the root, which corresponds to the resting or terminal phase of the individual hair growth cycle. Others exhibit a conical thickening, with a typical striation (narrow, parallel series of grooves) near the root. These are cadaver hairs with rootless hair-shafts, caused when growing hairs are suppressed in the skin. This happens, for instance, when the blood supply to the follicles fails owing to some trauma. This finding suggests that these 'cadaver hairs' are in fact hairs lost by the Iceman after his death. The normal process of slight decay attacking the epidermis after death is here confirmed and explains the hair loss.

Other hairs can be classified as body hair on the strength of their divergent form and recognizable unevenness of the hair-shaft surface, although for the time being it is not clear whether we are dealing with armpit, pubic or other strongly developed body hair.

A fairly substantial sample of human hair was recovered from the upper edge of the grass cloak, i.e. from the part of the garment which lay immediately under the corpse's chin, neck and upper body. This represents a dark-brown to black curly material and the individual hairs are clearly thicker than those on the head. At the widest point of the hair-shaft they are as much as 160 micrometres thick. This sample is provisionally regarded as beard hair.

Presently available methods do not make it possible to establish the Iceman's hairline and as this is a major factor in a person's appearance, we cannot make any detailed statements about the Iceman's face. All we know is that he had hair at least 9 centimetres long; also, in view of his age, we may assume that he was balding – either a receding hairline or a thinning pate.

It is highly likely that he had a beard. Of course, we have no indications regarding his hair style or shape of beard. The individual strands of head hair point to the fact that he wore his hair loose and not plaited into pigtails or into a knot.

An interesting point here is the result of a trace metal analysis of hair samples from the Iceman. The concentration of lead in them is very substantially less than that of a modern population. Though this finding is hardly surprising, the researches confirm, albeit at present only on hair samples and not verified by tissue samples, that today we live in a heavily polluted environment. The Iceman was still able to breathe clean air and eat uncontaminated food.

Like the hair, all the Iceman's finger- and toenails have fallen out. One nail was retrieved during the second examination of the site in the late summer of 1992.

6 Position at Death

As a result of desiccation during mummification the dead man's skin is now leathery and tough. Its colour varies from one part of the body to another, ranging from pale ivory through brownish beige to a deep, at times almost blackish brown. During the first few days, when it began to thaw, it transpired that the corpse had not hardened completely, as is usual with dry mummies. This was very evident when the upper body of the corpse was lifted up *in situ*, although the legs were still frozen in the ice. The corpse could be moved, at least to a limited extent. Likewise, when it was placed in the coffin in Vent, the projecting left arm was bent back; although the bone of the upper arm was probably broken then, the skin remained intact. Possibly this flexibility was due to a certain rehydration while the corpse lay in the meltwater. This would also explain the greater weight of 'between 20 and 30 kilograms', estimated after its delivery to the Forensic Medicine Institute. By way of contrast, under stabilized conditions, the Iceman now has a constant weight of 13.030 kilograms. In view of his almost total dehydration, his original weight can be calculated to have been about 50 kilograms, perhaps a little more, perhaps a little less.

The body's position when it was found is more or less in line with a natural dying posture. But there are a few anomalies. For one, there is the unnatural position of the left arm, which is extended to the right and appears to have been pressed flat against the neck directly under the chin. Likewise, the left outer ear is folded over, out of its natural position, obliquely towards the front. The fold runs in a strikingly straight line. On the face deformities of the soft tissues are very noticeable. Nose and upper lip are shifted upwards to the right. In this context it is worth mentioning a small area of skin next to the left eye, which shows fine granular depressions caused after death. This suggests that after the Iceman's death there must have been small dislocations of the body due to ice pressure and ice movement. And so we can attempt a reconstruction.

Evidently overtaken by a blizzard or sudden fog, or both, the Iceman was in a state of total exhaustion. In a gully in the rock, perhaps familiar to him from previous crossings of the pass, he sought what shelter he could from the bad weather. With his failing

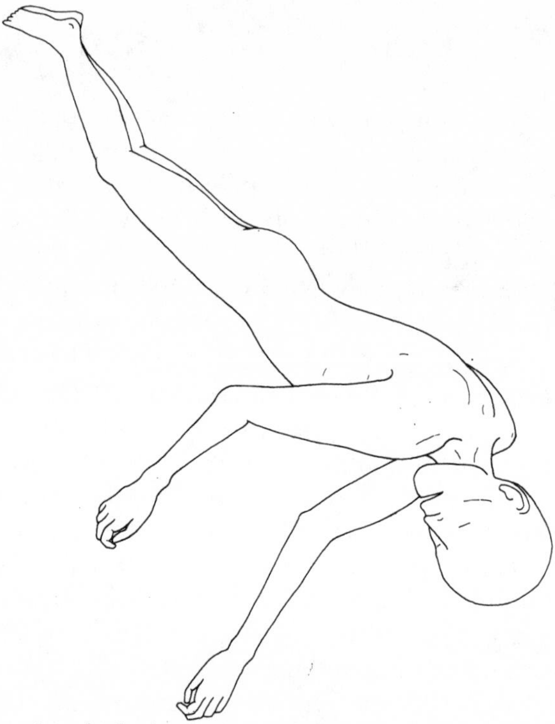

The position in which the Iceman died. He lay down on his left side. Given an unhealed serial rib fracture on his right side, this is the least painful position. The head was resting on a boulder. Both arms were extended forward, their muscles relaxed. The feet were lying one on top of the other. The left ear was folded forward. In this position the man probably froze to death within a few hours. There are no indications of distress. Death by freezing in a state of utter exhaustion is a gentle death, not perceived by the conscious mind. Originally the man was lying on the rocky ground fully clothed.

strength he settled down for the night. He deposited his axe, bow and backpack on the ledge of the rock. It is possible that he consumed here the last of his food store: a piece of tough dried ibex meat. Two bone splinters had inadvertently been left in the strip of meat as he cut it off: these he chewed off and spat out. Meanwhile it had grown dark. To press on might prove fatal. It was snowing ceaselessly, and in the gale the icy cold penetrated his clothes. A terrible fatigue engulfed his limbs. Between his will to survive and increasing indifference towards his physical danger he once more pulled himself together. He knew that to fall asleep meant death. He reeled forward a few more steps. He dropped his quiver. Before him he felt a towering rockface, barring the way. He turned back. Below him there was only loose scree. He tripped and fell heavily against a boulder. The container with the hot embers slipped from his hand; his cap fell off. Again pain pierced the right side of his chest. He only wanted a short rest, but his need for sleep was stronger than his willpower. For a very definite reason, which we shall discuss later, he turned on to his left side to dull the pain. He laid his head on the rock. His senses numbed, he no longer noticed the awkward position of his folded ear. His left arm, its muscles relaxed and probably slightly bent at the elbow, lay in front of him. His right arm was almost extended and was hanging down forward. His feet rested one on the other; the left shoe under the right. Soon his clothes froze to the rough ground. He was no longer aware that he was freezing to death. Overnight the body froze stiff.

Whether this fictitious reconstruction of the Iceman's death accords with what really happened, we shall never know. It is merely one way of visualizing the event.

The glaciologist Patzelt told us that at that altitude it is quite possible for 2 metres of snow to fall in a single night. Wind action produces even deeper snow cover through drifts and cornices. But there was no need for the snow blanket to have been that deep by the time the next day dawned. Certainly it protected the body; otherwise it would have been discovered immediately and attacked by vultures or raven-type birds. (At that time of year insect swarms can be ruled out.) Such snow remains air-permeable for several years. It was during this phase that combined freeze-drying resulted in the body's mummification.

Snow takes about ten to twenty years to turn into ice. As soon as the ice above that spot reached a certain thickness the glacier began to flow downhill. This releases enormous forces. According to the glaciologist's estimate the maximum ice cover at the Hauslabjoch would have reached 20 to 25 metres. However, as the site in question was in a protected gully, the main mass of the glacier flowed over the dip, while the basal ice in the gully remained virtually static. Nevertheless, modified ice pressure movements and a slight basal slipping of the ice mass may be expected even in the depth. The overlay pressure, which varies according to ice

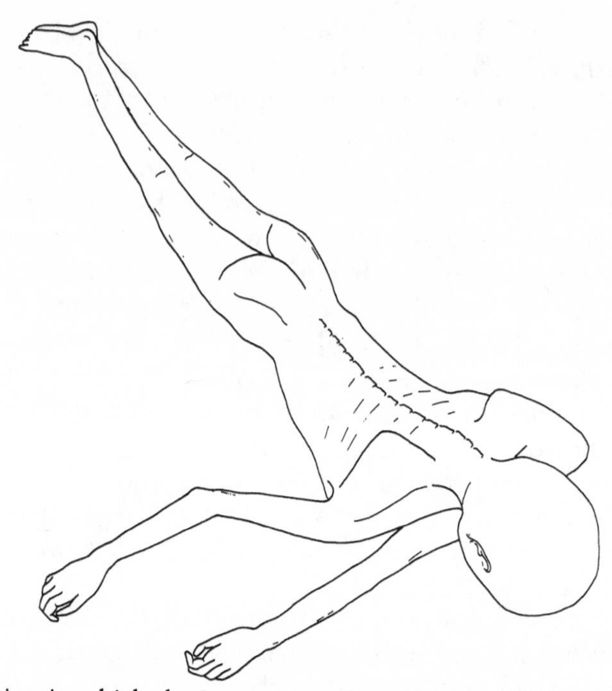

The position in which the Iceman was found in September 1991. Compared to the position in which he died (see previous page) minor changes have taken place. As a result of ice pressure and a slight sliding movement of the ice mass in the lower part of the gully the body was rotated by about 90 degrees, from a position on its side to one on its stomach. This caused the soft tissue displacements in the face, which was then lying directly on the rock. The nose and upper lips were displaced towards the right and upwards.

thickness, is partially diverted downward. This results in shifts, though these are within the decimetre range. At the bottom of the gully the ice movement is unlikely to have exceeded 30 to 50 centimetres.

It was to such slight moving forces that the Iceman's body was exposed. From its original position on its left side it was turned – a process which might have taken tens or even hundreds of years – into something like a position on its stomach, which was how it was discovered in September 1991. As the left arm was frozen on to the rock, the body was, as it were, pulled across it, producing the unnatural position of the left arm. During this axial rotation by a little less than 90 degrees the right arm hung down beside the boulder on which the dead man was resting – which is why the freeing of the right hand caused so much trouble. It was only finally released by Rainer Henn during the official recovery operation on Monday, 23 September 1991. Originally the left temple was lying on the rock, which is when the granular impressions over and beside the left eye were caused. The ice then moved the head in the rotary direction of the face, causing the soft tissues and the upper lip, as the most prominent areas, to be flattened and pulled upwards towards the right. The left ear, already mummified during the corpse's rotation in the ice, retained its forward-folded position with its sharp folding line. The corpse's feet remained in their original position, unaffected by the forces in the ice.

The direction of flow of the glacier's main bulk at the Hauslabjoch was at right angles to the line of the gully, down towards the southwest. Admittedly, the body on the ground was rotated in the opposite direction. This accords with the moving forces of the ice, which, in sliding over the gully, pressed diagonally forward against the rib of rock bounding the gully towards the southwest. Thus the ice, albeit minimally, was forced in a circular motion, pushing in the opposite direction at the bottom of the gully.

Right from the start everyone who saw the corpse pointed out that its left hand was closed in a way that suggested it had been gripping something. That was why rumours sprang up during the first few days that the dead man had held the axe or the bow in his hand. But this is not borne out by the facts. Both hands are in a natural death position; the subsequent freezing of the body preserved the position of the fingers at the moment of dying. This is not a case of the clawing produced by rigor mortis, a phenomenon often observed with other corpses.

7 The Tattoos

When Markus Pirpamer and Blaz Kulis first visited the site, immediately after the discovery of the body by the Simons, they noticed marks on the dead man's skin. Initially there was a suggestion of branding or even lash-weals, which is why the public prosecutor and the investigating magistrate were first involved. As the skin has darkened in many places, some of these marks are difficult to make out. The following details have been observed so far:

two parallel stripes around the left wrist;
four groups of lines to the left of the lumbar spine;
one group of lines to the right of the lumbar spine;
a cruciform mark on the inside of the right knee;
three groups of lines on the left calf;
a small cruciform mark to the left of the left Achilles tendon;
a group of lines on the back of the right foot;
a group of lines next to the right outer ankle;
a group of lines above the right inner ankle.

Medical examinations all agreed that the two stripes on the left wrist were caused by a rather tight cord – not, of course, tight enough to impair circulation. Even so, any external pressure causes a slight congestion of the blood, which can show up after death as a dark discoloration of the skin. Forensic scientists are familiar with this phenomenon. In cases of strangulation it provides the first clue. We can therefore conclude that the Iceman had a doubly wound band round his wrist. Unfortunately, when the left hand was freed from the ice, no attention was paid to whether the dead man was carrying anything tied to his wrist, and if so what. Nevertheless, two observations can be made.

To start with, he could have been carrying one or more objects attached to this cord round his wrist; among the objects found with the body only the two birch fungi, which were strung on long fur strips, would fit the bill. The precise spot where they were found is not certain. So while this may be a very probable and satisfactory solution, a different explanation should not be entirely ruled out.

The second possibility is that the doubly wound band represents a protective device of the kind occasionally used by archers. Archers from historical and modern periods frequently wear a coarse leather gauntlet which protects the lower arm and/or the thumb of the hand holding the bow against the backlash of the bowstring. If they are right-handed they will, of course, wear it on their left hand. From prehistoric times we know of stone discs to protect the arm and thumb; these were made to curve slightly to follow the shape of the limbs and could be tied to the forearm and thumb by means of holes punched into the narrow edge. The Bell Beaker folk at the end of the Neolithic period were fond of using such forearm and thumb protectors ground very carefully out of soft stone. In one exceptional instance such a plate was actually made of gold; no doubt its owner was a person of high standing, perhaps the chief of a tribe. Needless to say, the possibility exists of a leather arm protector which has not survived but which would likewise have been fastened with strings or straps to the forearm.

If the Iceman, whom we know to have been an archer, had worn such protection it may have been lost during the unsystematic recovery operations and blown away by the wind, or perhaps it can simply no longer be reconstructed from among the numerous scraps of leather and fur which have been handed in.

Of the two possible explanations for the stripes on the left wrist, carrying the birch fungi seems to me the more likely, particularly because the length and thickness of the two strips of fur on which the fungi are strung fit very well into that picture.

All the other marks on the Iceman's skin are evidently tattoos. Normally the skin is punctured, scored or cut with a very sharp or pointed instrument. A coloured paste is then rubbed into the wounds. The pigment most frequently used is powdered charcoal, stirred up with saliva or tepid freshwater, which produces the familiar blue tint. As the Iceman's tattoos have a decidedly blue colouring, charcoal was probably used to make them. The precise draughtsmanship of the tattoos on his back clearly shows that he cannot have made them himself. As for the other marks, on his legs and feet, it is possible that he pricked those out with his own hand. Indeed, the bone awl with its needle-sharp point, one of the tools he carried in his belt-pouch, would have been ideal for the task. However, we know from experience that self-tattooing is practised only in very exceptional situations, and we have no reason to suspect such circumstances for the Iceman. We may therefore assume that all the tattoos were done by someone else.

No system is discernible in the way the marks are distributed over the body. The groups of lines to the left of the lumbar spine are arranged vertically. From the top down, they start with a group of four, followed by two groups of three, followed in turn by a gap, and finally by another group of four. Any further continuation has not been preserved because of the damage done to the body in the region of the left hip and buttock. But it may be worth noting that a group of four is applied to the right of the lumbar spine at exactly the same height as the gap on the left-hand side.

Below the left knee, on the calf, a group of seven marks can be made out, followed further down, roughly 13 centimetres away, by a group of three, which in turn is followed, off-line towards the inside, by a single line. On the left foot there is a small cruciform mark tattooed beside the Achilles tendon a little way above the point of the outer ankle. Equally haphazardly distributed are the tattoos on the right leg, yet here they are confined to the knee and ankle. The cruciform mark on the inside of the knee is striking, deep blue at a point where the surface of the skin is a very pale beige. Grouped around the right ankle are three groups of three lines each, parallel to the axis of the body. One of these is placed slightly outwards to the back of the foot, the other laterally at the outer ankle, and the third above the inner ankle. It is not impossible that, on the very dark areas of the skin, there are further marks which have not been detected with the methods used so far.

Tattooing and other forms of skin decoration, for instance by painting or scar patterns, are found at all times and among nearly all the peoples on earth. I would like to make special mention of the elaborate pictures tattooed onto the skin of the nomadic ruler of Pazyryk in the Siberian Altai Mountains, dating from around 400 BC, and to the tattooed facial decoration of the Greenland Eskimos of Qilakitsoq from about AD 1500. Even in our highly civilized contemporary culture tattooing has by no means gone out of fashion. In the past it was chiefly sailors who had themselves tattooed to ensure that, in the event of shipwreck, their bodies could be identified and would be given a Christian burial. Prostitutes sometimes use tattoos to indicate their profession. Prisoners use thematic tattoos to express their protest at the society which puts them behind bars. As members of an élite male association, SS men identified themselves by having their blood group tattooed on the inside of the upper arm. In a corresponding measure concentration camp inmates, victims of the Nazis, had a number forcibly tattooed on their arms.

This brings us to the question of why the Iceman had himself tattooed. Application of marks to the human skin has a psychological or practical purpose, the two sometimes converging. Tattoos may answer the search for beauty, in cases of artistic representation or striking patterns. Occasionally ritually prescribed marks proclaim certain phases in life, such as, very often, sexual maturity. At times sexual signals are reinforced by tattoos or emotional states generally expressed, such as love, hate, sorrow or thirst for vengeance. Visible or concealed marks may represent a political or religious belief. Another variety of tattoos indicates rank and dignity within a society, property rights, executive power, descent or identity, membership of a group or, last but not least, personal achievements, such as the number of enemies slain, animals killed or women subdued. Finally tattooing can serve a therapeutic purpose, in much the same way as similar procedures, such as cauterization.

Professor Renate Rolle of the Institute for Pre- and Protohistory at the University of Hamburg has drawn attention to the use of skin marks in folk medicine. Particular attention should be focused here on hot branding, a practice found down to the present day among the nomads of the Eurasian steppes as well as in Tibetan medicine. As treatment for rheumatic and arthritic complaints the affected joints are burned; this is also supposed to achieve a hardening against the effects of winter. A variety of instruments are used, from hot pipe-bowls to cauterization tools such as punching and branding irons. To this day Tibetan medicine burns herbs into the skin to cure aching joints, sterility and back trouble. The tip of a glowing branding iron is filled with herbs and placed on the skin – a method also successfully practised by numerous primitive communities. Even in ancient medical literature, in the *Corpus Hippocraticum*, reference is made to the fact that the Scythians had brands all over their skin as a consequence of therapeutic cauterization.

In addition to his other magnificent tattoos, the man from Kurgan 2 of the Pazyryk necropolis has skin marks which are believed to represent the traces of such therapeutic treatment. A vertical string of eleven pinpoints to the left of the lumbar spine is matched by a similar line of only three points to the right; he also has an arc-shaped line of six pinpoints on the right ankle. Apart from the fact that the Pazyryk man has pinpoints while the Hauslabjoch man has lines, the correspondence between the marks of the nomad ruler and the Iceman is so striking that one feels reluctant to believe in coincidence, even though the two are separated by thousands of kilometres and years. Unfortunately, no

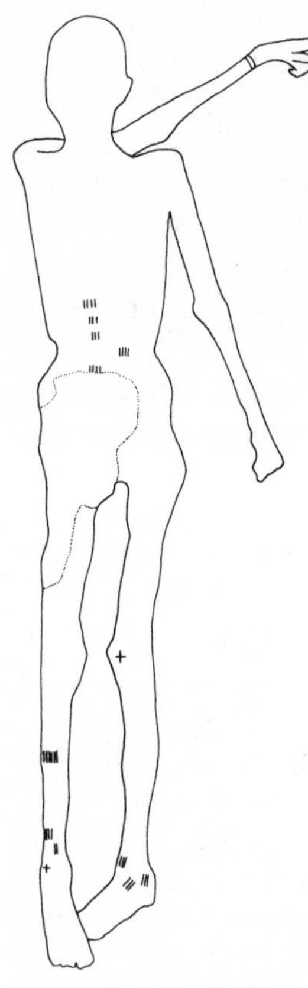

Overall pattern of the tattoos so far discovered on the body of the Iceman. The damage caused by the pneumatic chisel to the left buttock and thigh is outlined. The tattoos are concentrated at the lumbar spine, the right knee and the ankles, as well as the left calf. The two lines on the left wrist are pressure marks.

X-rays were taken at the time of the discovery of the Altai mummies and computer tomography did not yet exist. Hence external observations could not be related to the condition of the bones. But in the case of the Iceman, we can discover whether the bones adjacent to the tattoo marks show any unusual features, or deviations from the norms of a healthy joint.

First of all, the lumbar spine, the right knee and the two ankles were examined, as it is there that tattoos have been applied in proximity to joints. With considerable suspense we awaited the first X-ray pictures. What we had hardly dared hope for came true. Professor zur Nedden described his findings thus: 'In the lumbar spine we were able to establish discrete to medium degenerative changes (osteochondrosis and also slight spondylosis). Likewise there are medium-degree wear-and-tear phenomena at the knee joints and especially in both ankle joints.' It seems likely that there is a connection between the tattoos on the Iceman's skin and

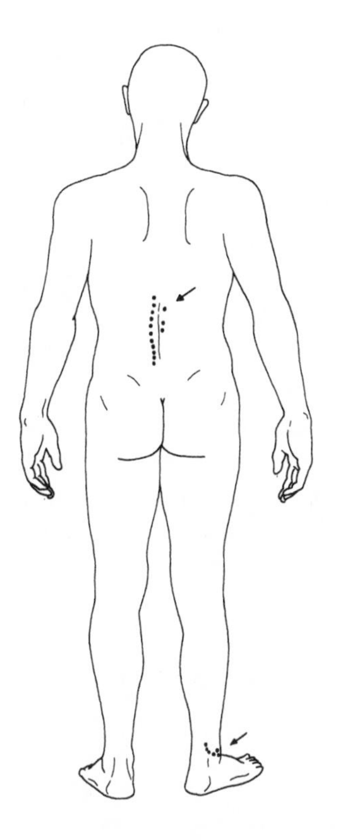

A sketch reconstruction of the preserved male body from Kurgan 2 at Pazyryk in the Altai Mountains in Siberia. As with the Hauslabjoch man, his dot-shaped tattoos at the lumbar spine and the right ankle (arrows) served a therapeutic purpose. For the sake of clarity the other tattoos on the skin of the Scythian nomad leader have been omitted here. They are representations of interlaced fabulous creatures of high artistic quality.

the age-conditioned or strain-induced degeneration found by the radiologist, and that the bluish marks had the purpose, at least partially, of relieving pain in the joints – a method practised by folk medicine to the present day. An explanation remains to be found for the bundle of lines on the left calf, which does not relate to any bone condition. The calf muscles, more than any others, are prone to ache under heavy strain, for instance in mountaineering – a condition popularly described as *Muskelkater* ('muscular hangover'). Perhaps the Iceman occasionally suffered from it and submitted to the medical treatment of the day.

8 The Genital Region

Every spectacular find instantly triggers a host of rumours. Half-truths and untruths are spread as established facts, and certain media and self-appointed experts eagerly seize on those bubbles even when they have long since burst. Things take on a life of their own and get out of control. Although I had made up my mind to confine myself to the presentation of scientific facts, and to discuss the assumptions, interpretations and hypotheses deriving from those facts, there are two press reports I feel obliged to deal with.

The reputable German weekly *Der Spiegel* (No. 10, 1993) took up the rumour that 'sperm was found in Ötzi's anus', a story said to have appeared in gay periodicals. This is a complete fabrication, particularly obvious since the anal region of the corpse, through to the bony pelvic area, was destroyed by the first recovery team and their pneumatic chisel. No examinations were conducted on the preserved soft tissue in the pelvis, nor were samples taken. The rectum has not even been identified in the computer tomograph of the body's internal organs, greatly shrunk as they were through dehydration. Much more important would be the analysis of any contents of the rectum, i.e. the faeces, which might provide interesting information on the Iceman's diet and digestion, the presence of parasites and possible intestinal diseases.

The most persistent rumour was that the Iceman had been castrated – that his penis had been amputated and his testicles removed in his lifetime. This story evidently also stems from an article in *Der Spiegel* (No. 5, 1992), which claimed that Platzer, the anatomist, revealed the existence of a further mystery. The scrotum, it seemed, was present, but not the testicles. 'It looks as if they had been scraped out', Platzer was reported as saying. The prehistoric man's penis, too, was said to have disappeared without trace. I countered the idea of castration by pointing out that the recovery workers' brutal methods had damaged the corpse long after the Iceman's death.

The castration theory, which started in *Der Spiegel* as a cautiously formulated hypothesis based on statements allegedly made by Platzer and myself, soon took on a life of its own. Surmises became facts. Matters reached such a pitch that a private individual in Munich accused me of having stolen the Iceman's penis. The most absurd ideas circulated, culminating in the claim that the corpse could not have originated in the Alps. Instead, the Iceman must have come from a culture in which castration rites were practised for religious reasons. And this could not have been in the Alps, but must have been in the Mediterranean region or the Middle East. There the priests had sacrificed their genitals in order to be allowed to officiate in the temples of the Great Mother.

What is true is that the schedule for examination of the body left the genital region until relatively late in the proceedings. For a variety of good scientific reasons other related work, such as computer tomography and photogrammetry, had to be undertaken first and the results awaited. The photogrammetrical investigations, in particular, were very time-consuming because a new procedure had to be developed to avoid fitting markers to the body. Otherwise there would have been a serious risk of dangerous contamination.

The first external, initially purely visual, examination of the corpse's genital region took place on 1 April 1993. Before that date, no authoritative statement could be made on that set of problems. The stories disseminated by the media and the self-appointed Iceman experts, which had begun to appear almost a year earlier, were at best frivolous or at worst irresponsible flights of fancy by science journalists.

The examination began at 2 p.m. Apart from myself, Professor Othmar Gaber and Dr Karl-Heinz Künzel were present, as well as auxiliary staff. As with all the other investigations, Professor Herbert Maurer filmed the entire procedure to ensure that records of all the Iceman research were kept complete. Important individual observations were recorded in close-up. Platzer dictated his findings onto cassette, and contributions by the other scientists were also tape-recorded. Before we started, we had all put on sterile, disposable surgical clothing – rubber gloves, caps and face-masks, as well as surgical gowns. No tissue samples of any kind were taken, so the results given below must be regarded as provisional pending later histological verification.

The condition of the sex organs had already been described as 'foliate, desiccated' by the forensic experts when the corpse was delivered to the dissection room. The organs had not undergone any changes since, but were stored frozen at minus 6° Centigrade. Nevertheless, despite extensive dehydration, the genitals could be moved to a certain extent and without doing any damage. This made examination considerably easier.

The major part of the foliate, initially compacted, corrugated structure proved to be the scrotum, to which the equally flattened penis had frozen. Once the outline of the penis had been more or less established, with rubber-gloved fingertips it was possible, millimetre by millimetre, to separate it slightly from the scrotum.

Because of the almost total loss of body fluid during mummification the scrotum now appears, seen from in front between the inguinal flexures, as a structure of irregular semicircular outline about 6 centimetres wide and about as long. Its thickness decreases from the top down. Where it emerges below the *symphysis pubis* of the pelvis it amounts to over 1 centimetre; towards its lower edge it thins down to barely 2 millimetres.

One must keep in mind that the measurements given here are only rough estimates. To our dismay we discovered that the sterile electronic measuring devices specially prepared for the job did not respond at temperatures below 0° Centigrade. In view of the short time available it was not possible to set up conventional sterilized mechanical measuring instruments, which would have given more precise data. This incident illustrates the problems with which a unique research project is, time and again, confronted and also explains why it takes so long to obtain scientifically verifiable results.

On the front side of the scrotum the wrinkled vertical middle line, the *raphe scroti*, was clearly visible; this is the seam at which, in the male, the two main sections of the genitals grow together in the embryo. In the female they do not close up but form the external lips of the vulva, the *labia majora*.

The two testicles could not be identified by probing. Only histological examination can provide more exact information here. One possibility is an extreme flattening, caused by loss of body fluid, thus making it impossible to feel them through the skin of the scrotum. Another possibility is that they have reduced through autolysis. Being endocrine glands, the testicles can be subject to disintegration by the body's own enzymes after death. One result of this first examination of the Iceman's genital region is that castration before or after his death can be ruled out. Only at the lower edge of the scrotum were slight postmortal lesions established; these had been observed from the outset and have resulted in a lamellar, barely perceptible, foliation of the basal scrotal skins. These slight injuries were certainly caused in the course of the various attempts at recovery. Nevertheless, it is astonishing how well such a delicate organ has survived over a span of more than 5,000 years.

In front of the scrotum the root of the penis emerges below the *symphysis pubis*, and like the scrotum it is flattened. At its root it is 3 to 4 centimetres wide and slightly narrows downwards, to a somewhat irregularly rounded end. Its length is about 5 centimetres. While the urethra could be clearly felt, from the root to the glans, the erectile tissue could not. A sickle-shaped fold a short way above the tip of the penis is probably the foreskin. No damage or injury, suffered either before or after death, was observed on the penis during that first preliminary examination.

There would normally have been no need to dwell at such length on the more intimate parts of the Iceman's body – but wild speculation must be refuted with scientifically corroborated facts. Thanks to an almost incredible concatenation of favourable circumstances a Neolithic man is available as an object of research to us, and, I hope, to future generations of scientists. Nothing of the kind could have been expected by the man himself, and it probably would not have been compatible with his religious sentiments. He did not, as the governor of Tyrol once put it, voluntarily 'give himself as a present to science, and indeed the whole of mankind'. Thus, we are all the more obliged to apply the ethical and moral standards of our society to his dead body. In a paper given at the first international symposium on the subject of 'The Man in the Ice', held in Innsbruck in 1992, Dr Hans Rotter, SJ, of the Institute for Moral Theology and Social Sciences, urged us, notwithstanding the scientific interest of the case, not to overstep the limits of piety and to preserve the human dignity of the Iceman even beyond his death.

To conclude this section we might consider an idea voiced by the Swedish anthropologist Sjøvold, concerning the link between the Neolithic man from the Ötztal glacier and ourselves, and whether we can regard him as a direct ancestor. Sjøvold quotes a well-known example from population genetics, one that elucidates the concept of kinship.

What degree of kinship exists between us and persons who lived centuries before us?

A simple consideration: everybody has two parents, four grandparents, eight great-grandparents, sixteen great-great-grandparents, and so on. He has 2^n ancestors n generations back, if there was no kinship between them. Assuming that on average a generation covers twenty-five years, then the span of a century is determined by four generations. Assuming that there is no kinship at all, the number of our ancestors a thousand years ago would be

2^{40}. That number is greater than the total number of people who have ever been born.

For the period 5,100 years ago we arrive, according to Sjøvold, at the astronomical, unimaginably large number of 15 followed by fifty zeros: 1,500,000,000,000,000,000,000,000,000,000, 000,000,000,000,000,000,000. (The number may even be an under-estimate, as Sjøvold used a generation lap of 30 years.) This suggests that the Hauslabjoch man must be distantly related to us all.

9 Internal Organs and Injuries

We have now discussed some of the important findings from the external examination of the Iceman's body; they are as reliable as the state of research at the time of writing allows. Since examination of the internal soft tissues has only just begun, as yet there is little of importance to impart.

The skin structure showed that the subcutaneous fatty tissue is very weakly developed, which gives an indication of the Iceman's state of nourishment at the time of death. This should not be taken to mean that he was undernourished, let alone anorexic. It would be more accurate to say that he carried not an ounce of surplus fat on his body. Calorie intake and calorie expenditure were balanced. His mode of life, which presumably involved more than just one temporary sojourn in the high mountains, prevented him from putting on fat.

Histological samples, on the other hand, all revealed a marked reduction of the body's fat system, which consists of stored fat, for instance in the subcutaneous fatty tissue or in the ventral region, and the structural fat which is indispensable for the proper functioning of all the organs of the human body. The reason for this reduction is not yet quite clear. Possibly, given that he had no stored fat reserves, the Iceman had been subjected before his death to an involuntary starvation diet. However, we cannot rule out degradation of the body fats after death.

We already have some information on the state of his brain, so far obtained exclusively from computer tomographical analysis. The tomography was conducted both in 4-millimetre layers and in spirals. The skull was differently placed for the two scans which

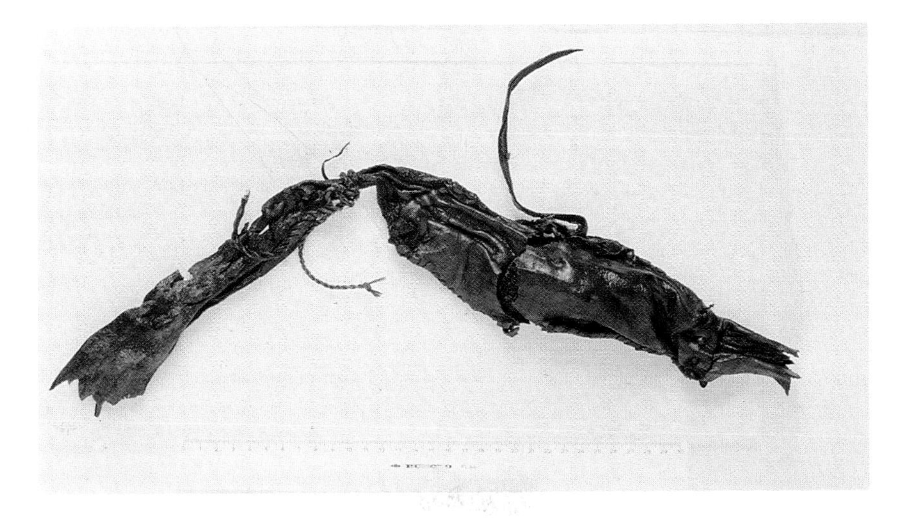

The belt pouch of calf leather. Both belt ends are torn off. The item had been used for a long time and shows several repairs. It opened at the short upper side. The gatherings on each side were caused by the leather straps of the leggings, which were attached there. The way in which it was worn can be seen on page 108.

Parts of a 'true tinder' fungus from the belt pouch. Adhering particles of pyrites indicate that the Iceman carried fire-lighting equipment with him. However, the spark-striking flint and the lumps of pyrites have been lost – whether in the Iceman's lifetime or during the various recovery attempts has not yet been established.

A sharp-edged scraper made of flint from the belt pouch, a multipurpose tool suitable for scraping, cutting, carving, planing and smoothing. Fine 'sickle-gloss' on one edge reveals that the Iceman also used it for harvesting grass and/or grain.

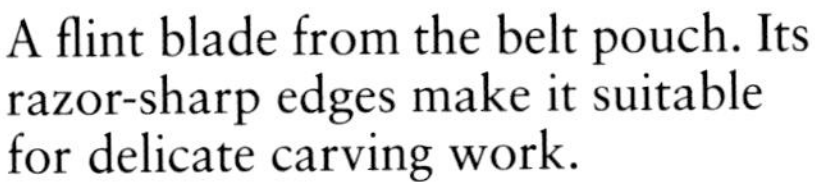

A flint blade from the belt pouch. Its razor-sharp edges make it suitable for delicate carving work.

A drill-like flint implement from the belt pouch. The small end was used for fine drilling, the broad end for planing and producing shavings.

An awl made from the metatarsal bone of a small ruminant. Both ends have secondary fractures. It is not clear whether they were caused during the Iceman's lifetime or during the recovery operation. The needle-point could be used for delicate sewing and also for tattooing. The flat end could be used for loosening tinder.

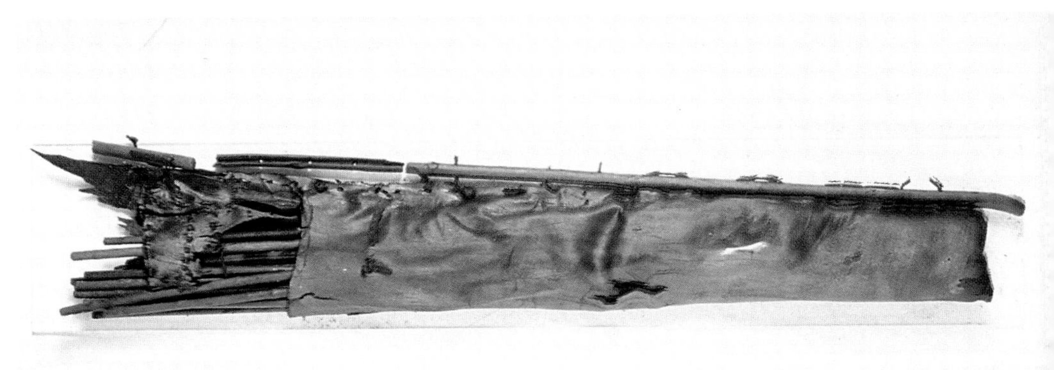

The quiver. It consists of a fur bag, stiffened by a hazel rod knotted to the side. As with the body itself, the hairs have almost completely fallen out, so that the material now looks like leather. Projecting at the top are the ends of the fourteen arrow-shafts. All damage to the quiver occurred during the Iceman's lifetime.

The upper part of the quiver with its semi-circular flap. The contents have already been taken out. The front closure cap and the carrying strap of the quiver had been torn off and lost before the Iceman's death.

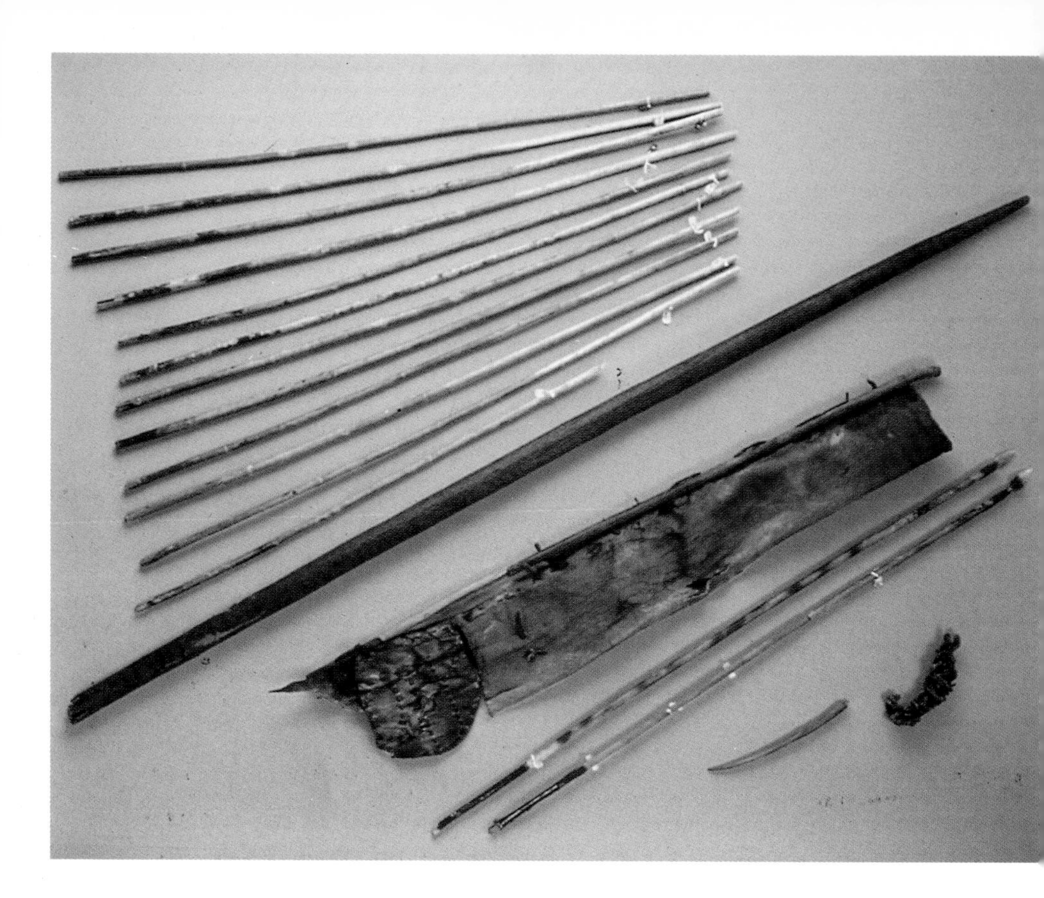

In the middle, the broken-off bow-stave, above it the arrow-shafts, one shortened; below it the two finished arrows, the stag-antler spike and the bundled bast cord, which presumably was intended as the bowstring for the unfinished bow.

The bundled bast cord from the quiver; one end is knotted. Its length is difficult to estimate, but could be about 1.9 to 2.1 metres.

Four fragments of stag antler, tied into a bundle with long strips of bast, were inside the quiver. Only rough-worked, they were presumably 'spare parts' for making arrowheads and other things.

A long, slightly bent spike, carved from the integument of a stag antler whose curve it follows. This multi-purpose tool was found in the quiver.

Crude sinew, tied into a bundle with strips of bast, found in the quiver. The two sinews come from the heels of an animal the size of a cow or a stag.

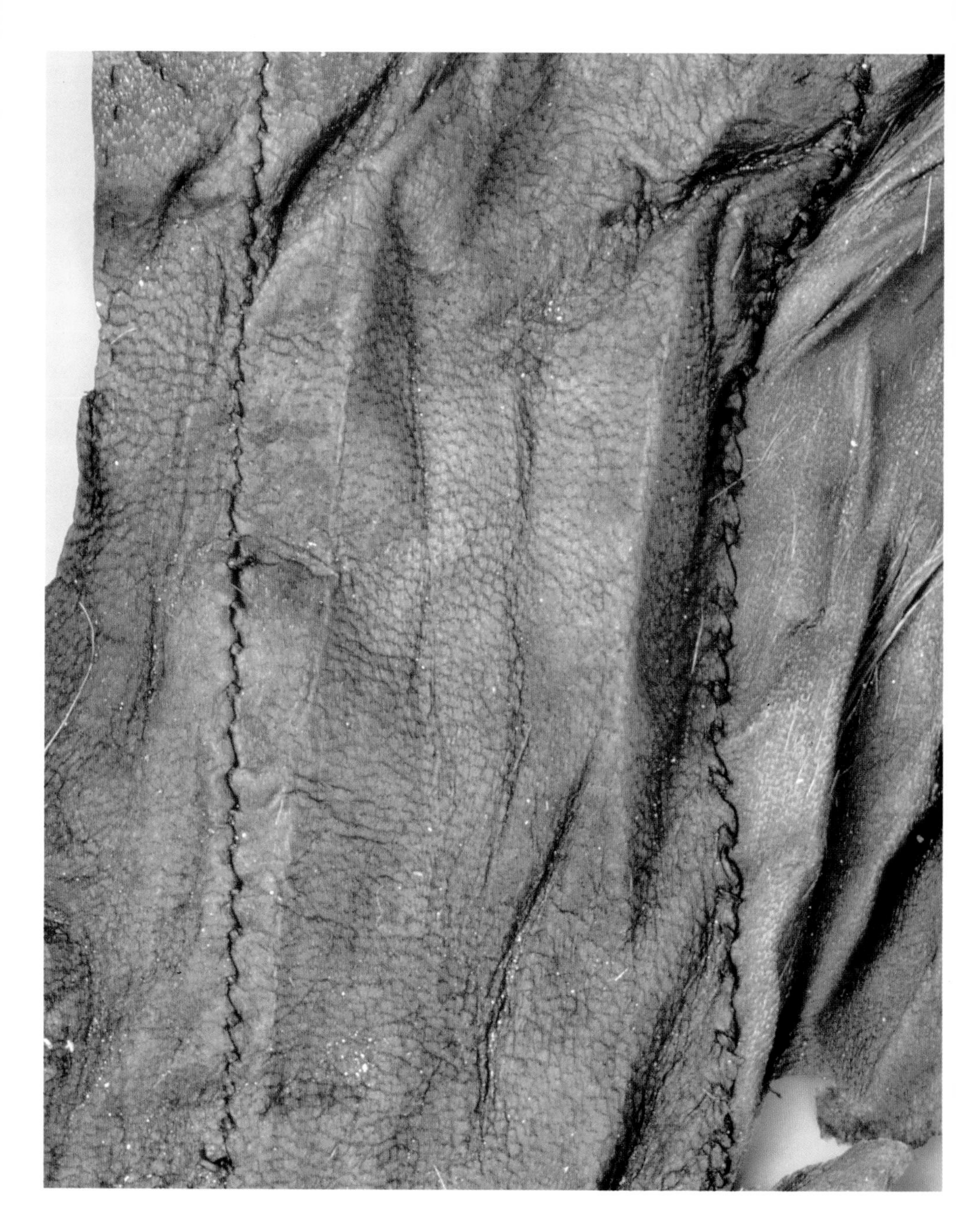

Detail of the Iceman's upper garment. The fur has fallen out, so that the garment now looks like leather. The picture shows the precision of the seams, which were sewn with frayed and twined animal sinews.

Fragment of the cloak plaited from grasses about one metre long and interwoven with horizontal strands of thread at intervals of 6–7 cm. The cloak was about 90 cm long and reached down to the Iceman's knees.

The Iceman's clothes show a few crudely made seams with threads of loosely twisted grasses. These are evidently temporary repairs made by the Iceman himself when he was away from his native settlement for any length of time.

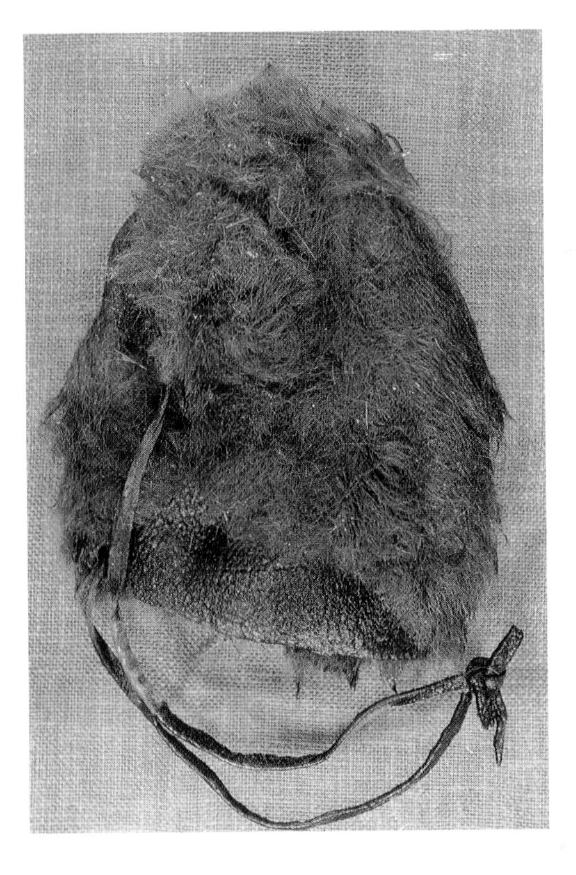

The Iceman's footwear. The better preserved right shoe in its original state and the inner netting of the less well preserved left shoe, which held the grass stuffing (used for heat insulation) in place. Only scraps of its leather uppers have survived. The shoes were fastened with grass cords wound round the ankles, and two leather straps knotted under the sole leather were designed to ensure a better grip.

The Iceman's cap. It was found when the site was re-examined in 1992, about 70 cm away from the Iceman's head, at the base of the boulder against which he had rested.

made it possible to establish that the brain was changing position in the cranial cavity according to the position of the head. These changes were particularly marked when the corpse was turned from lying on its back to lying on its stomach. The section revealed two lighter lines running through the brain mass at right angles to the direction of the frontal and occipital brain. They suggest that the brain, which had been frozen and was therefore rigid, was fractured in these regions by movements of the corpse before, during and after recovery, and perhaps especially during the journey to Innsbruck.

The high-resolution method even revealed slight hardening of the arteries in the area of the base of the brain. Such changes, while varying from one individual to another, normally only occur with advanced age. Even if the Iceman's age at death is taken to have been thirty-five to forty, the changes must have occurred relatively early in his life. A daring hypothesis would be that he had a metabolic susceptibility to early arteriosclerosis, possibly due to a high blood cholesterol level.

The description of the Iceman's equipment in the third part of this book, especially the bow-stave and the quiver with its contents, provided very specific clues that, shortly before his death, the man must have suffered some disaster, an event associated with the action of considerable force. I also mentioned that he had a pain in the right side of his chest, and therefore, as he was dying, lay down on his left side for relief. Let me explain this.

An X-ray taken in the prone position revealed that of the normally present twelve pairs of ribs only eleven were fully developed. No joint surfaces are to be seen on the twelfth pectoral vertebra. Statistically speaking, the absence of a twelfth rib on both sides is a rare anomaly. Physiologically it is of no importance and, remaining unnoticed, in no way curtailed the man's range of activity.

Far more significant are several rib fractures, because they testify to events in the man's life. The fifth, sixth, seventh, eighth and ninth ribs on the left show healed fractures, suggesting a one-time multiple trauma. All five ribs have healed well, even though some distortion remains. Serial rib fractures occur mainly in people at risk of falling: they are diagnosed chiefly in drunks, sportsmen and mountaineers. The Iceman, therefore, long before he died, had an accident which crushed the left side of his thorax. Fractured ribs generally heal well and without further treatment, provided the arm is kept more or less immobilized. After about three months the healing process is complete.

A totally different picture, however, is presented by further rib fractures on the right. Here the X-ray shows that the third, fourth, fifth and sixth ribs are broken and are somewhat out of position. In this case there is no callus formation, no trace of the bones having healed, a fact which limits the time-frame in which the break happened to no more than two months before the Iceman's death. Fracture after death is also a possibility because, even after the most detailed evaluation of the X-ray, it cannot be entirely ruled out that it was caused by ice pressure or during recovery. The most likely explanation, however, is an accident shortly before the Iceman's death.

If this injury is indeed viewed as the result of a violent conflict, this would support a scenario which explains the Iceman's inadequate equipment. His physical impairment due to the pain must have been a serious handicap. Nevertheless, he set off for an inhospitable region, and without sufficient food. All this can be regarded as an indication of 'flight with his last remaining strength'.

The unfinished weapons he had with him, the bow and the arrow-shafts, also fit into that picture. The time needed to obtain the raw materials and for starting on the trimming can be estimated at a few days, at the most one to two weeks. The disaster could well have happened before then. To return to his dying position on his left side, this would be the most comfortable position for an unhealed serial rib fracture at the right side of the thorax, far more comfortable than lying on his back, stomach or right hip. It was the position, in short, which caused the least pain.

Mention should also be made of another interesting observation on the Iceman's skeleton. The X-ray of the two upper arms shows differing degrees of calcification: the left humerus appears normal, but the right is slightly decalcified. This shows up on the X-ray because the left bone has absorbed more radiation and therefore is a lighter shade. The radiologist identifies this as the result of inactivity, which characteristically appears after about two to three weeks. It seems, therefore, that the Iceman had 'gone easy' on his right arm for that amount of time, no doubt to keep the pain in his right thorax bearable. Besides, some time in the past he had suffered the same injury to the other side of his chest and so knew that immobilizing the arm promotes a rapid and complication-free healing of rib fractures.

Considering the evidence as a whole, the 'disaster' probably happened from two or three weeks to two months before his death. It is not surprising that, thus handicapped, he did not survive the cruel weather at the Hauslabjoch.

10 The Teeth

The first thing one notices about the Iceman's teeth is the gap, about 4 millimetres wide, in the upper jaw between the two first incisors. Such a gap is called, in anthropology, a *diastema*. It is a variant, and not a very rare one at that, well within the norm. Only a few studies exist on the frequency of the *diastema* in prehistoric skeletons. In Austria, 16 per cent of cases examined from the Early Bronze Age had this feature.

As for the anatomical cause of a *diastema*, the literature lists a large variety of possibilities, ranging from direct heredity to a particularly low frenulum of the upper lip which forces the teeth apart. The Iceman has an unusually broad *canalis incisivus*, which leads onto the palate behind the incisors. This exceptionally large opening (*foramen incisivum*) could also be responsible for the *diastema*. Another feature observed directly on the Iceman's jaw, and one that might favour *diastema* formation of such proportions, is the relatively wide dental arc in the upper jaw and the related absence of wisdom teeth.

This, then, is another peculiarity in the arrangement of the Iceman's teeth and in his jaw structure. X-rays and computer tomograms show that all four wisdom teeth are missing. Although partially present in the jaw, not one of them has broken through.

The decrease in the total number of teeth in humans, mainly due to the absence of the rear-most molars or the second incisors, is regarded as an evolutionary trend. The reduction of the jawbone in the course of man's evolution, from an ape-like snout to the present-day flat lower face, should result in a reduction in the number of teeth. Statistical studies of teeth arrangement from various prehistoric, historical and modern populations reveal a frequency of missing wisdom teeth from 1.5 to 34 per cent. For

Australians, for instance, the factor is only 1.5 and for Greenland Eskimos, whose teeth are subject to particularly heavy wear, it is as high as 31.1 per cent. In consequence the absence of all four wisdom teeth in the Iceman has to be viewed, also for his time, merely as a deviation and certainly not as an abnormality when compared with a fully developed set of teeth. Nor does the absence of those four teeth run counter to the evolutionary trend. The 5,000 years that have elapsed are a mere second in the history of human evolution. So short a time span does not justify the assumption that physiological differences existed between Neolithic and modern man.

Thirdly, it is instantly noticeable that the teeth are extremely worn, with an exceptional level of abrasion. A deceptive impression is gained by merely looking: due to shrinkage the gums have receded somewhat, exposing more of the teeth and making them look longer than they really are.

Abrasion continues throughout a person's life. The degree of wear therefore provides a clue to the Iceman's age at death. The formula applied suggests, for the Iceman, a range from thirty-five to forty years. One possible cause for the heavy abrasion of his teeth may be the habitual consumption of dried meat (pemmican), as evidenced by North American Indians. The main cause of abrasion – and not only in the Neolithic period – was probably the consumption of cereals ground in querns made of quartz sandstone.

One might have thought that the abrasion caused by excessive demands made on the teeth would cause the Iceman's upper and lower teeth to close firmly one upon the other. But this is not the case. Instead, great wear and tear is evident in the left frontal upper jaw. As a result, a flat-curved horizontal opening remains when the jaws are closed one upon the other, comparable to the irregular tooth wear seen in pipe-smokers today. Such phenomena are known to dentists as functional wear and tear, unrelated to normal chewing action. The dental diagnosis of the Iceman suggests that his functional tooth wear is due to some habitual action which, unfortunately, we cannot ascertain. In recent primitive societies similar wear and tear has been observed as a result of the frequent use of blowpipes.

Also observed in the frontal region, i.e. on the incisors and canines, are polished facets a few millimetres long. Perhaps these features may also be explained by some routine daily activity. Excavations in Neolithic pile-settlements have produced a number of elongated birch tar lumps bearing very clear impressions of human teeth. For that reason they were instantly labelled 'prehistoric chewing gum'. The real function of these artefacts, however, remains uncertain – they may have served as stimulants, as hygienic aids, or as working material. Chewing of such lumps – after all, the Iceman used birch tar for his haftings and as an adhesive – might easily have produced those polished facets on his teeth. The warming associated with chewing would have made the material more ductile and easier to work.

Finally, we should mention a negative diagnosis. The Iceman's teeth are entirely free from caries. An important cause of tooth decay is the fermentation in the mouth of carbohydrates, such as sugar and finely ground flour, through the action of microorganisms. This sets free acids which dissolve the enamel and dentine. In Europe a progressive incidence of caries has been noted since the Neolithic, more dramatically so since the Middle Ages. Its incidence today is more than 99 per cent. With his healthy set of teeth the Iceman pre-dates the onset of this deplorable development.

11 The Skull Model

In the scientific study of preserved corpses the anthropologist finds himself in the awkward position that his usual comparative standards are not available. The parameters used in human biology for individual and comparative studies are all derived from living persons or from skeletons. A mummy, especially a dry mummy, occupies an in-between position. Even though radiology and computer tomography can nowadays overcome many of these problems, they do not meet all the demands made by a complex investigation - for example, measurement of a mummy's skull. Added to this is the fact that, because of the risk of contamination and gradual damage, the mummy should be touched as little as possible. Preference is therefore given to procedures which operate largely, or better still entirely, without contact.

From the outset, it was decided to produce a model of the Iceman's skull. Methods for the reconstruction of skull models on the basis of computer tomographical data have been in use for some time and have proved effective. But they are confined to two-dimensional models, i.e. models that reproduce the outward shape of a skull but are solid inside.

The Innsbruck radiologist Dieter zur Nedden took up the challenge of building the world's first three-dimensional skull model from the computer tomographical data of the Iceman's head. To this end he made use of stereolithography, hitherto applied mainly in aviation and astronautics, but also in the motor industry and in architecture.

The stereolithography system consists of an ultraviolet laser, mounted above a basin filled with photosensitive liquid acrylic resin. A perforated base plate is dipped into this container, accurately layer by layer. The laser beam, controlled by the computer tomographical data, scans the plate point by point, in layers of 0.25-millimetre thickness. Where the ultraviolet beam strikes, the liquid resin hardens through polymerization. The hardened layer is allowed to drip dry and is again dipped into the bath. Again the laser beam travels through the liquid, hardening the next layer. In this way, after nearly forty hours of work, an exact three-dimensional model of the Iceman's head was produced, a model in which not only the cranial cavity but also the sinuses and the lesser cavities of the skull are formed as empty spaces.

The first skull model built by this method contained some annoyingly unclear structures. There were, for instance, perforations in the frontal region; it was initially unclear whether they represented genuine, disease-related changes or computer errors. The method was subsequently refined and a second scan showed that these perforations had been eliminated. Now we had a reproduction of the Iceman's skull that was accurate to within fractions of a millimetre.

Even the first assessment showed deformities of the skull. These were mechanically produced asymmetries, almost certainly caused by pressure and movement of the ice, which resulted, among other things, in a fracture of the right zygomatic arch with a displacement of the fracture faces by just over 2 millimetres. The right bony eye-socket is also slightly sheered. Minor asymmetries are found on the occiput, which can only be interpreted as the results of deformities occurring after death. By means of a specially developed computer programme, we are now trying to eliminate these changes. Only when we possess a corrected skull model will it be possible, using the method developed by the Russian Mikhail M. Gerasimov and since repeatedly improved, to attempt a three-dimensional reconstruction of the Iceman's face.

The stereolithographic method first successfully developed for our purposes was very soon applied in medical practice. A young Libyan girl was brought to Innsbruck, her head disfigured by a median facial split which would have condemned the child to a wretched, marginalized existence. Such defects can nowadays be surgically eliminated. However, intervention in the cerebral region is always an exceedingly risky undertaking with often unpredictable complications. It was decided to make a stereolithographic three-dimensional model, on which zur Nedden prescribed the line of the cut. Throughout the lengthy operation performed by Professor Hans Anderl the sequence of interventions was continually checked against the skull model. The operation was successful and the girl can now lead a normal life. Some plastic surgery will still be necessary, but when it is completed there will be nothing to show that the face was once hideously deformed.

This poignant case demonstrates that the research into the Iceman is not an exclusive objective. Numerous innovative ideas are being developed which, extending far beyond the information supplied by the Iceman, will prove beneficial to mankind.

V
The Iceman and his World

1 The Neolithic Age in the Circum-Alpine Region

The cultural period of the Stone Age is divided by prehistorians into three sections – the Early, the Middle and the Late Stone Age, or, in scientific language, the Palaeolithic, the Mesolithic and the Neolithic. In the southern Central European region the most ancient human settlement confirmed so far dates from the middle of the Palaeolithic Age, some 250,000 years ago. These people lived as hunters-gatherers, not so differently from many primitive tribes today. For approximately 240,000 years until the end of the Mesolithic, in the sixth millennium BC, human lifestyle did not change. As for the climate, several warm periods, interrupted by glaciations, were followed first by the Würm glaciation at the end of the Palaeolithic about 10,000 BC and then by the moderate climate of the present (the Holocene). Heat-sensitive animals, such as the mammoth, the woolly rhinoceros and the cave bear became extinct in our regions or else, like the reindeer and lemming, followed the edges of the glaciers retreating to the north in the direction of Scandinavia. Aurochs, bison, elk, bear, wolf, otter, beaver, lynx, wildcat and tortoise would still inhabit our fields and forests if man had not been so careless about animal life. Many of these species are already extinct in this part of Europe, or at least endangered. Few of them have a chance of survival.

For the humans of the Palaeolithic and Mesolithic periods wild animals provided the basis of their food and hence their existence. Finds of animal bones at resting and dwelling places, where the hunters, in tents or behind screens, would seek shelter from bad weather, testify that anything, from reptiles, birds, fish and small mammals up to the dangerous large animals such as mammoth, aurochs and bear, served them for food. Gathered herbs, roots, fruit and seeds supplemented the diet of the few families or small kinship groups that inhabited our region.

Food was obtained by means of spears, composite harpoons and other specialized hunting weapons, for the manufacture of which highly developed flint tools were used. Pictorial representations of men and animals, painted on rockfaces or carved in ivory, as well as scarce finds of human burials point to magic beliefs (hunting magic) and ritualized practices.

In the course of the sixth millennium BC a fundamental change took place in the economy, made possible by the development of farming and stockbreeding. Modern scientists agree that the transition from food gatherer to food producer was not developed by the indigenous population but was adopted from outside. The wave, originating from the ancient eastern centre of cultural development, the Fertile Crescent, passed over Anatolia and the Balkans to reach the Central European settlement areas.

It is a contentious point whether this process, described as 'Neolithization' or the 'Neolithic revolution', took place as a result of the immigration of alien peoples, or whether the new achievements spread step by step, by way of cultural transfer from adjacent populations. There is a third option, a combination of the two – which is the one I favour.

For the native Mesolithic population of hunters, whom I believe were integrated into this Neolithization process in our region, the process brought almost unimaginable changes in lifestyle. Of course, the traditional economy was not simply abandoned and certainly not abruptly. Hunting and gathering continued to be important methods of food acquisition in the Neolithic. Indeed, even at its height, there were village communities that met the bulk of their meat requirements by hunting deer.

The switch to a producing economy led to a total transformation in settlement patterns. Temporary resting-places such as tents, windbreaks and rocky overhangs were no longer sufficient as homes: livestock had to be accommodated under cover during the winter, storerooms had to be established. Fields were cultivated for several months together, and if land was needed forests had to be cleared. All this required permanent dwellings with winter-proof dwelling places, occasionally protected by ditches and palisades.

In the agricultural society of the Neolithic an individual would possess the total knowledge of his period, as well as all the skills necessary to perform his daily tasks. He would know how to hunt, fish and gather food, how to till his field, look after his livestock and build huts. He would make his own pots, prepare his food and manufacture the tools and implements needed for all these activities. He would submit to the social behaviour of his community and he would be acquainted with its customs and rites, including the burial rites which, even in the Neolithic, were surprisingly diversified and typical of the various cultural groups. We shall need to remember this multiplicity of skills and knowledge when, in a subsequent section, we consider the likely activities of the Iceman in his lifetime.

Needless to say, there were frequent innovations, changes, progressive developments and surely also reverses over the nearly 4,000 years of the Neolithic period. Man was constantly on the road of progress – whatever his goal. The archaeologist can confirm such changes even on the strength of his usually scant source material, albeit only intermittently and by inference. On the basis of identifiable stages of change it is possible to subdivide the Neolithic into three cultural phases. These are known as Early Neolithic, Middle Neolithic and Late Neolithic. At about 2200 BC the Early Bronze Age began.

The Early Neolithic was the period of land seizure. The first agricultural settlers still had a wide choice and were able to appropriate the best soils. Crop farming and stockbreeding spread throughout all the major loess regions of Central Europe. Nowhere did settlement penetrate to the foot of the Alps.

By the Middle Neolithic unused loess areas had already become rare, but there was still enough good soil for farming. Neolithic culture first advanced close to the Alps from the north, when the Swiss Mittelland was settled. The advance continued up the Isar and Middle Neolithic man also established himself in the Salzburg region.

At that period, in the fifth millennium BC, the settlement areas were also approaching the Alps from the south. The piedmont was occupied, and the valleys running down from the north were also followed up, almost to within sight of the main ridge. The favourable climate of South Tyrol lured early man far deeper up the valleys than was the case in the north.

The real opening up of the Alps took place in the Late Neolithic, which probably started no later than about 4000 BC. The first extensive forest clearances were carried out in the valley landscapes favouring settlement, and new forms of agriculture were developed specifically for the high mountains. We shall return to this topic when we discuss the living space and the lifestyle of the Iceman.

About that time, however, something crucial happened to promote the settlement of the Alpine valleys: man in that region learnt how to use metal. There are abundant deposits of copper in the Alps. In their search for ore men moved up the valleys, crossed ridges and adapted their lifestyle and economic methods to the mountains. The typical cultures of the Late Neolithic developed. But none of them crossed the passes or saddles of the main Alpine ridge. A breathing space intervened.

Not until the final centuries of the Late Stone Age, at the beginning of the third millennium BC, did cultural phenomena emerge – albeit still very hesitantly – which left comparable traces on both sides of the Alps.

The Iceman lived during the decades between 3300 and 3200 BC. He therefore belongs to the Late Neolithic period, indeed to its final phase. Whether by then the first manifestations of later cultures were in evidence is a question which present-day archaeology is not yet clear about – but it cannot be entirely ruled out. Irrespective of this, however, the question arises whether, from the archaeological point of view, his cultural sphere can be narrowed down further: whether, to put it more simply, he can be assigned to one of the cultural groups already identified by research. This obviously also includes the question of whether his home settlement was north or south of the Alps.

2 Neolithic Cultures North and South of the Alps

Classification of Neolithic finds and their assignment to individual cultures and groups is based almost exclusively on pottery. The shapes and ornaments of ceramic items – pots, jugs, ewers, basins, bowls, dishes and plates – were subject to rapid changes of fashion in time and space. In that respect the Neolithic period did not differ from ours. Neolithic research also benefits from the fact that pottery items could be manufactured locally anywhere with simple tools and that, because of their fragility, they were not transported over great distances. The Iceman was evidently equipped for distant travel – at least the birch-bark containers he carried would seem to suggest that.

The more or less isolated cultural groups are mostly named after a famous location, or after the first location mentioned in the literature, such as Pfyn, Cortaillod and Horgen in Switzerland, Michelsberg, Altheim and Cham in southern Germany, Mondsee and Baden in Austria, or Lagozza and Remedello in Italy. In addition there are designations based on a culture's particular ceramic characteristics, such as the Square-Mouthed Pot culture, the Corded Ware culture or the Bell Beaker culture. All these cultures are to be found around the central Alpine region during the Late Neolithic period, with Horgen, Cham, Corded Ware, Bell Beaker and Remedello largely regarded as belonging to its latest stage. Lesser local groups will not be considered here. As for Remedello, there has been some suggestion recently that its oldest phase may have begun a little earlier. This is of considerable importance for the cultural chronology of the Hauslabjoch find, as it would make it possible to establish a chronological connection between it and the grave finds of Remedello near Brescia in Lombardy.

If the Iceman had carried a pottery vessel, his assignment to a definite Neolithic cultural group would be much easier. Even so, his native village will presumably have to be sought in the Neolithic settlement areas nearest to where he was found. These are the Inn valley in North Tyrol and the Adige valley with its side-valleys on the south flank of the Alps. From the site at the Hauslabjoch down through the Ötztal to the valley lands of the Inn is a distance of some 60 to 70 kilometres. So far no confirmed Neolithic find is known from the Ötztal itself. This may be due to the fact that the Ötztal was not yet settled then; on the other hand, as in all high-altitude Alpine valleys, the soil relief of the valley is subject to considerable changes from the forces of nature. Slope erosion, landslides and stone avalanches fill up the often narrow valley floors. If a narrow ravine is blocked by a rockfall, water backs up behind it to form a lake and it often takes centuries for it to eat its way through and drain again. The streams are continually seeking new beds. There are numerous reports from historical times of houses and farms being buried by mountain disasters. The resulting gravel cover on the valley floor could easily have buried existing Neolithic settlements and burial grounds of past millennia under many metres of detritus, making them inaccessible to archaeology. No systematic survey, combined with appropriate fieldwork, has so far been conducted in the Tyrol.

Conditions are much the same in the Inn valley. What Neolithic finds are known both up and down the Inn would fit into a shoebox – a few flint implements, including arrowheads, a few stone axes and barely a handful of shards. These are all individual, scattered and accidental finds, but they are enough to prove that the Inn valley was covered by a network of Neolithic settlements, however sparse. So far we have no evidence of any penetration up the side valleys.

Attribution of the scant remains to specific Neolithic cultures is highly problematic. At least we can rule out that any of them come from the earlier phases, the Early or Middle Neolithic. The situation is different where the more recent phases of the Neolithic are concerned, the ones covering the time of the Iceman. Some very few shards found in Innsbruck-Hötting could, albeit with considerable reservations, be attributed to the Late Neolithic Altheim culture. Others, though coming from the same sandpit, as well as from Angath near Wörgl, exhibit certain similarities with the Cham culture, which probably began towards the end of the Late Neolithic.

In order to determine the native village and hence the cultural stage of the Iceman it is helpful to rule out all the cultures to which he cannot have belonged. Let us start with those that follow the arc of the Alps along the north, proceeding from west to east. We will deal first with the earlier cultures of the Late Neolithic, and afterwards with those of its later stage.

The Cortaillod culture is a purely central Swiss affair. Thanks to forty-five carbon-14 datings so far we have an excellent picture of its duration. It peaked between 3800 and 3400 BC, with the date-range extending into the next millennium. From the point of view of time, the Iceman would fit into it. However, its main distribution is in western Switzerland. Its easterly ramifications in the north reach just into the Zürich area; in the south it extends, up the Rhône, a little beyond Lake Geneva. That is well over 200 kilometres, as the crow flies, from the Hauslabjoch, and a lot further on foot and over difficult passes. As there are contemporaneous cultures closer to the Hauslabjoch, Cortaillod may be ruled out.

The Pfyn culture, located principally around Lake Constance, comes closer to the location of the Iceman's discovery. However, the entire span of its carbon-14 spectrum from 4200 to 3500 BC lies so clearly before the Iceman's time that it, too, can be eliminated.

The same applies to an even greater degree to the Michelsberg culture. Although its influence reaches into southern Bavaria and the Salzburg area, so coming very close to the foot of the Alps where the Inn emerges, its carbon-14 range from 4400 to 3800 BC clearly places it – with a slight overlap – even before Pfyn.

The Altheim culture fits better into the picture. Its main distribution area is Bavaria south of the Danube, from where it extends towards the Alpine piedmont, reaching as far as the area around Salzburg. So far we only have ten carbon-14 datings for it, though it is by no means certain that these cover the whole duration of the Altheim culture with its many important sites. Its narrow spectrum starts at 3700 and ends about 3400 BC, thus just about touching the lower limit of the Iceman dating. Despite this reservation we shall, for the time being, list it as a possibility.

The Altheim culture is joined in the east by the Mondsee culture, which penetrates noticeably into the Alpine valleys from the north, and in the west radiates as far as the Salzach. Its twenty-two carbon-14 datings range from 3700 to 2900 BC and thus cover the Iceman's dates. However, its westernmost extensions are still more than 200 kilometres away, as the crow flies, from the Hauslabjoch. For that reason the Mondsee culture may also be ruled out.

The Baden culture can be disregarded with even more confidence. Although with its carbon-14 spectrum from 3400 to 2900 BC it fits very neatly into our period, it does not in the west, where it follows the Danube, reach even as far as Linz. It must be eliminated on geographical grounds.

So much for the earlier Late Neolithic cultures of the northern Alpine arc. In the later stage, conditions become more complicated, mainly because cultures often overlap a lot more, both in time and topography, than during the preceding period. Let us start in the west, with Switzerland.

Here the Horgen culture is one of the most important phenomena of the final phase of the Neolithic. Its territory extends from western Switzerland through the Mittelland, along both sides of the Upper Rhine, around Lake Constance and the Alpine Rhine up to Liechtenstein. Its twenty-one carbon-14 datings provide a good overview of its duration. The core range lies between 3400 and 2900 BC. Although this timespan accords well with the Hauslabjoch dates, the Horgen culture ought to be eliminated because of the great distances involved.

The Cham culture is found in Bavaria, with extensions into Bohemia and Austria. Its western limit is the Lech, making a clear boundary between it and the Horgen culture. In the south it pushes across the Munich gravel plain to the Alpine piedmont and even thrusts up the Salzach into the Pongau area. Members of this culture had the courage to penetrate deep into the mountains. The sixteen carbon-14 datings we have so far cover the narrow range from 3000 to 2700 BC, with standard deviations from 3200 to 2500 BC, so contact, and possibly overlap, with the Hauslabjoch date spectrum is just possible.

The Corded Ware culture is found over large regions of Central Europe, following a later trend for cultures to cover extensive territories. But it kept away from the mountains. In Switzerland a broad band extends from Lake Geneva through the Mittelland to Lake Constance; this band gets narrower towards the east and, with a few scattered sites, advances only to the lower reaches of the Inn in Upper Bavaria. The thirteen carbon-14 datings we have place the flowering of the Corded Ware culture in the period from 2900 to 2500 BC, with a fluctuation range from 3200 to 2300 BC, which makes it decidedly more recent than the Cham culture. However, its unquestionable inclination to avoid the mountains eliminates it at once from our inquiry.

The Bell Beaker culture with its pan-European distribution – from the North African coast in the south to Jutland in the north, and from Portugal in the west to the Hungarian plain in the east – is one of the most fascinating phenomena of the end of the Late

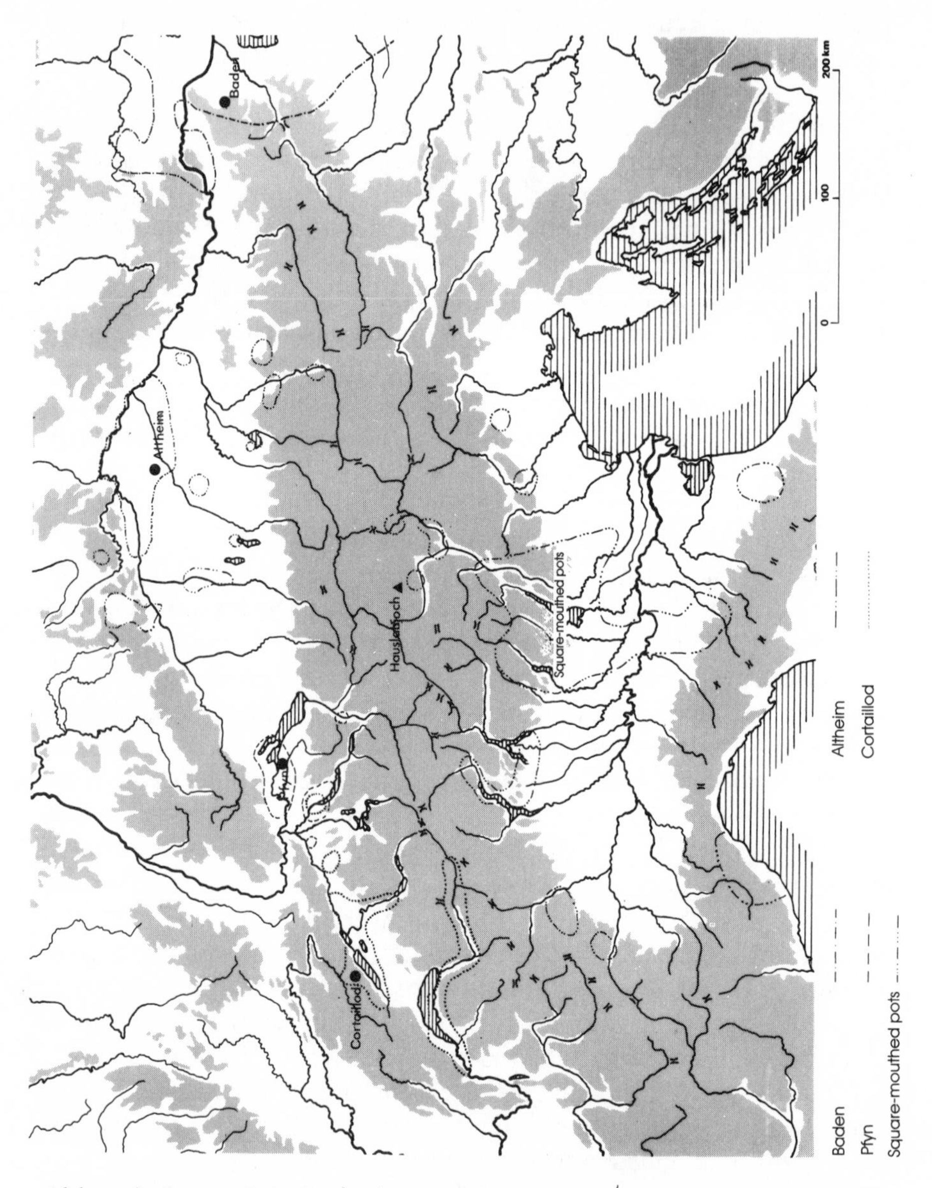

Although the north Italian culture of the square-mouthed pots advanced northwards far into the Alpine valleys, even settling in the Val Venosta, it came to an end several centuries before the Iceman lost his life at the Hauslabjoch. The map also shows the remoteness of the Hauslabjoch from the settlement areas along the northern arc of the Alps.

The Michelsberg, Mondee, and Lagozza cultures. This map, too, shows the remoteness of the Hauslabjoch.

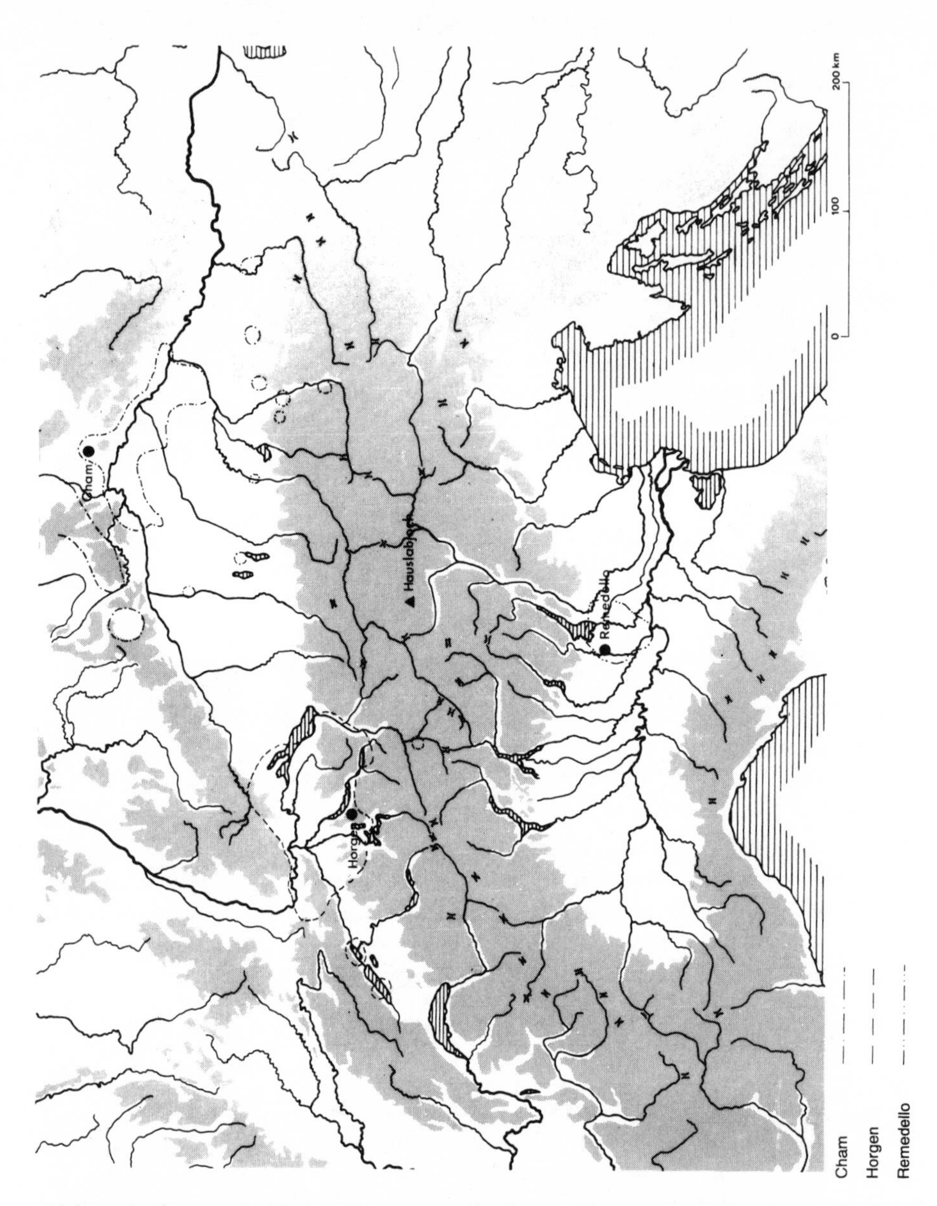

Although the north-Alpine Horgen and Cham cultures sporadically penetrate into the Alpine valleys, and although some are at least partially contemporaneous, they are nevertheless at a good distance from the Hauslabjoch. South of the Alps is the core area of the Remedello culture, contemporaneous in an early phase with the Hauslabjoch. Cultural phenomena related to Remedello penetrate far into the Alpine valleys and even reach the Val Venosta, thereby bridging the relatively great distance between Remedello and the Hauslabjoch. They have not yet been sufficiently researched to be described as evidence of an independent culture.

Neolithic period. Sites exist both north and south of the Alps. Flexible and mobile, with sites such as Sion-Petit Chasseur in the Swiss Valais, it thrusts deep into the heart of the central Alps. It would be the perfect setting for our Iceman if only the dates would match. However, measured on fifty-seven samples, the Bell Beaker culture existed from 2800 to 2200 BC. Standard deviation would stretch only as far as 3000 BC, so there can be no link with the Hauslabjoch find.

Thus of all the Neolithic cultures north of the Alps, the only possible contenders are Altheim and Cham.

Extending in an east-westerly direction along both sides of the Upper Adige, and barely 20 kilometres south of the Hauslabjoch, is the Val Venosta, the Vinschgau. With a length of roughly 50 kilometres between Mallès Venosta and Merano and a wide valley floor, it provides favourable conditions for settlement. So it is not surprising that it was densely inhabited in the Neolithic. It even appears possible to differentiate between several distinct Neolithic groups. It is clear, however, that, from Lake Garda and up the Adige, the great Neolithic cultures of northern Italy radiate into this small settlement area between the glaciers of the Ötztal Alps and the Ortles (Ortler) group.

Along with a multiplicity of local variations, particular attention needs to be paid to the Square-Mouthed Pot culture and the Lagozza culture in the early Late Neolithic and to the Remedello culture in its final centuries. However, synchronizations and date limitations of the kind possible for the zone north of the Alps can only be carried out here on a greatly reduced scale, as a relatively small number of carbon-14 datings are available.

The late Square-Mouthed Pot culture has its principal area of distribution in the eastern part of North Padania, i.e. in the Trentino and the western Veneto. Its area of influence extends far up into the Alpine valleys and is also evident in the Val Venosta. However, the spectrum of its carbon-14 datings ends in the middle of the fourth millennium BC, to be more accurate about 3600 to 3500 BC, well before our Iceman was born.

The western neighbour of the Square-Mouthed Pot culture is the Lagozza culture of the north Italian Po plain. Its northern boundary runs along the foot of the southern Alpine arc. Its influence can be observed as far as the Lake Garda area and up the Adige beyond Trento, though it misses the Val Venosta itself. The carbon-14 datings range from 3800 to 3400 BC. There are also two more recent datings (3300 to 2800 BC), but these are both said to come from uncertain sites and to be unreliable. Strictly speaking, this means that Lagozza could just about border on the widest date range of the Iceman. One would probably have to take this culture into consideration if no more suitable alternative were available.

However, such an alternative does offer itself, despite exceedingly complicated conditions, in the Remedello culture, which, in the region we are looking at and alongside some less important groups, represents the main manifestation of the later Late Neolithic. This culture is widespread north of the Apennines, mainly in the eastern Po plain. Offshoots of this culture, admittedly under locally changed conditions, can be identified well into the southern Alpine valleys and even into the Val Venosta itself. The few carbon-14 datings so far available are predominantly from the first half of the second millennium BC and are therefore too recent for the Hauslabjoch find. There exist, however, six carbon-14 datings of related cultural manifestations, whose fluctuation range from 3300 to 3000 BC clearly overlaps the Iceman's date. It is therefore not unreasonable to assume a connection, if initially only a chronological one, between the Hauslabjoch and Remedello.

What, then, is left of the Late Neolithic circum-Alpine cultures after our elimination process? In the north we have the late Altheim and early Cham cultures, and in the south early Remedello. What these three cultures have in common is that their main distribution areas extend north and south of the foot of the Alps and that all three advance into the Alpine valleys. There they inspire existing or developing local Neolithic groups, transmit to them various ideas and help them perfect local practices. The cultures named thus radiate far beyond their core regions, initiating new developments. Not infrequently their specific influence becomes blurred in the process and so cannot easily be demonstrated by archaeologists. Especially in the presence of innovations from elsewhere, perhaps even from across the main ridge of the Alps, it becomes almost impossible to untangle the finely woven net of intercultural links.

We are still looking for the Iceman's native village. We have established that in his day the nearest Neolithic settlement north of the Alpine main ridge was in the Upper Inn valley, and south of the watershed in the Val Venosta. The North Tyrolean Upper Inn valley is some 50 kilometres from the Hauslabjoch, and the South Tyrolean Val Venosta about 20 kilometres as the crow flies – substantially more, of course, on foot. What could be more natural than to place the Iceman in the Val Venosta? But this would be too easy. Considered dispassionately, we are dealing with an archaeological find whose age we have determined but which we have not yet assigned to any culture. The absence of pottery compels us to resort for clues to the other objects found.

Our interest focuses primarily on the axe. The Iceman must have had contacts with a cultural environment with a considerable mastery of copper technology. I am uneasy about the assertion, made time and again but totally unproven, that copper was regarded as particularly valuable in the Late Neolithic, that it was exceedingly rare and that it conferred a high status on its owner. It is a commonplace that for the archaeologist knowledge of the metal inventory of a prehistoric cultural group derives predominantly, often exclusively, from grave finds. As mentioned earlier, grave goods were chosen in accordance with rituals we do not understand. On the whole, it seems unlikely that a dead man invariably had all the metal objects he owned buried with him.

This brings us to the crucial point. We do not know any graves of the cultures located near the Hauslabjoch in the north. Altheim and Cham practised the elimination of dead bodies, which of course means they are lost to archaeological study. This deprives us of an essential source for the study of their metallurgy. So we are reduced to analysing what has survived at their settlements. As copper can be remelted at any time, nothing much is left to us.

To repeat, we know the Altheim culture only from its settlements. They are, for preference, on wide fertile valley floors and on gentle slopes. Hill or spur positions that favour defence were not a first choice and only three bog settlements are known. In nearly all Altheim villages the original surface and levels of use have been eroded, so excavations mostly reveal only artificial interventions in the ground, such as post-holes, pits and ditches. Complete plans of dwellings are known from the few valley-floor settlements. It is interesting that many dwellings are surrounded by defensive ditches or palisades, suggesting a need to protect the settlement communities.

The Altheim culture, in the usual way, is characterized by the typical features of its pottery. Its lack of decoration follows a trend of the period, and there is nothing more boring than sorting Altheim shards. Often the openings of vessels are reinforced by a clay bulge which is pressed down by the fingertips at close intervals, producing the so-called Arcade (*Arakaden*) rims. In the absence of other ornamentation this is described as the decorative technique typical of the Altheim culture.

For the manufacture of flint implements the favourite material is Jurassic slab flint. This was also used for the well-known Altheim sickles, which lend the culture a marked agricultural character. It is important to note that Altheim flint arrowheads are always equipped with a straight or narrowed base, quite unlike the Iceman's arrowheads which have hafting tangs.

Altheim metallurgy is well documented. We have flat axes, awls and ornamental pendants made of copper. There is also a heavy, massive cast-copper axe with a shaft hole, which is attributed to the Altheim culture and which was copied, within that culture, for axe-hammers of identical shape made of stone.

An advanced metallurgy is observed also in the Pfyn culture adjacent to the west and in the Mondsee culture adjacent to the east. Pfyn has a particular predilection for lakeshore and lake dwellings, which is why its cultural layers have survived far better than those on mineral soils. Evidence of copper working was discovered in the form of thick-walled coarse-clay melting pots, encrusted with slag on the inside. Copper finds as such are rare as no earth graves were dug. In the bog village of Schorrenried near Ravensburg, however, a very beautiful 12-centimetre-long copper dagger was found. It has a trapezoidal rounded tang with rivets and a narrow triangular blade with no central rib.

The manufacture of copper objects was also known to the Mondsee culture. Here, along with melting pots, spiral pendants and a riveted dagger with a slight suggestion of a central rib, mention need be made mainly of the flat axes. In form they are close to the Altheim axes; the absence of hammered flanges distinguishes them clearly from the Hauslabjoch axe. While it can therefore be proved that the Late Neolithic cultures north of the Alps had a well-established copper metallurgy, no specific links exist between Altheim and the Hauslabjoch.

More pronouncedly than the Altheim, the Cham culture preferred to settle on hills and spurs and such settlements were invariably protected against the hinterland by a system of trenches. Cham produced outstanding high-quality pottery, abandoning the monotonous products of its predecessors and creating new forms and motifs. One might say that the Cham people were the avant-garde cubists among uninspired neo-realists. Their pots, vases, bowls and cups are covered with applied strips arranged in grids, bundles and groups. All the decorative elements deliberately accentuate the vessels' shape, a trait which lends the Cham culture its individual and unmistakable mark. Even with seemingly unimportant peripheral items, such as spinning implements, Cham ingenuity challenges its neighbouring cultures. The outsize spinning whorls on the wooden spindles are often attractively decorated.

The partially contemporaneous Swiss Horgen culture displays a similar extravagance in the shapes of its pots. The Horgen moulded their vessels with a restrained sense of balance and a deliberate and practical simplicity. By avoiding excessive surface treatment and by applying sparse but effective decoration, they endow their pots, bowls and basins with an attractive rustic appearance, pleasing even to modern taste. At the same time they made a useful invention by very deliberately lowering the centre of gravity of their ware; it is almost impossible to knock over one of their massive solid-bottomed vessels. In this they contrast with the unstable shapes produced by neighbouring cultures, large-bellied items with often minute bases. Holes often had to be dug into the hut floors to ensure that beakers and flasks did not topple over.

As the Cham people had no archaeologically traceable burial customs, and as their hilltop settlements were often badly affected by erosion, evidence of their metallurgy is scant. That they nevertheless did practise metal-working is attested by a copper fishing hook from the Cham settlement of Riekhofen and a crucible from Köfering, also near Regensburg, which was used for melting gold.

Cham flint arrowheads with a straight and narrowing base are plentiful and resemble those of the preceding Altheim culture. In addition, though very rarely, tanged arrowheads have been found – so far at five sites, two of them in the Alpine piedmont. One of these is near Dobl, 40 kilometres north of Kufstein, where a fine flint dagger, worked on both faces with a retoucheur, has been dug up.

Tanged arrowheads and flint daggers of the kind carried by the Iceman began to appear north of the Alpine arc in the Late Neolithic and were fashioned during the later part of the period by the Corded Ware and Bell Beaker cultures. Their appearance in conjunction with evidence of metallurgical skills cannot be ignored. Cross-influence is conceivable, particularly as the Bell Beaker people of the Iberian Peninsula, with its great wealth of metal, actually possessed long-tanged arrowheads made of copper.

From what we have said, connections, however faint, between the Hauslabjoch find and an early – still Late Neolithic – phase of the Cham culture seem much more likely than any connections with the older Altheim culture.

Yet even if the few – possibly Cham-culture – shards found at Angath and Innsbruck-Hötting are included, we are still left with a very long journey from the most southerly genuine Cham settlement at Dobl, all the way up the Inn valley and the Ötztal to the Hauslabjoch. On foot this would be more than 250 kilometres over often very difficult ground.

Of the Neolithic cultures north of the Alps, the Cham best fits the Hauslabjoch finds in terms of date. But the Iceman certainly did not belong to the Cham cultural community. At that period, just before the beginning of the third millennium BC, the linking elements are only part of a general cultural trend. Specific connections between Cham and the Hauslabjoch do not exist.

3 The Remedello Culture and the Val Venosta

Unlike the north-Alpine cultures examined in the preceding section, the Remedello culture is attested not only by settlements but also by burial grounds. Hence we have a much better idea of its metal objects than was the case with Pfyn, Horgen, Altheim, Cham or Mondsee. Both flat and flanged copper axes are known from the Remedello culture. In addition, there are awls and armlets of the same metal. Particularly interesting is a special type of dagger with a narrow triangular blade and central ribs. At the haft there is a small rectangular tang with a hole to take the rivet. The very high standard of metal-working is confirmed by a magnificent hammer-head pin and a pectoral ornament in the shape of a sickle moon decorated with rows of repoussé work, both made of silver.

In addition to metal objects Remedello, of course, also produced a wealth of articles from traditional materials. For our purposes the flint implements deserve attention. There is a fairly large number of large and small daggers, worked by retoucheur on both surfaces. The types of arrowhead vary, but most are tanged in shape.

The Remedello burial ground has also supplied the closest comparable item to the Hauslabjoch axe so far. Apart from the fact that the Remedello axe, measuring 11.8 centimetres, is 2.5 centimetres longer than that from the Hauslabjoch, there are no formal differences whatever. The neck surfaces are elongated and rectangular, the outline is a narrow trapezoid, the sides are faceted, the delicately wrought flanges are identical, and the cutting edge of both axes is slightly curved. The Remedello axe was found in Grave 102. With it were a dagger and four tanged arrowheads, all of flint. Admittedly, the blade of the dagger, worked on both sides with a retouching tool, measures 13 centimetres and is thus twice as large as its counterpart from the Hauslabjoch. It also has a rounded haft-end. But there are, also, from the same necropolis, smaller flint daggers with a recessed tang – for instance one, not associated with any particular grave, of 9.2 centimetres and another, from Grave 13, with a length of 9.6 centimetres. These two come very close to the Hauslabjoch dagger, in both shape and size. The blade of the Hauslabjoch dagger, because of its broken tip, is now only 6.4 centimetres long. Originally it probably measured some 7.2 to 7.3 centimetres, and would fit among the longest arrowheads of the Remedello burial ground, which reach a length of up to 8.8 centimetres. It would, as I have said, have been classified as an arrowhead if the blade had not been found set in its handle. With this in mind, one can find among the stock of Remedello flint objects hitherto labelled arrowheads some items which deviate noticeably from the classic arrowhead shape with a pointed tang, and whose outlines match those of the Iceman's dagger blade. It is clear, therefore, that large arrowheads and small dagger blades have a certain overlap. In an individual case attribution must be doubtful unless a handle or arrow-shaft is

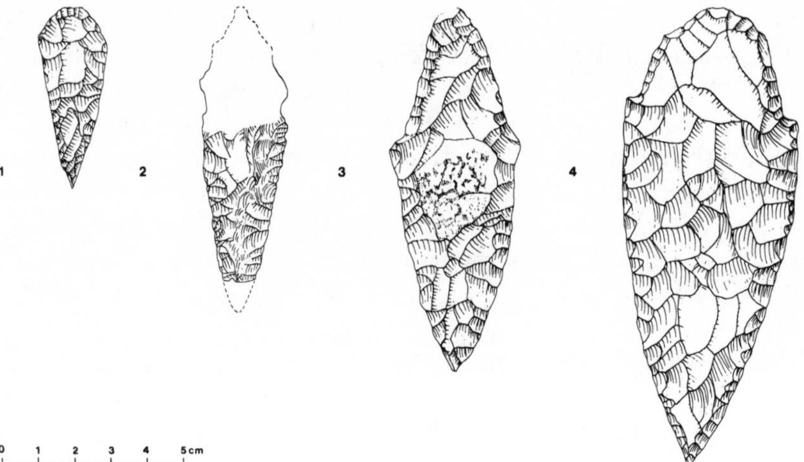

The blade of the Hauslabjoch dagger (No. 2) compared with a point (No. 1) and two dagger blades (Nos 3 and 4) from the Remedello necropolis near Brescia in northern Italy. All the implements are made of flint. The drawing fills in the point of the Hauslabjoch blade, which was broken off. The outline of the area within the shaft is visible by X-ray, but because of splintering at the upper end the exact shape of the tang could not be clearly established and is given here as a dotted line.

preserved. The minute Hauslabjoch dagger matches formally identical items from Remedello, some of whose flint implements should evidently be re-analysed in the light of these findings.

The agreement between the Hauslabjoch finds and those from Remedello Grave 102 is striking. In both cases the man's equipment includes a copper axe, a flint dagger and tanged arrowheads, moreover in virtually identical form. The contents of Remedello Grave 102 and the Ötztal glacier finds could be exchanged without anyone noticing the difference.

This unexpected parallel would unequivocally assign the Iceman to the south – if only the Remedello graves were in the Val Venosta, at a distance of 20 kilometres, and not between Cremona and Lake Garda, some 150 kilometres away as the crow flies. Along the upper course of the Adige the Late Neolithic cultural manifestations are different.

What we need to know is whether traces of Remedello culture have been detected across the Trentino right up into the Late Neolithic settlement areas of South Tyrol. They have. The flint

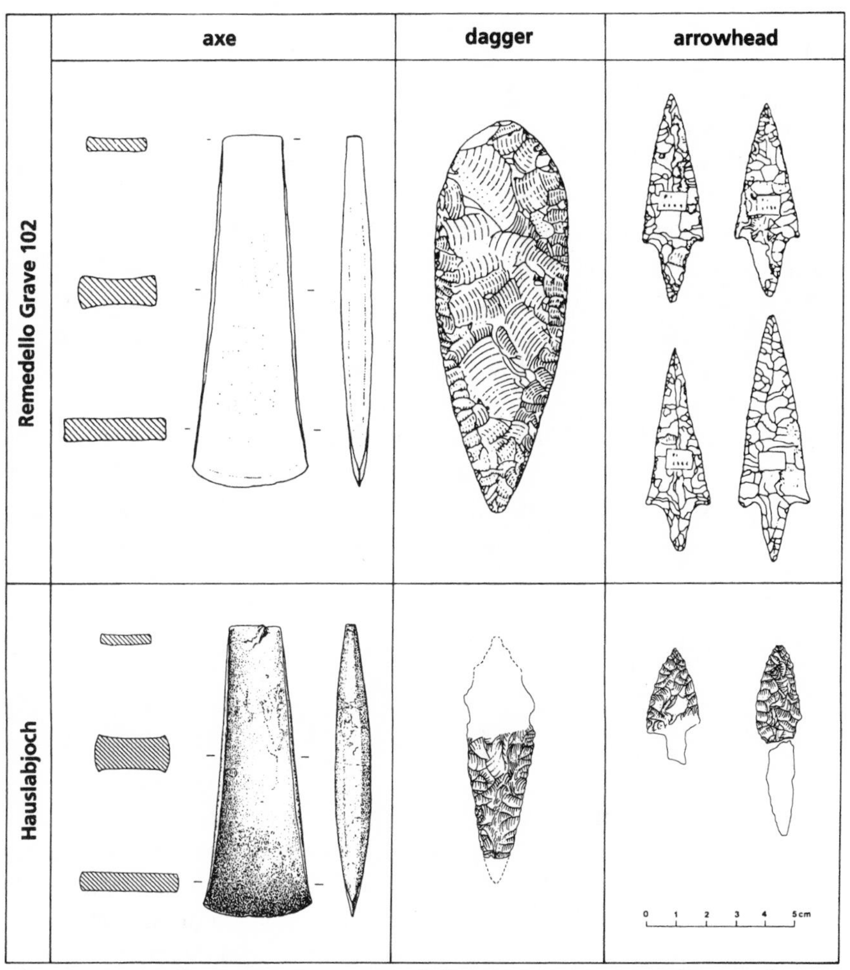

The contents of Grave 102 at Remedello near Brescia, northern Italy (top), compared with the corresponding implements found at the Hauslabjoch (bottom). The axes are of copper, the other implements of flint. The outlines of the tangs of the Hauslabjoch arrowheads were drawn on the basis of X-rays. The drawings of the Remedello arrowheads include the rectangular adhesive labels with the museum's reference numbers.

daggers of Eppan, Reutsch and Kaltern, all of them near Bolzano, have always rightly been linked to Remedello. Similarly, the flat copper axes of Hirzlsteig near the Rosengarten, Kollmann in the Isarco valley and Lana near Merano can be connected with relevant influences, which means that – with Merano situated at its eastern exit – we have come fairly close to the Val Venosta.

In this attractive valley basin only a few traces of Neolithic settlements have so far been found. Worth mentioning, taking the sites from west to east, are Müstair with a leaf-point and a stalked arrowhead, Tartscher Bühel with a shard and a stone cudgel, Eyrs with an axe-hammer, Laas with a tanged arrowhead, Kortsch also with a tanged arrowhead and other flint implements, Schlanders with a shaft-hole axe, Goldrain with a leaf-point, and Schloss Juval, where the mountaineer Messner lives, with numerous finds of shards and flints. At the eastern opening of the Val Venosta, around Merano, the sites become more frequent, with a stone axe and a flint blade at Algund and two further pointed-neck stone axes from Dorf Tirol. Deserving of special attention are the stone chamber tombs of Algund and Gratsch, though these probably date from well before the Iceman's time. The stones of the Gratsch tomb, with their 'soul holes', were re-used towards the end of the Late Neolithic for the construction of an individual grave with body burial. Grave goods included two boar tusks and a stag antler tine, which have been lost. These finds, though so far rather scanty, prove clearly that the Val Venosta was inhabited during the Late Neolithic. As in the Inn valley in North Tyrol, unfavourable conditions for archaeological excavation and insufficient research prevent us from gaining a comprehensive overview.

These finds alone are not enough to prove convincing links with Remedello. Some help, however, is provided from a rather different source, the 'picture stones', also described as 'statue menhirs'. These have been found in two locations in the Val Venosta, one in secondary use as an altar stone in the Bühel church at Latsch, and the others as a group of four found during work in a vineyard at the breaking edge of a steep gravel fan descending to the bed of the Adige near Algund.

These picture stones are of essential importance for judging the relationship between Hauslabjoch and Remedello. Latsch is situated almost exactly due south of the Hauslabjoch at a distance of only 18 kilometres; Algund is another 20 kilometres or so east of Latsch down the Adige at the eastern opening of the Val Venosta. If it proved possible to link these five picture stones with Remedello, then the location of the native village of the Iceman could be greatly narrowed down. Interestingly, similar 'statue menhirs' are found sporadically towards the south through the Trentino, across the main distribution region of the Remedello

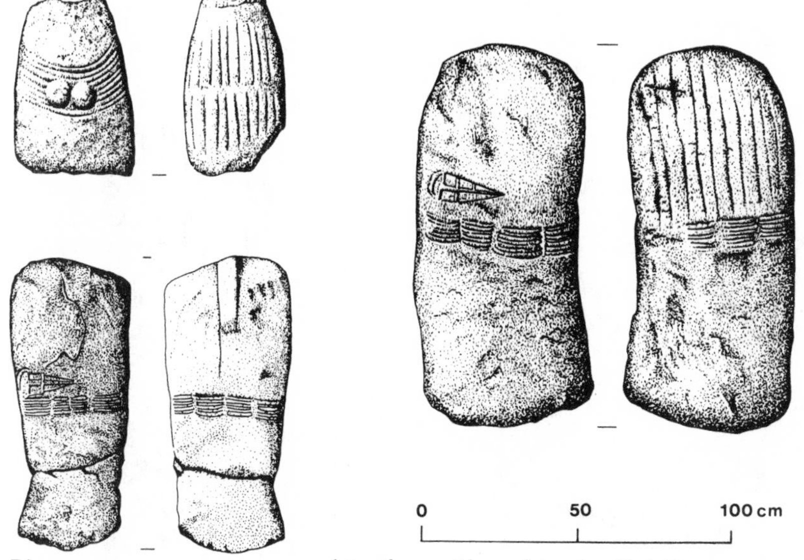

Picture stones or statue menhirs from Algund in the Val Venosta, each shown from the front and the back. Top left: Stone 1; bottom left: Stone 3; right: Stone 4.

culture, all the way to Liguria. North of the main ridge of the Alps they are totally absent.

Weighty tomes have been written about the meaning of these stones, and there is no need for us to add to their interpretations. Certainly they cannot be dated on the grounds of particular aspects of their discovery, nor from accompanying objects. Their dating has to be based solely on the pictures carved on them.

The four Algund stones were found lying close to each other and were undoubtedly part of a common design. They differ in size, state of preservation and pictorial representations. For the most part the pictures are engraved; evidently the outlines of the drawings were first carefully pricked out and then ground in.

Stone 1 was discovered on 11 February 1932. It emerged at a depth of about 1 metre at the foot of a scree when the foundations were being dug for a retaining wall. It is a torso, broken off at the top and the bottom, whose missing parts have not been found. The central part, a rounded rectangle in cross-section, narrowing like a pyramid towards the top, still has a height of 1.1 metres. Like the other stones, the fragment consists of locally found coarse-grained Töll marble. The stone is very carefully worked into

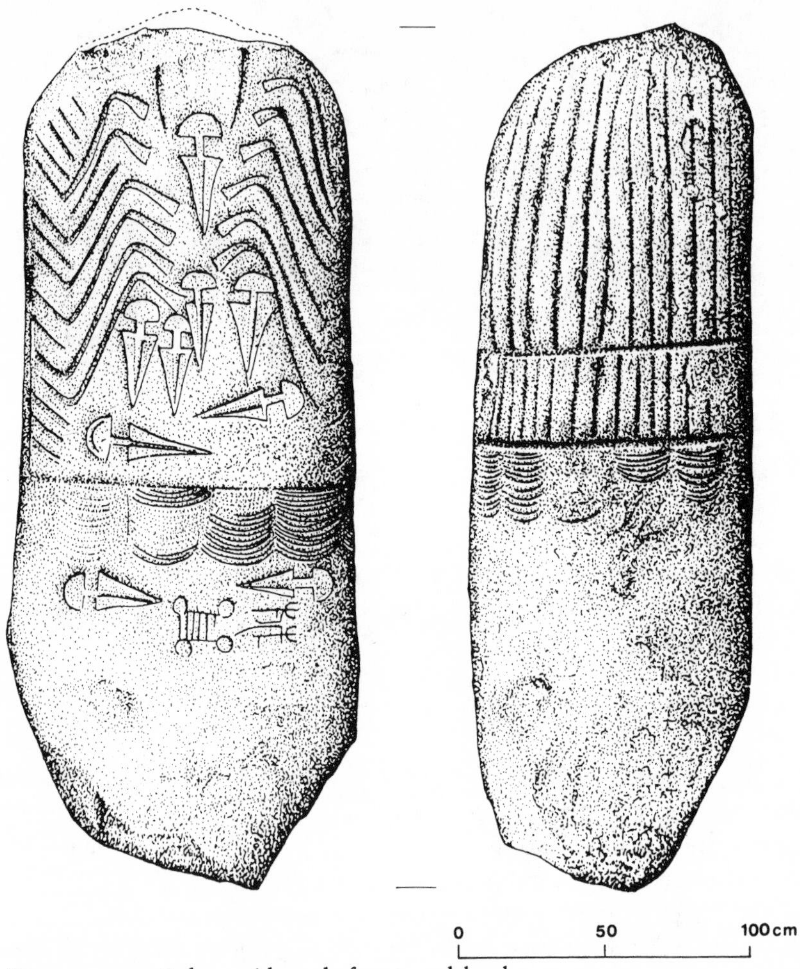

Picture stone 2 from Algund, front and back.

smooth surfaces all round. Two hemispheres projecting on the front side are thought to represent female breasts. They are framed by a broad bundle of grooves, gathered on the sides like garlands. On the back face the grooves run vertically.

On the following day the workmen struck Stone 2. This is a huge slab of Töll marble, 2.75 metres high, 1 metre wide, 0.35 metres thick and weighing 2,420 kilograms. It was also in the position in which it had crashed, about 8 metres from Stone 1, 0.5 to 0.7 metres deep in the ground, at an angle, and covered by slipped gravel. The front side of the block is flat, the back slightly curved. Working and design match those of Stone 1. A bundle of lines, gathered up at intervals under a horizontal groove, run around the stone like a belt. In the upper part there seems to be a hint of a face or a dagger-holster. Underneath is a group of eight daggers. To the right and left the picture is decorated with a group of seven daggers on each side. Under the belt are another two daggers and a four-wheeled chariot drawn by two large animals with horns. On the rear of the slab vertical grooves are visible down to the belt.

Ten years later, on 29 March 1942, Stone 3 was discovered. It was lying not at the foot of the scree, but on the slope itself, some 20 metres higher and 5.5 metres below the eroding upper edge, flat upon an ancient ground level. This had been covered by later erosion detritus. A Töll marble slab, it is worked like the other stones. It has a height of 0.95 metres, a width of 0.38 metres and a thickness of 0.11 metres; its weight is 62 kilograms. As an ornament this stone exhibits only the horizontal band of grooves, repeatedly gathered, as a belt; above it, on the front face, is a dagger.

Stone 4 was found a mere 20 centimetres away from Stone 3, but went unnoticed until 11 August 1942. This slab remained standing, albeit a little off-kilter, when the old ground level was overrun by a stone avalanche. It is 1.15 metres high, 0.5 metres wide, and weighs 170 kilograms. Material and treatment are the same as for the other stones. Stone 4 shows a similar decoration to Stone 3, except that on the rear face it has eleven grooves arranged vertically and covering the whole upper half.

From the observations made by the people who discovered the stones the following conclusions can be drawn. On a fan-shaped erosion slope, gently descending from the north, the stones were erected close to one another some 20 metres above the valley floor; their bases were let into the ground deep enough to provide sufficient stability but not so deep as to cover any of the pictures. As that soil level was densely permeated with fragments and flakes of charcoal, fires must have been lit there. This is why scientists speak of a cult site. At an unknown date the area was buried to a depth of more than 5 metres by one or more stone avalanches. At some time the meandering Adige forced its way towards the north and began to gnaw at the foot of the gravel fan. A steep bluff was formed, increasingly receding and growing in height, until one day the first picture stones crashed down. Two of them – there may originally have been more – were found again in 1932. Two more remained *in situ* higher up and were discovered in 1942.

Less than a year after the Iceman was found Hans Nothdurfter, in the course of archaeological work in the Bühel church of Latsch, discovered yet another richly decorated picture stone of local unworked marble. It emerged, as if on cue, from under the baroque wood shuttering of the altar table, thus still playing a 'cultic' role. A certain syncretism is conceivable: pagan symbols were thus accommodated to Christianity. The original location of the stone is unknown, but it was no doubt in the immediate neighbourhood, possibly in the direction of Laas, where the marble came from. This new find has not yet been definitively studied. But the same typical gathered belt found on the Algund stones can be made out, along with a jumble of daggers, sun symbols, axes and grooves. One gains the impression that several layers of decoration are superimposed one upon the other.

The same applies to the large Stone 2 from Algund. Scholars believe that the daggers and the belt represent the first phase of working, while in the second phase the upper part of the stone was fairly symmetrically decorated with axes. The wagon drawn by two horned animals was engraved last.

Such overlapping pictures begin to make sense if one assumes that the engraved designs were filled with colour. Once the first 'painting layer' had weathered and the colours been washed away, new designs were created, whose fresh colouring rendered the earlier designs scarcely noticeable.

For us the axes and daggers in the paintings are of utmost importance; on all the Val Venosta picture stones they both invariably represent an absolutely uniform type.

The axes exhibit elbow-shafting with a slightly curved handle. The hafting branch projects almost at a right angle. The narrow blade, widening towards the cutting edge in a trapezium shape, indicates that a metal axe-head is here represented. If one takes the Hauslabjoch axe and places it on the stone slab one might think that the former had been used for tracing the latter.

The daggers on the picture stones all have a slender triangular blade with a raised central rib. At the haft-end the blade has a wide base, perpendicular to the axis of the blade. The short haft ends in a massively shaped semicircular pommel. Here, too, there

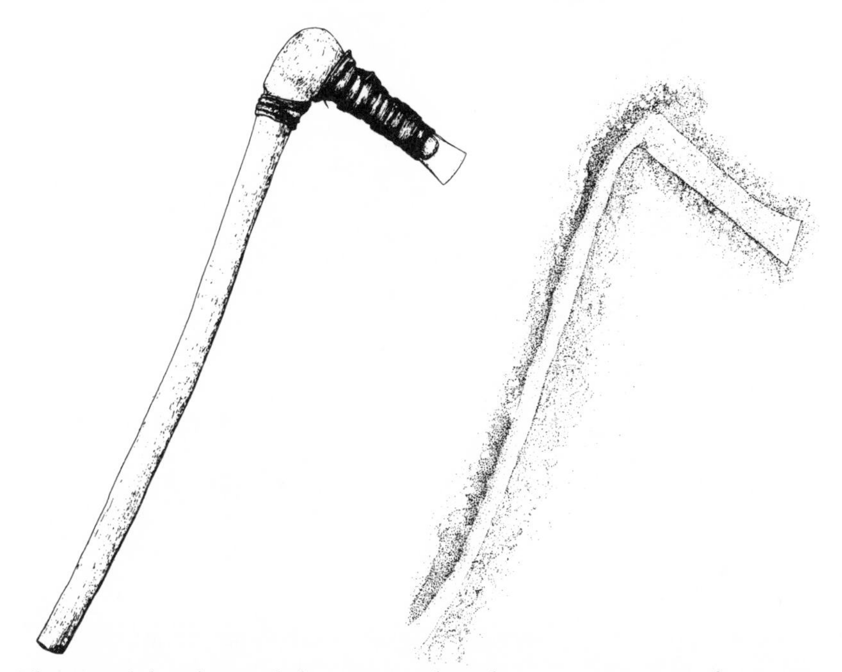

The Hauslabjoch axe (left) compared to the representation of an axe on Stone 2 from Algund in the Val Venosta. This shows that the sculptor of the Algund stones intended to represent an axe with a copper blade.

is no doubt that these are drawings of metal blades. The only comparable – and moreover very specific – companion pieces of these Val Venosta daggers are the copper daggers of the Remedello culture. Needless to say, the handles of these, made of wood or horn, have not survived. But the blades can be placed on top of the stone pictures without the slightest deviation. Moreover, the incidence of these unique copper daggers is confined to the Remedello culture. What few metal daggers are known from other copper-using Neolithic cultures north and south of the Alps invariably exhibit different characteristics of shape and outline.

I have already pointed out that the Iceman's equipment – flanged axe, flint dagger and tanged arrowheads – shows definite affinities with the south. In particular the male Grave 102 of the Remedello necropolis, with its flanged axe, flint dagger and tanged arrowheads, shows unambiguous agreement with the Hauslabjoch. On the other hand, because of unfavourable conditions for archaeological excavation and insufficient research, we have no know-

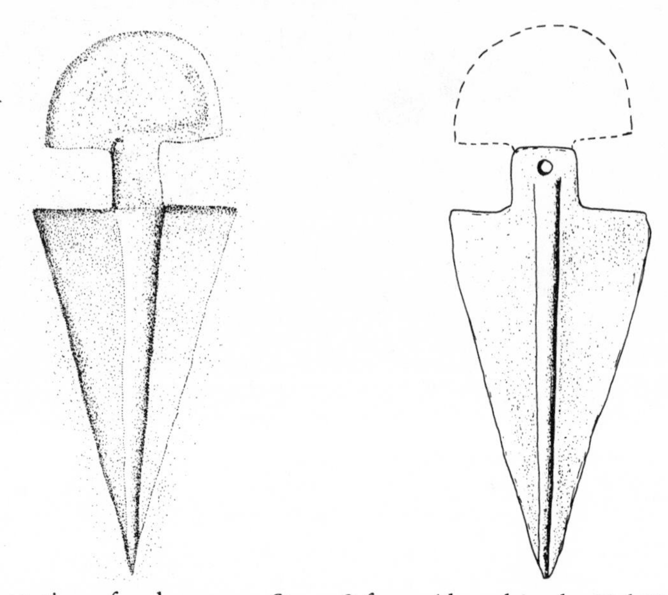

Representation of a dagger on Stone 2 from Algund in the Val Venosta (left) and a copper dagger blade from the Remedello necropolis near Brescia in northern Italy (right), with the pommel drawn in. Clearly the Algund sculptor used a Remedello-type dagger as a model. This proves the closeness between the Val Venosta and the north Italian Remedello culture.

ledge of the Late Neolithic cultures which inhabited the Val Venosta at the time. Not even the Late Neolithic finds of shards and flints at Schloss Juval at the mouth of the Val di Senales (Schnalstal) have so far been made available to researchers. No doubt most of the remains are still in the ground and, with a little luck, may be dug up in the future.

The picture stones of Algund and Latsch clearly suggest that the artist who created these monuments was familiar with Remedello daggers and axes of the kind found at the Hauslabjoch and at the centre of the Remedello culture and used only those for his models. This indicates that during the Iceman's lifetime, the Val Venosta was settled by a Late Neolithic population that maintained close contacts with the core area of the Remedello culture.

If, bearing all this in mind, we look once again at the location of the Hauslabjoch, we find that from the Val Venosta, through the Val di Senales, it can be reached quite comfortably in a few hours, or at most in a short day's ascent. From the Inn valley it would take at least three, if not four, strenuous days over difficult, at times pathless, terrain. The cultural connections implied by the Iceman's equipment very largely, and probably unambiguously, point to the south. To avoid being accused of bias I have extensively discussed the contemporaneous Late Neolithic cultures north of the Alps. But there was nothing to be discovered there, beyond a general spirit of the age, to provide links to the Hauslabjoch.

In terms of scholarship the find at the Hauslabjoch, however sensational and spectacular it may have been, is nothing more than a new site on the archaeological map. A peculiarity of it, of course, is that it is high among the Alpine glaciers, away from known settlement regions. Nonetheless, well-tried methods of scientific research have to be applied here too. In line with modern procedure we first investigated its chronological attribution. This was convincingly accomplished by means of carbon-14 dating. The next task was its assignment to one of the cultures described and defined by scholarship, unless such an assignment was provided from the outset by culture-specific objects found at the site. The most successful method for attributing Neolithic finds is, as a rule, via the pottery – but at the Hauslabjoch there was none. We therefore had to resort to other objects, all of which pointed to the south. It has proved possible, albeit by the roundabout route via Remedello, to demonstrate that the Val Venosta had late Neolithic settlements which agree with the Hauslabjoch find.

The ritual sites with their picture stones, where people gathered around a blazing fire, argue convincingly against the kind of sparse settlement found in the Inn valley. In the Val Venosta, therefore, at the time in question, there lived a population with established ritual practices, firmly settled and using the land, in line with Neolithic economic practice, for growing crops and raising livestock. To assign a new complex of finds to a culture far removed from the nearest settlement region would run counter to all archaeological methodology.

Thus it seems appropriate to seek the native village of the Iceman in the Val Venosta. I would go even further and say that perhaps the Iceman saw the sacred stone of Latsch with his own eyes.

4 Anthropological Problems

The conclusion, in the preceding section, that the Iceman came from an as yet insufficiently defined Late Neolithic settlement community of the Val Venosta and that his range of movement extended from there at least as far as the main Alpine ridge, and possibly beyond, was reached on purely archaeological grounds. This will now have to be confronted with other scientific findings.

We are often asked why we do not simply determine, by anthropological examination, whether the Iceman belonged to a north-Alpine or south-Alpine 'tribe' and so establish where his home was. But this is the wrong question. With all the attempts to describe the Iceman by the methods of classical anthropology and to assign him to a specific contemporaneous population, it should not be forgotten that, from the viewpoint of scientific research, he is merely one individual. We have no adequate information on the population to which, in an anthropological respect, he might belong, because suitable comparable samples are lacking over extensive areas of his cultural background. But quite regardless of that lack of data, we should avoid burdening the Iceman with the long-outdated and discarded racial labels of the late nineteenth and early twentieth century, which might have assigned him to some Mediterranean, Dinaric or Nordic racial type.

Serious anthropology is above all a comparative discipline, whose real objective consists in demonstrating differences of anthropological characteristics over space and time. For prehistoric anthropology this means that measured data of individual skulls or mean values of series of skulls are compared with one another in order to demonstrate similarities or differences of individual characteristics or combinations of characteristics. Results so far have shown that major morphological similarities of measured data frequently reflect genetic or ethnological relations and hence may be useful in solving the question of an individual's or a group's origins. Only in this sense can an attempt be made to determine the anthropological position of the Hauslabjoch glacier corpse within the Late Neolithic populations of the circum-Alpine region.

For such a comparative analysis we possess data on several thousand skulls and – admittedly on a lesser scale – of long bones from various Neolithic cultural groups in Europe. Unfortunately, only a few anthropological finds are available, with the exception of Remedello, from the cultures directly adjacent to the Hauslabjoch. For the Horgen culture of Switzerland we have only the data of a single male individual from Meilen on Lake Zürich. The Baden culture on the eastern edge of the Alps is relatively well attested by skeletal finds. However, we have also included in our examination a considerable number of comparative series from more distant cultures. In this analysis the Hauslabjoch corpse exhibits the greatest similarities with the anthropological finds of the Remedello culture.

The height of the Iceman, compared to present-day populations, is relatively short. Even so, it is slightly above the average for the Neolithic population of Tuscany, which is 1.58 metres. The north Italian and South Tyrol finds show heights the same as, or only slightly greater than, the Iceman's. Swiss Neolithic body height data equally provide relatively low figures, some of them well under 1.60 metres. The Horgen skeleton also measured only 1.60 metres. The representatives of the Baden culture, with an average height of 1.63 metres, are likewise not much taller than the Iceman.

Comparative series of Neolithic population groups from North Africa, Sicily, Sardinia, France and eastern Europe, as well as those of the Corded Ware and Bell Beaker people of Central Europe, average about 1.70 metres. Thus, even inclusion of body measurements from remote cultural regions shows that the Iceman fits very well into the circle of Alpine and circum-Alpine Neolithic groups. Of course, the link to Remedello should not be overestimated on anthropological grounds, as, in the absence of north-Alpine data, this is, for the time being, no more than a conclusion *ex silentio*. It does not, however, contradict the cultural connections postulated here between Hauslabjoch, Val Venosta and Remedello.

5 The Wooden Items and other Vegetable Finds

To make his equipment the Iceman used a lot of organic, mostly vegetable, materials. These included wood for his tools and weapons, bast as a cording and binding material and bark for the containers. Grasses were used extensively for his clothing. Finally, even the contents of the ember-container consisted exclusively of vegetable matter: grasses, leaves, tree needles and charcoal. Even the pollen adhering to his clothes and deposited in the ice at the site of the find is of great importance.

Apart from the pollen all the materials were chosen deliberately. Having said that, we must not assume that the Iceman invariably gathered the materials he needed with his own hands. One could imagine, for instance, that he did not plait his grass coat himself, but that it was made by other members of his village community. They would also have collected the special metre-long grasses for it. The same could apply to the expertly plaited scabbard.

There is also the possibility that he did not make other parts of his equipment, and so did not have to find the material himself. Acquisition could have been by trade, exchange, gift, theft or pillage, and for the flint or the copper axe – raw materials not available locally – such options might apply. Bones, on the other hand, presented no problem and stag antlers could be readily collected from those shed naturally.

However, the fact that prior to his death the Iceman had worked on his bow and on several arrow-shafts surely suggests that he collected the wood for them himself. The same has to be assumed for the contents of the ember-container, and for the grass – which, after all, would have had to be changed frequently - in his shoes. Hence, we believe that the Iceman obtained a large part of the raw material for his equipment himself.

As for the wooden objects, the variety of tree species represented is astonishing. This is no accident. Every implement is subject to certain, often very disparate demands, so the chosen raw material must satisfy particular ergonomic and technological preconditions. For his equipment the Iceman invariably chose the best-suited materials. This calls for considerable experience of a kind virtually lost to our own civilization, but indispensable for survival in a Neolithic environment. Naturally, the right trees and shrubs are not found everywhere: some prefer the valley, others higher altitudes. Within the range of his wanderings, therefore, the Iceman deliberately searched out certain trees. If we can demonstrate where they grew, we will gain a good insight into his living space.

As late as the nineteenth century the Val Venosta was one of the bread-baskets of South Tyrol (growing rye and buckwheat). This indicates that the soil and climate favoured cereal growing – which is no doubt what attracted the Neolithic farmers to settle there. Nowadays the principal agricultural line is fruit farming, in the east also grapes for wine. Intensive soil cultivation has squeezed out the native vegetation, with the result that we can nowadays scarcely visualize what the landscape looked like 5,000 years ago. Only in inhospitable elevated positions is there a chance of the original flora surviving undamaged.

In order to imagine the settlement area of Neolithic man we must first eliminate from our minds all the modern traffic arteries which now cover the landscape in a fine net, such as railways, motorways and other roads, down to the (invariably surfaced) farm and forestry tracks. Likewise the ugly high-voltage pylons and cables, and indeed the often disastrous regulation of even small streams. Industrial zones, leisure complexes and cultivated land make it impossible even to surmise the original state.

Trodden paths were all that was then needed for communication between localities and across regions. Given the small demands for movement, the routes probably changed from time to time. Added to this, in the Val Venosta, path alignments have presumably changed anyway as a result of the changing soil pattern. The Upper Adige valley is characterized to an especial degree by the gigantic gravel cones of the mountain torrents, which are subject to continual movement because of rock falls and gravel avalanches.

The entire valley floor was swampy and covered with dense primeval forest, through which the Adige time and again carved a new bed. It was most unsuitable for settlement, but abundant in game for the hunter.

For his settlements Neolithic man chose hillocks and elevations on the edge of the valley; good examples in the Val Venosta are the castle hill of Juval and the Tartscher Bühel. From there he created small clearings: the cultivated fields were in the immediate neighbourhood of the farmsteads. Next came more or less open woodland, from which timber for building was cut and which served for wood pasture and cattle droving. Beyond lay impenetrable primeval forest, and up the hillside extended dense mountain forests, the area where the Iceman roamed, in which he found his raw materials. The table on the next page shows the different kinds of wood he used.

Yew is a superb commercial timber. In prehistoric times it was

Tree	Latin name	Part used by the Iceman	Use by the Iceman
Yew	*Taxus baccata*	Wood	Bow, axe-helve
Lime	*Tilia sp.*	Wood (branch), bast	Retouching tool, cord, binding material
Ash	*Fraxinus excelsior*	Wood	Dagger handle
Hazel	*Coryllus avellana*	Wood (stem)	U frame of backpack, quiver bracing
Larch	*Larix decidua*	Wood	Boards of backpack, fuel
Wayfaring tree	*Viburnum lantana*	Wood (shoot)	Arrow-shaft
Cornel tree	*Cornus sp.*	Wood (shoot)	Arrow-shaft
Birch	*Betula sp.*	Bark, sap	Container, tar
Reticulate willow	*Salix reticulata T.*	Wood	Fuel
cf. Amelanchier	cf. *Amelanchier ovalis*	Wood	Fuel
Alder	*Alnus viridis*	Wood	Fuel
Norway spruce	*Picea abies*	Wood, needles	Fuel, ?
Pine	*Pinus sp. (non cembra)*	Wood	Fuel
Elm	*Ulmus sp.*	Wood	Fuel
Juniper	*Juniperus sp.*	Needles	?
Norway maple	*Acer platanoides*	Leaves	Insulating material
Blackthorn	*Prunus spinosa*	Fruit	Food

used almost exclusively for making bows. That the Iceman also chose it for the handle of his axe is an exception. Contemporaneous axe-helves known from circum-Alpine pile-dwellings are made – in order of diminishing frequency – of ash, oak, beech and pine. Yew's evergreen conifer wood is resin-free, tough, highly elastic and heavy, non-rotting and subject to minimal attack by wood-boring beetles. Because of its hardness it is difficult to work. The sharp-ridged trimming facets on the Hauslabjoch bow suggest that the Iceman used his copper blade on it. Yew is widespread in the southern and northern Alpine regions. It does not grow abundantly in the plain, but prefers medium-high mountain environments up to an altitude of 1,400 metres, as well as protected locations on south-facing slopes, including ravine faces and steep inclines.

Lime is a soft, not very flexible, but tough wood which is easy to work. In prehistoric times it was hardly ever worked; only rarely was a dagger handle or a knife handle fashioned from it. The Iceman used a piece of lime branch as the pressure rod of his retoucheur. Lime bast can be used as a binding material: separated into fibres, it is twisted or twined into threads and cords which resist tearing. The Iceman used it to wrap his antler points and raw sinews into bundles to place in his quiver. The sewing material for the birch-bark container and what is presumed to be the bowstring also consist of lime bast. Lime has no preferred habitat: in the Central Alps it occurs sporadically up to an altitude of 1,500 metres. It frequently invades stands of beech and alder in the plain, but also climbs slopes along with oak, hornbeam and maple forests. Along with elm, maple and ash it also grows in forested ravines.

The wood of the ash was used for many purposes in prehistoric times. It is hard, fine-grained and long-grained, elastic, pliable and easy to work. In prehistoric times it was used extensively for tool handles, especially for hatchet- and axe-helves. It was also often used for wooden dishes and bowls. The Iceman made his dagger handle from it. Nowadays ash is widespread as a result of planting. It occurs naturally in the Alpine region, but is absent from the Upper Inn valley and the Alpine Lech valley, as it is also from the Bolzano area and other parts of South Tyrol. It prefers humid riverbank and forest meadow habitats, is found in riverbank areas and in the mixed deciduous forests of the plains and on fairly mountainous positions up to an altitude of 1,400 metres. It avoids dry soils, but appears to thrive on chalky subsoils.

Hazel forms a moderately hard, tough, flexible and easily-split wood, with trunk diameters of up to 20 centimetres. As a working timber it played a minor role in prehistoric times when it was occasionally used for hatchet- and axe-helves. For the clay-rendered wattle walls of dwellings and as pliable switches for baking-oven domes it was indispensable. The Iceman chose one hazel switch each for the U-shaped frame of his backpack and for the lateral bracing of his quiver. The hazel shrub, more rarely tree, is widespread throughout the Alpine region. It forms part of the undergrowth in well-lit broadleaved forests, along with alder, elm, lime and maple. It prefers the edges of woodlands, hedges and mixed habitats in the plain and in the mountains it is found at medium-altitude locations, up to 1,600 metres.

Apart from the yew, the most durable coniferous wood is provided by the deciduous larch which is pliable and easy to work. Larchwood implements are not otherwise known from the Neolithic; in this respect the Iceman is again an exceptional case. He used it to cut two boards for his backpack. The larch is found from the Maritime Alps to the Vienna Woods. It occurs in pure stands or mixed with Arolla pine and Norway spruce, less often with mountain pine, beech or fir. It prefers the uppermost forest belt of the subalpine level, to which it climbs from the plain (300–500 metres) up to a maximum altitude of 2,300 metres.

Of the wayfaring tree (*viburnum*) only the thin shoots were used. These are very hard and heavy, also particularly straight, so that they were mainly suitable for arrow-shafts and occasionally for fine combs. Most of the Iceman's arrows were cut from the shoots of this tree. Its preference is for the sunny edges of woodlands, hedges, and sparse oak and pine woods. It grows in the meadow landscapes of the plains and is found up to 1,600 metres on moderately dry chalky soil.

Like the wayfaring tree, the cornel tree – about whose properties in use nothing is known – was used almost exclusively for arrow-shafts. Very occasionally the handles of daggers and knives were cut from it and the Iceman fashioned the short foreshaft of his composite arrow from its shoots. Of the species (*Cornus sp.*), the varieties most likely to have been used are dogwood (*Cornus sanguinea*) and cornelian cherry (*Cornus mas*), which are indistinguishable in terms of wood anatomy. Dogwood is widespread both north and south of the main Alpine ridge, mainly in meadow copses, while the cornelian cherry, north of the Alps, grows only in the Vorarlberg, in the Lake Constance area. There is also said to be a stand, albeit a doubtful one, in the Zillertal. South of the watershed it occurs in the eastern Val Venosta around Merano and, elsewhere in South Tyrol, only in the Bolzano area. It favours sparse mixed deciduous forests with undergrowth and climbs from the plain to mountain levels around 1,400 metres.

Birch was used in prehistoric times for two reasons. Its wood as such is not used except for firewood. The bark, on the other hand, is easy to detach, especially in young trees, and is flexible and highly resistant even when dried. This made it an ideal material for the manufacture of containers and cases, as well as boats (so far not proved in Europe). The Iceman had with him two such cylindrical containers, one of them to carry hot embers. Its sap, tapped from fresh cuts, can be boiled down to birch tar and used as glue. With this cement the Iceman shafted his arrows and the axe, though not the dagger. The birch favours sparse deciduous and coniferous forests, areas cleared by fire, poor pastureland, and, at higher altitudes, also mountain bogs. It grows from the plain up to the treeline at approximately 1,900 metres, especially in cool, humid climatic conditions.

Willows, here represented by the reticulate willow, grow in the Alps as creeping shrubs, chiefly above the treeline between 1,900 and 2,300 metres. They are found on slightly humic chalk-rich stony and coarse detrital soils which are under snow for long periods of seven or eight months, and are hence well soaked. They also occur as pioneer plants on exposed patches of earth and thus are important for the re-greening of cirques, areas from which stone avalanches have descended. The reticulate willow is of no importance as a commercial timber. The Iceman used it to light his campfire and he filled a container with some of its glowing embers.

The *Amelanchier* type has not been previously identified among prehistoric woods. Its fruit is considered unsuitable for human consumption. It favours chalky soils and is therefore mostly not found in the Central Alps. It occurs, sporadically and in groups, on firm slopes, in rocky clefts or detritus and on stony soils. From the plain it climbs to altitudes of about 1,600 metres. Charcoal remains of *Amelanchier* were found in the ember-container.

Alder is widespread in the Alps, but is found more often in chalk-poor subsoil. It grows on meadow soils, among the subalpine scrub and in avalanche ditches. Along stream-beds it descends lower, but it is also found at altitudes of up to 2,000 metres. Its coarse-grained, soft, but rather firm and elastic, wood was used in prehistoric times almost exclusively for whittled wooden vessels. The Iceman carried it as fuel in his ember-container.

Nowadays spruce is the most common conifer in the Alps. Normally it forms pure stands only above 800 metres. Mixed with other conifers or broadleaves it is found at lower altitudes. Even in cold climatic locations it grows up to the treeline, in the Tyrol to about 1,900–2,300 metres. Its soft, elastic and easily split wood was very rarely used and only a few small vessels are known. Granules of spruce charcoal were found in the ember-carrier. A few unburnt needles adhering to the Norway maple leaves are probably present only by accident.

Pine was not worked in prehistoric times. It is widespread and thrives on all soils, even the poorest. In the Alps it is found from the plain right up to the treeline. The Iceman carried it as fuel in his ember-container.

Elm has a hard, elastic and tough wood, which is, even so, easily split. In Neolithic times it was only very rarely used – as for an axe-helve or a digging stick. The Iceman's ember-carrier contained crumbs of its charcoal. Scattered or in groups the elm is a typical tree of meadow and ravine woodlands, and along with other species from the plain is found up to medium altitudes, though not above 1,400 metres.

Like the spruce needles, the needles of the juniper found among the insulating material in the ember-carrier should be regarded as present only by chance. Nowadays tough juniper wood is used for small carved and lathe-turned items. Its black-brown blue-bloomed berries are used in medicine, as a spice and as a fumigant. It is possible that its needles got into the birch-bark container accidentally, when the Iceman was gathering its berries. This evergreen shrub or tree favours sunny hillsides and manages on even the most infertile soils. On poor pastures, in shrub habitats and with dwarf shrub willows it can also be found at altitudes above 2,500 metres in the Alps.

The leaves of the Norway maple are a special case. They were used, along with tufts of grass, as insulating material between the hot embers and the wall of the birch-bark container. The stalks are missing; only the leaf blades remain. Their chlorophyll content indicates that they were freshly picked: as they are fully grown this must have been between June and September. Maplewood was very frequently used in prehistoric times for containers, less often for knife handles, hatchet- or axe-helves, or digging sticks. Its wood is very hard, resistant to rubbing and elastic. It is easily worked but tends to be attacked by wood-boring beetles and does not tolerate humidity. While the Norway maple did not originally grow in the Inn valley, it is widespread in the mountains and grows, singly or in groups, in oak woods and beech woods. It is found from the lowlands up to the mountain, less often the sub-alpine, level at 1,300 to 1,600 metres.

There is no prehistoric evidence for the use of blackthorn wood. Nowadays it is occasionally used for making walking sticks and is said to be very tough. Its fruits were generally popular, ripening from early August to early October. One sloe was found at the Hauslabjoch. Its fruit acid content and tanning properties make the fruit edible only after the action of frost, so that it can only be used between late September and early October. Blackthorn forms impenetrable hedges along the edges of woodlands and banks and is widespread, especially on sunny slopes, and also in sparse deciduous, more rarely coniferous, forests. It grows from the plain up into medium altitudes, to about 1,500 metres.

The vegetable materials used by the Iceman for making tools and equipment, as fuel, as food or for insulation were, in so far as they came from trees or shrubs, examined also in respect of their favoured habitats. They present an interesting picture of the range of his wanderings as well as his living space – though questions on the structure and economy of his 'native village' will be discussed later. But we must work on the assumption that this village, as archaeological analysis suggests, was in the Val Venosta.

Under normal growing conditions, not yet perceptibly affected or impaired by Neolithic man, the plants identified here can be found from the meadow forests of the valley floor via the plains, the terraces, low-level erosion gravel cones, hillside and mountain forests right up to the high-mountain zone above the treeline. The charcoal content of the birch-bark container, in particular, includes elm (which must have come from lower altitudes) and ranges through the other identifiable remains to species growing right up to the treeline. The remains of the reticulate willow with its thin stem (judging from the radius of the annual rings) even come from the region above the treeline. The raw materials deliberately chosen for the wooden implements grow mainly between these two extremes. It seems reasonable to suppose that the Iceman obtained these different species himself.

As for the charcoal remains from the ember-carrier, it seems certain that he gathered his firewood with his own hands. He lit campfires from the valley floor (elm) up to the treeline (reticulate willow). As a rule fires are lit at nightfall and so testify to more prolonged sojourns rather than a brief stop in one place. For the reconstruction of his lifestyle it is necessary to develop a model which includes activities both at the valley floor and above the treeline.

The cornelian cherry does not occur naturally in the Upper Inn valley or its side-valleys, but does grow naturally south of the main Alpine ridge, in the eastern Val Venosta and in the neighbourhood of Bolzano. If the arrow-shaft was made of that wood, and not of dogwood, then this provides a vital clue that the Iceman's range of movement was oriented to the south of the Hauslabjoch. Thus botanical examination may well point in the same direction as the archaeological analysis.

Examination of the major plant remains has also yielded important clues about the Iceman's time of death. Glaciological considerations, as well as analysis of related mummification phenomena, point to the time of the first snowfalls at the onset of winter. In support of this conclusion are the facts that the leaves of Norway maple were picked between June and September and that the sloe was presumably gathered in the second half of September or the first half of October. This is further confirmed by the analysis of the pollen spectrum.

Botany has developed methods for determining the seasonal stratification of a glacier by means of the pollen encapsulated in the ice. In the immediate vicinity of where the quiver was found samples were taken of the (at that spot undisturbed) ice. From this, 2,222 pollen grains were isolated in the laboratory. The count produced high figures for pine (*Pinus*), alder (*Alnus*) and meadow grasses (*Poaceae*). A medium pollen count was given by mugwort

(*Artemisia*), while birch (*Betula*), hazel (*Corylus*), spruce (*Picea*), elm (*Ulmus*) and beech (*Fagus*), along with ivy (*Hedera*), fern (*Pteridium*), stinging nettle (*Urticaceae*) and plantain (*Plantago*), were confirmed only individually – indicating that their main dispersal was over. It follows, therefore, that the ice in which the Iceman rested was formed between late summer and early autumn (September/October).

6 The Cereal Finds and Neolithic Agriculture

Of outstanding importance is evidence of domesticated cereals, which represented the principal source of carbohydrates in the Neolithic diet. During the Alpine Neolithic four types were cultivated, of which three were varieties of wheat: einkorn (*Triticum monococcum*), emmer (*Triticum diococcum*) and durum (*Triticum durum*). The fourth was barley (*Hordeum vulgare*). Cultivation of these varied in intensity both regionally and over time. The wheat varieties provided grain for bread, while barley was eaten as groats or added to stews.

In the course of the restoration work in Mainz two barleycorns were discovered. Both were in their husks and had become caught in the fur tufts of the Iceman's clothing. They did not, therefore, come from the Iceman's food store, but had got to the Hauslabjoch quite by chance. Also found among the insulation material of the ember-container were two uncharred remains of husk and one fragment of an einkorn ear, as well as fragments of chaff from an unidentifiable variety of wheat (*Triticum sp.*). These finds prove conclusively that the Iceman was in touch with a grain-growing community – presumably the community of his 'native village'. Grain is harvested in the late summer or early autumn, then threshed and stored for the winter. The remains of ears of wheat represent threshing or winnowing waste. It follows that the Iceman was in his village during or shortly after threshing time; after all, the insulating material had to be replaced from time to time and could not, therefore, have been in his ember-carrier for long. As harvest and threshing time called for the participation of all the able-bodied members of a village, one may assume that the Iceman, if he routinely spent prolonged periods away from his settlement, would have returned in the late summer specially to help with harvesting and threshing.

7 Neolithic Farming

Neolithic man's opening up of settlement areas and the resulting economic development in the mountain valleys proceeded along different lines from that of the plains and lake shores. One should bear in mind, however, that, alongside hunting and gathering, the Neolithic economy was always based on two pillars - the growing of crops and the raising of livestock.

Research into the bog settlements in the foothills north and south of the Alps, which has intensified over the past few decades, has yielded remarkable insights into the economic structure of the Neolithic period, testifying to highly developed arable farming. These findings may be applied, with some caution and certain reservations, also to the Val Venosta. Study of the – still rather scanty – Neolithic finds from the valleys between the Ötztal Alps and the Ortles massif has shown that the high ground along the valley flanks was favoured for village sites. Mention has already been made of the castle hill of Juval and the Tartscher Bühel. The topography of settlement thus reveals, on the one hand, a link with the valley plains and terraces and, on the other, with the higher mountain altitudes, which were evidently integrated in the economic system. Location of settlements on high ground was very probably due to a need for security, as it allowed the inhabitants to survey the immediate neighbourhood, so that imminent danger, whether from humans or wild animals, could be spotted in good time.

Land had to be cleared both for dwellings and for fields and this was done by the 'slash and burn' method. The risk of resulting forest fires was gladly accepted: unless the cleared areas were put to agricultural use, the burnt land would quickly be reclaimed by nature.

In addition to the cereals Late Neolithic man also cultivated oil crops such as linseed (*Linum usitatissimum*) and opium poppy (*Papaver somniferum*) and, alone among legumes, peas (*Pisum sativum*). One can assume that these crops were also grown in

Neolithic times in the Val Venosta, even though the Iceman's clothes and ember-carrier have so far only confirmed the cultivation of einkorn.

Certain agricultural practices are best detected by analysis of weed seeds found in crop samples from Neolithic settlements. Studies of Swiss pile-dwellings have shown that in the earlier phases of the Neolithic, methods of soil cultivation were not yet very intensive. It seems possible that shallow troughs were scratched with digging sticks at rather wide intervals and that the seeds were then placed in them. In consequence there was enough room left for other plants to grow in between. From the Late Neolithic onwards – the time of the Iceman – cultivation techniques improved; samples from that time contain typical crop weeds previously absent, which suggests that the crops were planted more closely, possibly as a result of seed sown by being broadcast. Weed analysis also shows that both winter and summer crops were planted. The grain varieties presumably provided the winter crops, sown in the autumn; in the spring summer crops such as linseed, poppy and peas were sown. Both kinds of crop were harvested in September and October.

It also emerges that there was an extension of the area under crops in the Late Neolithic, which is explained by a significant population growth. Additional areas were cleared at that time, even on poor ground and that facing slopes. This is suggested by an increase of those crop weeds which favour flatter, drier and more alkaline soils, a factor which seems to apply also to the Val Venosta, where, compared to the older finds, a definite increase in the density of Late Neolithic sites has been recorded. In Switzerland this began primarily at the time of the Horgen culture; in the Val Venosta one might think of influences from a late Square-Mouthed Pot culture and the Lagozza culture. It is possible that this development was witnessed by the Iceman, whom we have assigned to the early phase of the Remedello culture, which immediately followed.

The various crops were planted in rotation in separate fields, a procedure that invariably started with the demanding durum wheat. Though there is no firm evidence, we may assume the existence of fallow areas, which, along with the harvested fields, were used to graze livestock. The grazing animals' droppings improved the soil, so that a field could be cultivated over a number of years. Grain was harvested not by gathering the ears, but invariably by cutting the stalks. In this way straw was obtained, albeit on a limited scale. Along with chopped leaves (the leaf-bearing twigs of deciduous trees), the straw was used as fodder for the livestock in winter. Haymaking has not been proved to have occurred anywhere during the Neolithic period.

To cut the grain, flint sickles and harvesting knives were used. In this context we may recall the 'sickle gloss' on the Iceman's blade tool, something which could lend further support to the idea that he helped with the grain harvest and the threshing. The Late Neolithic peasant should certainly be credited with considerable agricultural skills.

Agriculture was supplemented by food-gathering, but this played no more than a subordinate role compared to the economic dominance of grain-growing. Of importance as food for storage were crab apples (*Malus silvestris*), which were also dried, hazel-nuts (*Corylus avellana*), acorns (*Quercus sp.*) and beechmast (*Fagus silvatica*). Seasonal variety in the diet was provided by fruit and berries, such as raspberries (*Rubus idaeus*), blackberries (*Rubus fruticosus*), rose-hips (*Rosa sp.*), elderberries (*Sambucus sp.*) and sloes (*Prunus spinosa*).

We may also assume that Neolithic man ate vegetable and salad plants which grew naturally in the wild, but this cannot be proved because these plants are picked before their seeds ripen. Tree fungi were gathered as tinder and for medicinal purposes; they are frequently attested and were also used by the Iceman.

8 Animal Products and Neolithic Stockbreeding

For the manufacture of his equipment as well as for his food the Iceman, as one would expect, made considerable use of animal products. His sources were both domesticated and wild animals. Palaeo-zoological research has not identified all animal species, partly due to the fact that no find has so far been made where delicate organic material, such as fur, leather, feathers and sinews, has been fully preserved, and partly because further methods of differentiation are still needed. In the meantime the following picture may be given:

Species	Latin name	Material	Used by the Iceman
Ibex	*Capra ibex*	Bone (as found)	Meat
Red deer	*Cervus elaphus*	Antler	Punch for retoucheur, large spike, 4 points
Goat, sheep, chamois or female ibex	*Capra sp.* sive *Ovis aries*	Bone	Awl
Domestic or wild cattle	*Bos sp.*	Calf leather	Belt
Red or roe deer	*Cervus sp.*	Hide (fur)	Clothing
Goat, chamois or ibex	*Capra sp.*	Hide (fur)	Clothing
Cattle or red deer	*Bos sp.* sive *Cervus elaphus*	Sinew	Threads and cords
Bird	*Avis (fera)*	Feather	Arrow fletching

Among the wild animals the first to be identified with certainty was the ibex (*Capra ibex*) from cervical vertebra fragments. These are evidently food waste and testify to a successful hunt for that animal.

Attested with equal certainty is the red deer (*Cervus elaphus*), whose antlers played an important part in the manufacture of implements during the Late Neolithic. It is not clear whether the Iceman's antler implements – the punch of his retoucheur, the long curved spike and the bundle of four points – were made from shed antlers or from those of slain deer. Both possibilities – hunting and gathering – have to be considered.

The bone material of the small awl from the Iceman's belt pouch can, at present, be narrowed down no further than to domestic goat, domestic sheep, chamois or female ibex (*Capra hircus, Ovis aries, Rupicapra rupicapra* or *Capra ibex*).

Much the same applies to the calf-leather belt, which may have come from either domestic cattle or the wild aurochs (*Bos taurus* or *Bos primigenius*). Indeed even the European bison (*Bison bonasus*) cannot be entirely ruled out, though in the Alpine region it is one of the rarest quarries of Late Neolithic hunters, favouring as it does extensive forest landscapes and so being found more often in the plains. The aurochs, on the other hand, was equally at home in meadow woodlands and could well have lived in the Upper

Adige valley, even though only in small numbers.

Otherwise the Iceman's clothing consists of fur leather obtained by tanning the skins of deer and goat-type animals. On the one hand we have wild animals – red deer, roe deer (*Capreolus capreolus*), chamois and ibex – and on the other the domesticated goat. Red and roe deer would have had to be hunted. The goat-type fur could have come from wild or domesticated animals. The sinew material from the quiver can, because of its size, only have come from a stag or from wild or domesticated cattle. The feathers of the arrow fletching must have come from a wild bird, as Neolithic man did not keep poultry. The first domestic fowl introduced to Central Europe is the chicken, first found also in our region in the Early Iron Age (Hallstatt culture) in the sixth century BC. This does not, of course, exclude the possibility that captured wildfowl were occasionally kept in Neolithic settlements.

Moulted bird feathers could be gathered. For the fletching of the arrows, however, numerous and carefully selected feathers were needed, which makes a hunted bird more likely.

Analysis of the animal material from the Iceman's equipment proves, first of all, that he hunted ibex, as well as deer and birds. In the case of stags and birds one should also consider gathering (shed antlers, feathers). But as it is not possible to decide whether the other materials come from hunted or domestic animals, the find in itself offers no basis for deciding whether the Iceman had anything to do with the keeping of domestic animals.

Nevertheless, as we have pointed out, stockbreeding was one of the two pillars of Neolithic farming; this is true for the entire circum-Alpine region and, of course, for the Val Venosta. In Neolithic Central Europe the five domestic animals were the dog, the goat, the sheep, the pig and the cow. The domesticated horse, attested at least in the Late Neolithic, stands outside this group. We shall deal with this useful animal as well.

The dog (*Canis familiaris*) is descended from the wolf (*Canis lupus*) and is man's oldest domestic animal. Its domestication can be traced back to the Palaeolithic period when it was kept in settlements in relatively small numbers and used as a guard as well as a hunting companion. In addition, it made itself useful as a consumer of refuse. Ornamental pendants from pierced canine metacarpals from Neolithic Switzerland testify to the special relationship between man and his dog. Its meat, as shown by slaughter marks on some dog bones, was eaten only rarely.

The domestic pig (*Sus domesticus*) has the wild boar (*Sus scrofa*) as its ancestor. If pigs kept at the settlement were driven into the woods in the autumn to feed on acorns and beechmast, mating with a wild boar might easily have happened. But the owner would probably not have been bothered if, in spring, a sow gave birth to a litter with faint longitudinal stripes on their backs, so long as they were healthy and numerous. The pig is not very demanding in terms of care, and its sole purpose is as a supplier of meat and lard. Pigs accounted for a significant, though varying, part of domestic livestock in all Late Neolithic settlements where bone finds have been anatomically examined.

There is no doubt that cattle (*Bos taurus*) were the most important domestic animals in the Neolithic. They are descended from the aurochs (*Bos primigenius*). If they were left alone on a pasture, mating with an aurochs bull might have occurred – a problem that could be avoided by slaughtering the calves early. Cattle are fairly demanding domestic animals. When looking after them becomes difficult, for instance through prolonged winter feeding, the percentage of cattle might be reduced in favour of pigs, because they will eat anything. At times, however, cattle could account for up to 60 per cent of domestic animal stock. Where they proved difficult to keep, the percentage could go down to 20 per cent or less. As suppliers of meat, lard, milk, butter and cheese, as well as of hides, sinews and other raw materials, cattle are the most useful domestic animals. Evidently by the Late Neolithic they were also used as draught animals and beasts of burden.

The history of the domestication of the horse (*Equus sp.*) is in many respects still a mystery. For one thing, this is because, at least at an early stage of domestication, the bones of domesticated and wild horses are not easily distinguishable. The oldest European evidence of horse domestication comes from the Ukraine, and is dated to the end of the fifth millennium BC. It is therefore entirely possible – in some cases even certain – that the finds of horse bones from Late Neolithic settlements in Central Europe are already those of domesticated animals. Besides, in the opinion of many scientists, the wild horse of the steppes would not have found suitable habitats in the primeval forests of Central Europe.

Finds of double yokes and parts of wagons with disc wheels from bog dwellings indicate the use of draught animals in the

Alpine region in the Late Neolithic period. This, of course, brings to mind the representation of a hitched wagon on Stone 2 from Algund in the Val Venosta. The picture is reduced to an awkward outline drawing. We recognize a rectangular latticed wagon box as seen from above. Its wheels are shown laid flat in side-view, quite out of perspective. Between the four-legged draught animals is a central shaft; whether the short crossbar at its front end is meant to be a yoke remains an open question. The literature identifies the draught animals as oxen or cows: the sickle-shaped arcs at the animals' heads are interpreted as horns. They cannot convincingly be interpreted as horses' ears.

The representation of the wagon on the Algund stone is considered to be the third 'painting layer', which would make it more recent than the engraved symbols in the shape of Remedello daggers and axes. At a very rough estimate this third working of the stone could have been performed at any time between the Neolithic and the Iron Age, i.e. within a timespan of up to 3,000 years. With the new finds of wagon parts and yokes from levels of the Pfyn and Corded Ware cultures, the history of the wagon wheel and hence of draught animals goes back very much earlier than had hitherto been supposed. Thus the Algund representation of the wagon might still be Neolithic and hence contemporaneous with the Iceman. Admittedly, the picture is of a cattle-drawn vehicle. However, finds of horse bones at late Neolithic settlements in the circum-Alpine region indicate that the horse, too, was then used as a draught and working animal. In all cultures of the ancient world where the horse was domesticated, it was used first as a draught animal and only much later as a mount for a rider.

After the dog, man's oldest domestic animals are the goat (*Capra hircus*) and the sheep (*Ovis aries*). Their domestication took place about 10,000 years ago in the Middle East, in the area of the Fertile Crescent. Sheep and goats can easily be put out to pasture together because they compete for fodder only to a limited extent and thus make the best possible use of the grazing. As today, it was sheep that predominated in prehistoric small-livestock herds: they were kept not only for meat, fat and milk, but primarily for wool. The goat, on the other hand, supplies more milk.

Whereas cattle, if only because of their enormous consumption of fodder, have always been grazed near the settlements or on specially cleared good-quality soils, a particular form of grazing practice developed for the less demanding sheep and goats at a relatively early date – the pattern known as transhumance. This involved the extensive use of natural meadows, usually on poor soil, which are unsuitable for any other purpose. In consequence no improvement to the soil occurs through humus accretion, manure or the cutting down of thickets. Especially suitable for transhumance are fire-cleared areas where natural reforestation is prevented by animals removing any early new growth. Goats, in particular, are fond of nibbling buds and green shoots, literally nipping a return to nature in the bud. As such a method of grazing, even for relatively small herds, calls for large areas, which are often a long way from the settlements, a shepherd can be away with his animals for weeks or months. We know from historical times that this type of roaming grazing economy sometimes covers many hundreds of kilometres. Especially in areas which are under snow during the winter, a shepherd will choose grazing where the winter is mild. Thus in the sheep-farming areas of the Swabian and Franconian Alb in southern Germany the flocks used to make two major moves each year – in the spring they were driven up to mountain pastures and in the autumn they were driven as far as Burgundy and the adjacent regions, sometimes even further south, to return once more in the spring.

European transhumance should not be seen as a kind of intermediate stage of nomadism. It had an established place within a locally based agricultural system. Professor Wolfgang Dehn, a prehistorian from the University of Marburg, reconstructed the model something along these lines: the farmstead with its permanently cultivated fields and stock of cattle, horses and pigs represented the core; in addition, sheep and goats were useful suppliers of meat, skins, hides, suet, milk and wool. In summer a shepherd would look after them on grazing in the vicinity – in Swabia and Franconia these would be mountain pastures – and in winter he would move with them to better grazing, from where he would not return until spring. It may be assumed that in prehistoric times transhumance operated in a similar way.

In our context it is of paramount importance that a similar transhumance economy, hitherto attested only in historical times, was practised by the Val Venosta peasants and, albeit on a lesser scale, is still being practised today. From mid-June onward the flocks are driven up from the south, through the Val di Senales across the saddles of the main ridge (Hochjoch, Niederjoch, Eisjoch, Hauslabjoch, Tisenjoch) into the Ötztal with its side-valleys (Niedertal, Rofental, Venter Tal, Gurgler Tal). There the Val Venosta farmers have grazings to this day.

The flocks start to be driven down in early September, a task which begins with the difficult job of rounding up the animals and searching for stragglers. The move back to the Val Venosta, a distance of some 50 kilometres, takes sheep and shepherds one or two weeks.

An important feature is the twice-yearly shearing of the sheep, which is one of the reasons why the animals return to the settlements in spring and autumn. This is not a job the shepherds can do single-handed. Prior to the invention of shears in the Late Iron Age (La-Tène culture) the mature wool was plucked. At the same time any superfluous ram lambs and old animals no longer suitable for breeding were slaughtered.

As regards the ratio between large and small animals, we have significant data from Late Neolithic settlements in Switzerland. There it was established that the ratio changes in favour of sheep and goat husbandry the more distant the villages are from the lowlands and the closer, for instance, to the high ground along the rim of the Swiss Jura. If the Val Venosta, a mountain region, followed this model, then in Neolithic times small animals must have played a much greater role than cattle, horses or pigs. Unfortunately, we have no dated animal bones from Late Neolithic settlements in the Val Venosta. But we do have research findings for the Ötztal collated by glaciologists, meteorologists and botanists at Innsbruck University, based on pollen profiles from bogs. Slope moors and high moors are created even in the mountains wherever there is a build-up of water; these record the pollen of plants over many thousands of years as though on a calendar.

Numerous pollen analyses were carried out in the Upper Ötztal in the course of glaciological, climatological and vegetation research. For that purpose the highest peat profile in the Alps, the peat bog at the Hinterer Rofenberg at an altitude of 2,760 metres, was evaluated. This bog is situated exactly opposite the Hauslabjoch. As the area lies well above the natural treeline, environmental influences could only be determined through analysis of grassy vegetation. It was found that changes in the plant communities, caused by human activity, date back as far as 4300 BC, i.e. to a thousand years before the time of the Iceman. At that period we observe a distinct restructuring in the vegetation cover, from natural scrub heath to herb-rich Alpine grass cover in which 'grazing indicators' predominate, especially *Ligusticum mutellina*, a plant closely related to lovage. (Indicators are plants growing mainly where herds of animals are driven for grazing.) We therefore have proof that high-Alpine pasture use, which occurred in the Ötztal in historical times, must have begun as early as the end of the Middle Neolithic. It should be noted that transhumance into the Ötztal in the Middle Ages and in modern times was practised from the Val Venosta in the south, and not from the north, from the Inn valley or the Lower Ötztal.

Pollen profiles from different altitudes in the Ötztal reveal a surprising fact. Normally one would expect economic expansion to proceed from the valleys upwards, in other words, that settlement would first be established at the valley floor, with higher areas being progressively integrated into the economy through clearance, initially near the villages, with new grazings being opened up gradually up the hillsides and mountains. In fact the exact opposite happened. The pollen profiles prove without a doubt that the extension of pasture started at the naturally tree-free mountain grasslands above the treeline – which, in climatically favourable areas here can be as high as 2,300 metres - and from there proceeded down towards the valley.

Yet another finding derives from the pollen analyses of the Upper Ötztal. The pollen of plants indicating human settlement - farmsteads and villages – does not appear until about AD 1000. This accords well with the first mentions of Ötztal localities in legal documents: these occur not earlier than the twelfth century. Moreover, the Upper Ötztal was used exclusively by South Tyrol farmers in the Middle Ages: one document from AD 1300 refers to 'Vent in the Schnalstal' (the Val di Senales). Indeed, until 1919 Vent, for ecclesiastical purposes, actually came under the parish of Tschars in the Val Venosta. This shows that the economic use of the Upper Ötztal, starting in the Late Neolithic and continuing well into the Middle Ages, was invariably effected from the south, more precisely from the Val Venosta, across the saddles of the main ridge of the Alps.

Two vital facts emerge:

- in the Iceman's day the mountain grasslands of the Upper Ötztal were used for grazing;
- his agricultural activity originated in the Val Venosta.

A few more aspects might be summed up here:

- in Neolithic times the keeping of sheep and goats was given increasing preference over that of large animals the more mountainous the settlement regions were;
- during the summer a shepherd would leave his native settlement with his flock for several months on end;
- the economic hub was the village in the valley, with arable farming and extensive stock-keeping.

9 The Iceman's 'Occupation'

The time has come to ask ourselves how the Iceman was integrated into the life of his Late Neolithic village. Even before I started writing this book there was a lively discussion as to his daily routine: an attempt was made to give him a definite occupation almost in the sense of a modern guild system. The following suggestions have been made: he was an outlaw; he was a shaman or priest; he was a prospector for metal ores; he was travelling on business; he was a hunter; he was a farmer; he was a shepherd.

Before discussing the various theories and possibly deciding in favour of one or the other, we shall briefly examine the question of his social status. In view of the fact that, by comparison with excavated contemporaneous graves, his equipment was fairly extensive, and especially in view of the copper axe he had with him, there has been a tendency to credit him with high social standing. 'He must have been a rich farmer,' someone said. To start with, we should bear in mind that the Iceman is an isolated individual, whom, moreover, we did not encounter in a grave. Hence we cannot confront him with equipment-based sociological researches such as have, sometimes successfully, been conducted at major prehistoric or early historical burial grounds. We have no adequate basis for comparison, the less so as grave goods and status during life can be two totally different things. Whether in the Late Neolithic a copper axe represented a particularly valuable piece of property is, as mentioned before, very much a matter of doubt and should certainly not be accepted without reservation. Besides, the huge number of representations of axes and daggers on the picture stones of Algund and Latsch suggests that copper artefacts were entirely commonplace in the Val Venosta at the time in question. They are probably more indicative of the man's background than of his standing or wealth. In the necropolises of the Remedello culture metal articles are likewise not particularly rare. Those buried with such articles certainly do not appear to belong to a particular social class – a view shared, no doubt rightly, by scientists. But as, with one single exception, we do not know any graves from the closer vicinity of the Hauslabjoch, such judgements are virtually impossible to make. We probably would not be far wrong if we assumed that he was neither strikingly poor nor significantly rich – in fact, a perfectly ordinary person.

The idea that the Iceman was an outsider, a person expelled from his community, is based on the remoteness of the Hauslabjoch site. But as we have established that the site is in the immediate vicinity of the passes (only 80 metres from the Tisenjoch saddle) over which, then as now, the grazings of the Ötztal were managed from the Val Venosta, this argument has no force. His 'wood scout' equipment and fur clothes, with an absence of textiles, is also given as proof of this theory. Furthermore, it is claimed, the tools, weapons and other implements that he carried, including materials for repairs, suggest that he had left his community to lead, and end, a lonely existence in the wilderness.

Quite apart from the fact that these are all conclusions *ex silentio*, a number of weighty arguments can be mustered against them. It is certainly correct that equipped as he was he could have survived for some time without regular supplies. We shall return to this point later. On the other hand, his medical tattoos show that he was undergoing treatment, which means that he was in touch, at least, with one other person with healing skills. Next, the threshing residue in his clothes and his ember-carrier testify that at harvest time in the year of his death he was still in a crop-farming village. The manufacture of his clothes and perhaps some of his equipment – the dagger scabbard especially – is also likely to have been someone else's work. This is corroborated by the very variable quality of the repairs to his clothing. He himself

was probably only responsible for the clumsy work with twisted grasses. All in all, there is no evidence of his having been an outsider; on the contrary, the Iceman was, at least temporarily, in close contact with a settlement community.

Another theory credits the Iceman with being a shaman. Linked with this theory is the idea that he had gone among the mountain peaks in order, in a priestly function, to be as near as possible to his gods or indeed to sacrifice himself up there. Shamanism is predominantly a phenomenon among primitive societies and is not infrequently linked to hunting populations. Although many books have been written about the nature of shamanism, most scholars of the civilized world face this phenomenon helplessly and uncomprehendingly, so that their studies are usually confined to a description of outward aspects. Accordingly, a shaman is masked and disguised as an animal and hung with numerous amulets, magical images and items which are used to produce noise. His head is hidden by a mask representing a stag, a bird,

Representations of hybrid creatures, interpreted as shamans. These pictures clearly demonstrate that the Iceman cannot have been a shaman, as has been repeatedly speculated. Left: 'Bison-man' from Gabillon, Dordogne; centre: 'Stag-man' from Les Trois Frères, French Pyrenees; right: 'Aurochs-man', also from Les Trois Frères. The date of these cave paintings is estimated at approximately 20,000 to 15,000 BC.

a bear, a bison or some other animal. Typically he dances to monotonous music, mainly from rhythmical instruments; this puts the shaman into a trance and can lead to exhaustion or fainting. Not infrequently, the action seems to be a submission to the animal to which he feels spiritually linked, but which nevertheless has to be killed to provide food for the clan. That is why defused hunting weapons are said to belong to a shaman's equipment. However, a shaman's activity would seem to be considerably more complex, to reach down to the depths of the human and animal soul and to embrace ghostly manifestations at various transcendental levels.

The idea that the Iceman was a shaman or a priest is ultimately a reflection of a widespread human tendency to mystify things that are difficult or impossible to understand. Reasons given for this interpretation include the loneliness and high altitude of the site, the fact that the Iceman carried an amulet-like object, the unfinished hunting implements, the tattoos and his alleged castration.

It is correct that the spot where the Iceman was found, at 3,210 metres above sea level, is the highest archaeological site in the Alps. But we have already shown that the nearby passes were used for economic reasons as long ago as prehistoric times. There is a rational, everyday explanation for the presence of a person who suffered an accident on that mountain track.

To interpret the marble bead with the fur-strip tassel as a shamanic device would imply that we should elevate every prehistoric grave containing at least one bead or similar object to the status of a shaman's grave. This would leave us with hardly any non-shaman graves. So far scholars have not developed any model for identifying shamans' graves – which would certainly make an attractive subject for research. On the other hand, a few rock drawings are known to us, mostly from the Palaeolithic period, which evidently portray persons dancing with animal masks. These are usually interpreted as shamans or animal deities.

Exact analysis of the traces left by working refutes the claim that the unfinished bow and arrow-shafts are shamanic make-believe weapons. There is no doubt that they were intended to make effective hunting gear. The Iceman's death prevented their completion.

The tattoos were unequivocally shown to be therapeutic aids within the framework of one or several treatments, designed to ease the Iceman's joint pains. They are therefore susceptible to rational explanation and need have nothing in common with spiritualism or mystical number games.

The castration theory is refuted by the anatomical findings and should be dismissed as nonsense.

There is thus nothing left which might connect the Iceman with any religious occupation.

It has been suggested that the Iceman may have been looking for or was indeed a 'professional' prospector for metals. This is based on the fact that he carried a copper axe with him. He must therefore, it is argued, have had something to do with the metallurgy of his day. Moreover, the site near a pass high up in the Alps suggests that he was on his way to look for ore deposits in the mountains. The difficulties associated with this set of hypotheses can only be inadequately touched upon and, as yet, cannot be satisfactorily resolved.

It will be useful, therefore, to set out what we know about metallurgy in the Late Neolithic, i.e. the time of the Iceman. Equally, we must set out what we do not know.

There is no question that metallurgy existed in the second half of the fourth millennium BC – indeed, the Iceman's copper axe confirms it. Our survey of Late Neolithic cultures north and south of the Alpine arc has shown us that metallurgical skills were common and very widespread at that time. We have also seen that metallurgy, at least in the region under consideration, was confined to the working of copper and exceptionally of gold (Cham culture). By way of contrast, further east, for instance in the Carpathian basin, there existed a highly developed gold metallurgy by the Late Neolithic.

In the cultures we have considered, those relevant to the time of the Iceman, there is not a single indication that other metals, such as silver or tin, were used. The two silver finds from the Remedello culture, the hammer-headed pin and the moon-sickle-shaped ornament, will have to be dated to the last centuries of the Late Neolithic, i.e. a time clearly later than that of the Iceman. In the Alpine rim cultures considered by us the first silver finds date only from the Bell Beaker culture and, as mentioned, probably an advanced phase of the Remedello culture. This entitles us to assume with certainty that, in our context, all metals in early use other than copper and gold can be eliminated.

Another important observation is based on the settlements of the Pfyn, Cham and Mondsee cultures. There crucibles were found, indicating clearly that copper or gold was worked. This is evidence that, even before and during the period of the Iceman, copper and gold articles were cast on the northern edge of the Alps. The melting and casting pots, of course, prove no more than that the people in the settlements melted copper and gold and cast them into desired shapes. They do not tell us from which deposits the metal came nor whether the ore was smelted.

The third source of information is represented by the copper articles themselves, such as the Iceman's axe-blade. As explained in the second part of this book, when the axe was described, the blade was not cast from pure copper, such as is occasionally found in the form of nuggets at outcrops of ore. Instead a copper ore was used, most probably malachite or azurite, which had first to be smelted. The crude copper was then cast or cold-worked into the desired shape. For the casting technique, at least, we have positive evidence from Pfyn and Mondsee.

The three paragraphs above sum up our hard knowledge of Late

Neolithic metallurgy. The catalogue of what we do not know is far more extensive.

North and south of the main ridge of the Alps there are numerous major and minor copper deposits which could have been reached by the Iceman in a few days' walk. The search for and discovery of surface outcrops of ore would have been no problem for Late Neolithic man and was no doubt practised by him with success. Mining archaeology as a rule proceeds from the assumption that the metallurgy of copper (and likewise of gold) began with the accidental finding of pure nuggets. *Homo faber* experimented with the gleaming objects and discovered their usefulness and potential applications. The next step, then, was the realization that the same metal can also be obtained from ores; the ability to provide the necessary high smelting temperatures was something Neolithic man had acquired from clay kilning.

But already we encounter considerable gaps in our knowledge. We do not know where Neolithic man mined his ores, what techniques he employed nor what implements he used. Dr Gerhard Sperl of the Solid State Physics Institute of the Austrian Academy of Sciences in Leoben has pointed out that the simplest method of ore extraction is the scraping off of outcropped malachite crusts. The trace metal content of the Hauslabjoch axe would be approximately that of ore which had been extracted in this way and smelted. So far, however, no Neolithic extraction or smelting sites are known in the Alpine region and the earliest evidence of prehistoric mining in the Alps dates back only to the Early Bronze Age. The question is whether older traces could ever be identified by archaeology, as most of them would be overlaid by more recent, and especially historical, mining work.

Another, no less important question is whether the persons engaged in finding and smelting ore, producing copper and manufacturing implements were specialized in these activities and were making their livelihood by trading their finished products. If so, this would mean that the Iceman had acquired his copper axe in exchange, for instance, for foodstuffs. Such an idea would be supported by the fact that in many cultures the miner and metalworker occupied a special position in his community, guarded the procedures of his craft as a secret, and because of his skills enjoyed particular respect.

In the circumstances, and given the exceedingly scant infor-

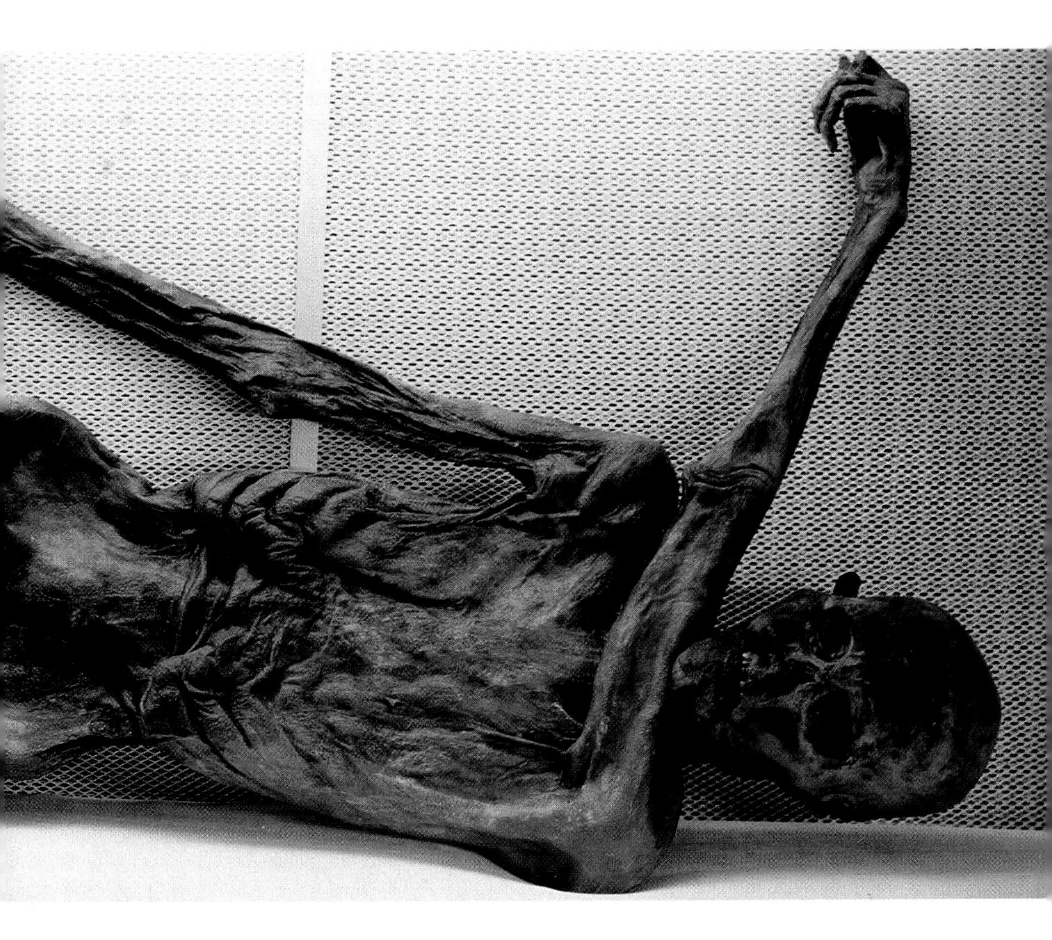

Front view of the Iceman's upper body. It clearly shows the unnatural position of the left arm, which was forced under the chin by slight movements of the glacier ice.

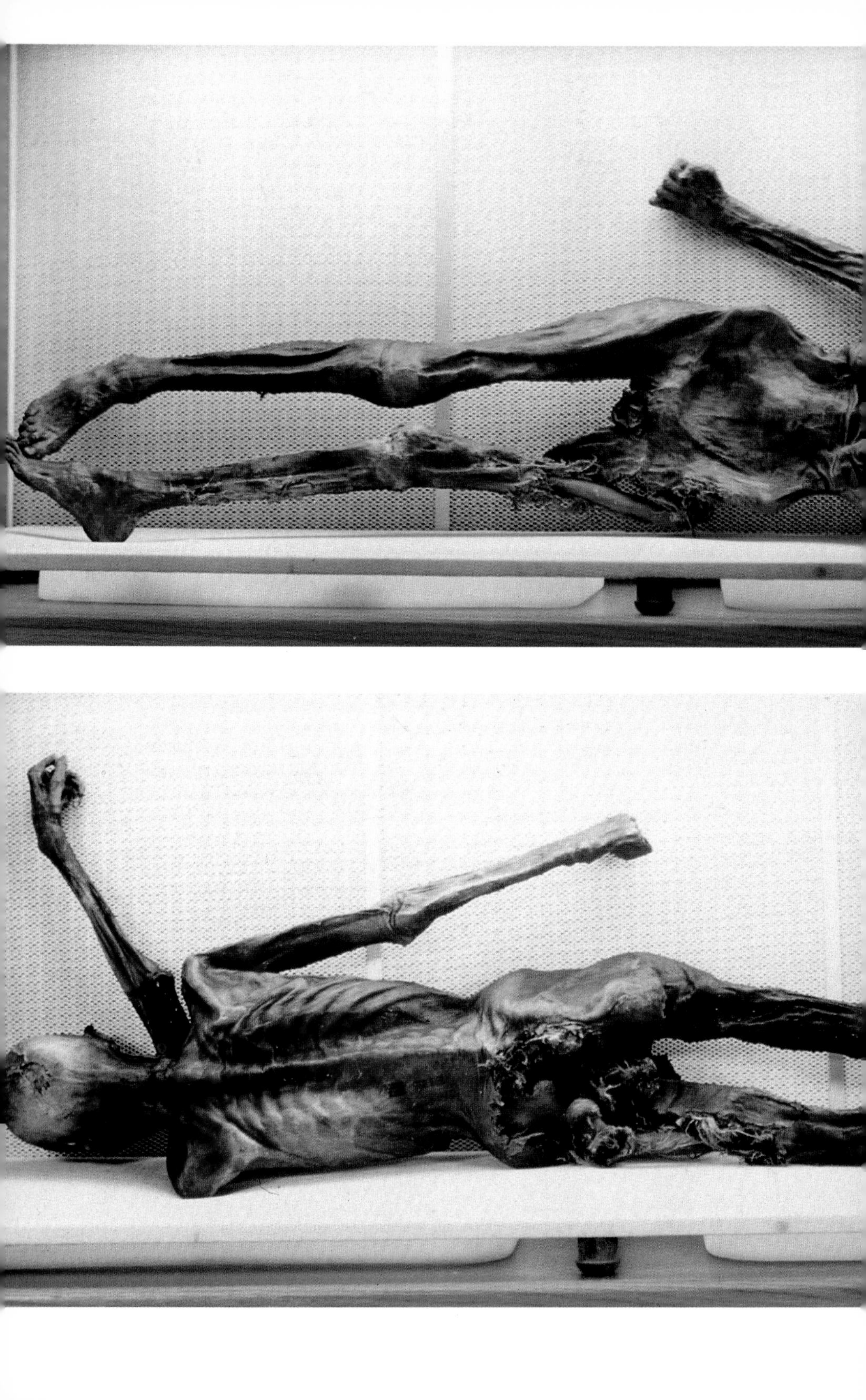

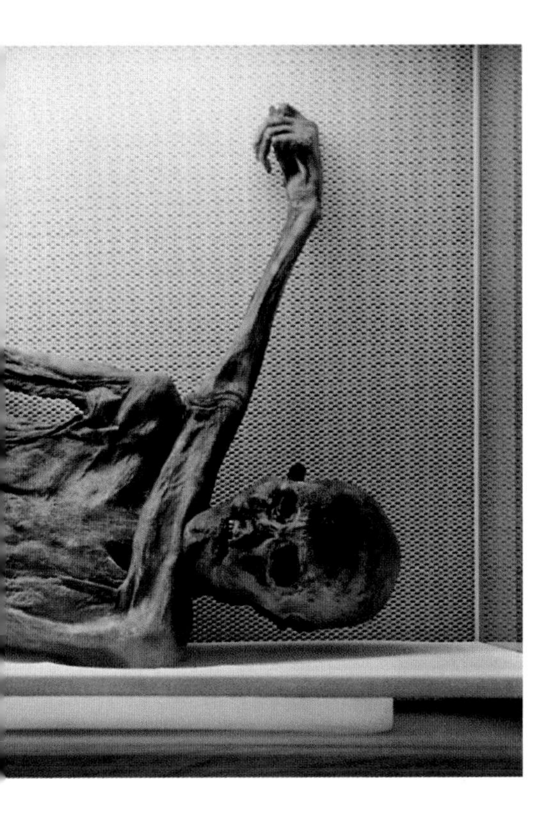

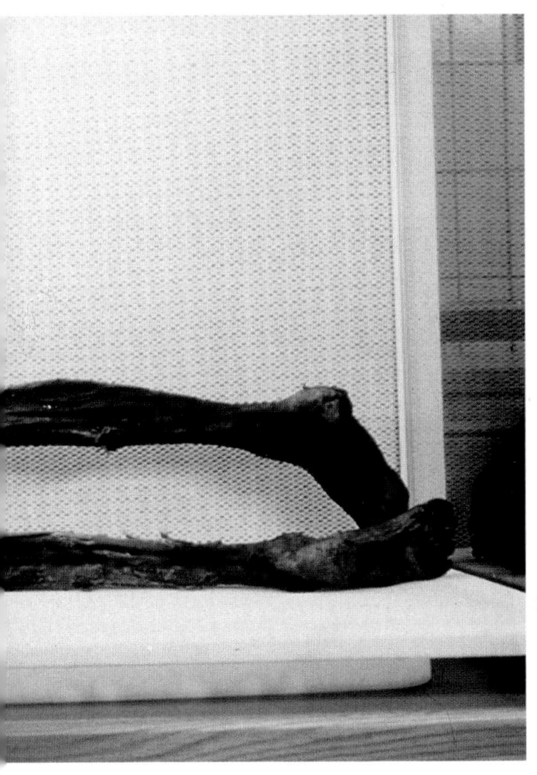

Front and back views of the Iceman. The left buttock and thigh show the serious soft tissue damage caused by the pneumatic chisel during the first recovery attempt.

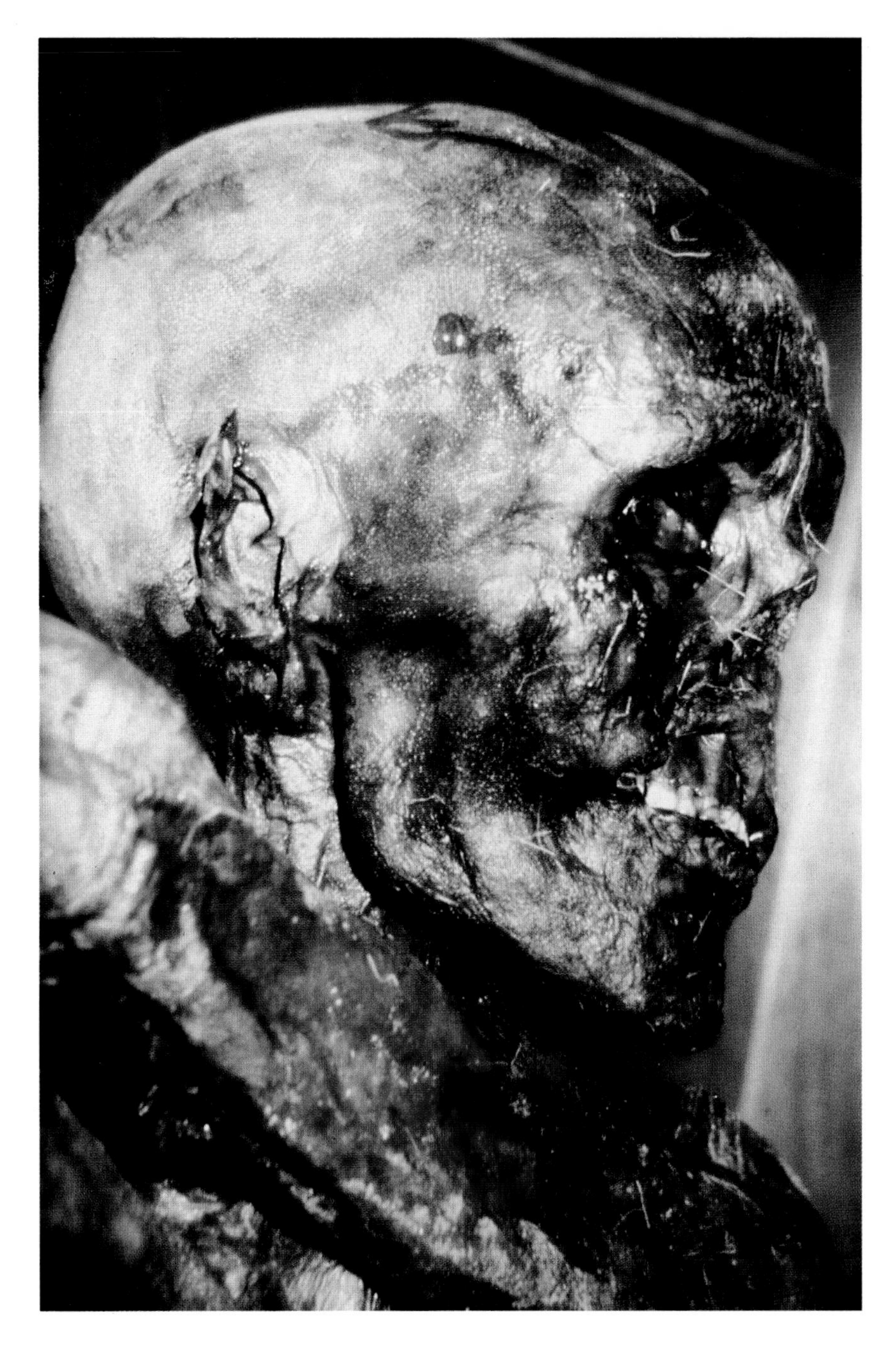

Right profile of the head, photographed immediately after the body's admission to the Institute of Forensic Medicine, before it was washed, on 23 September 1991.

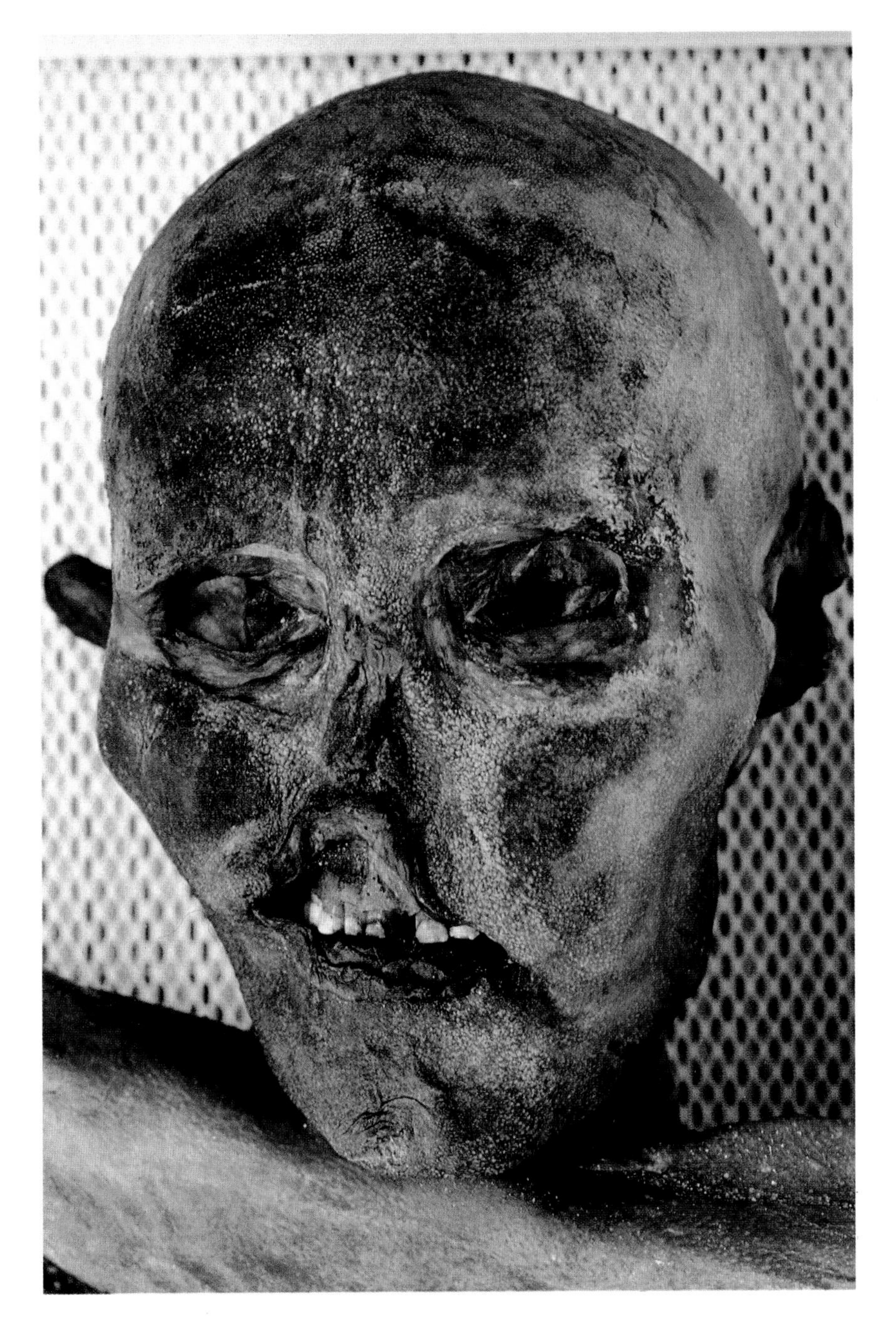

The face of the Iceman.

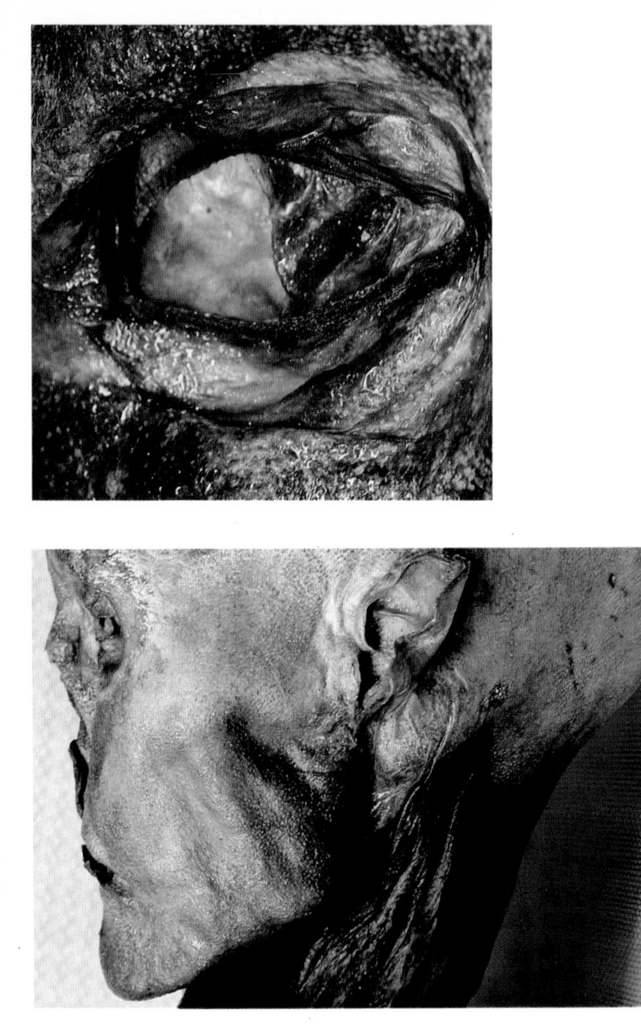

The Iceman's right eye.

The Iceman's left ear. Notice the vertical crease where the outer ear is folded forward.

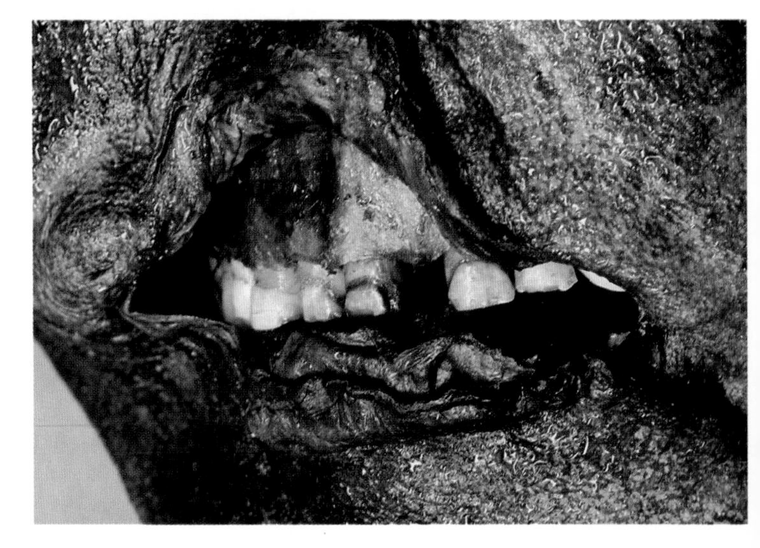

The Iceman's mouth. The congenital gap between the teeth in the upper jaw, the *diastema*, is clearly visible.

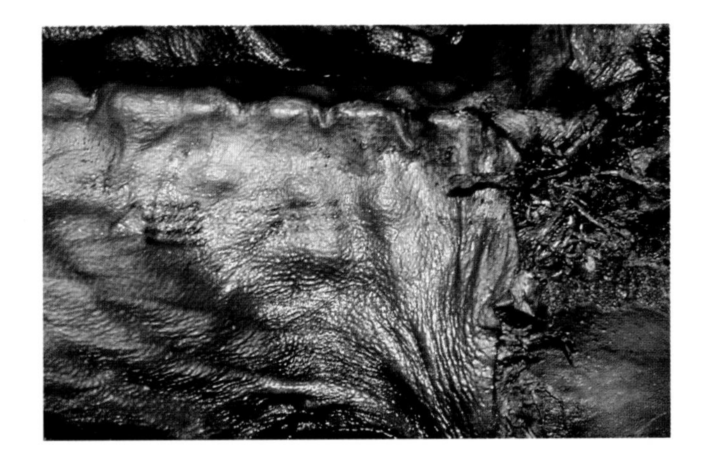

Groups of linear tattoos on both sides of the lumbar spine.

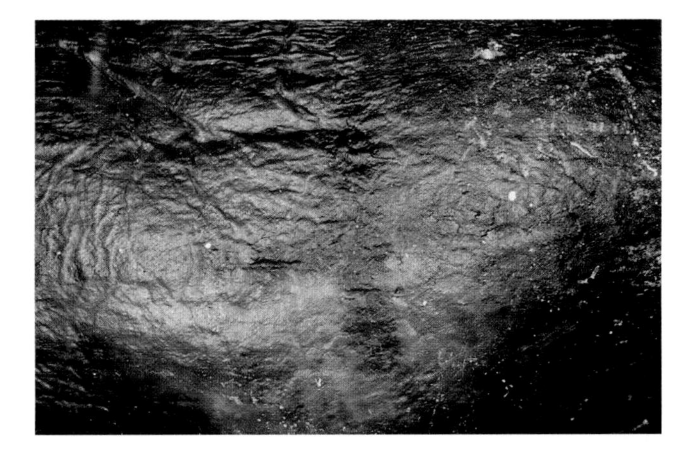

Cruciform tattoo on the inside of the right knee.

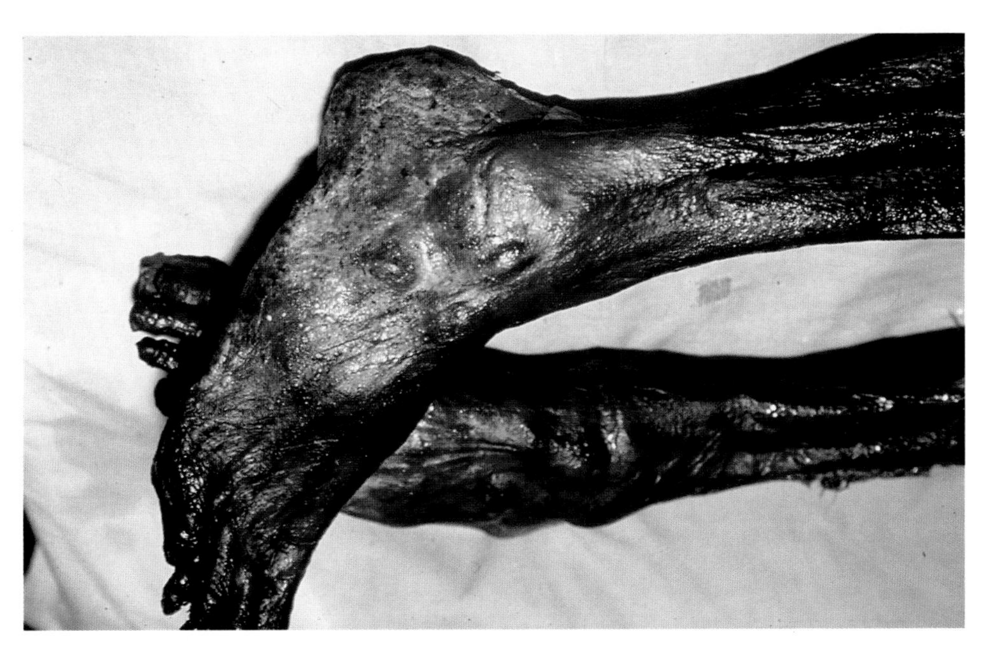

Small group of linear tattoos on the outside of the right ankle.

Statue menhir from the hilltop church of Latsch in the Val Venosta. The picture shows the back of the marble stone with its belt of garlands (right), groups of lines and sun symbols (left).

The front of the statue menhir from Latsch in the Val Venosta shows the belt with its repeated pinch-gathered effect (below) and, above it, symbolic representations of daggers, circles, axes, etc. For its secondary use as an altar slab the stone was narrowed (below left).

mation we have on Late Neolithic metalworking techniques, it would be pure speculation to assume that the Iceman was, either full-time or partially, involved in this activity. Equally, although there is no indication other than the axe itself, the idea cannot be ruled out.

Without listing possible barter goods in detail, it has repeatedly been asked whether the Iceman was not perhaps a trader or a merchant. This interpretation is based mainly on the quality and quantity of trans-Alpine traffic, especially in prehistoric times. The first artificial roads across the Alps, though not always usable all the year round, were built by the Romans, in order to facilitate the military administration of their northern provinces. The soldiers were followed by traders, merchants and private travellers. During the time that followed, up to the introduction of modern transport systems, such as tunnels, railways, motorways and aircraft, neither the principle nor indeed the routing of trans-Alpine communications underwent any marked change.

It is assumed that the Romans aligned their roads along ancient, even prehistoric, tracks. Consequently, Roman mountain-pass altars, where travellers made offerings for the success of their enterprise and, in the event of success, thanked the gods for a happy return, were based on prehistoric roots. It is certainly a fact that, alongside the major passes, such as the Great and Little St Bernhard, the Reschen, the Brenner, the Radstädter Tauern, the Septimer and Julier, there were also less important crossings which, at best, were developed as mule-tracks. The countless small saddles of purely local, or at most regional, significance deserve equal consideration here, because it is in this category that the crossings connecting the Ötztal with the Val di Senales belong, even though they played no part in trans-Alpine traffic.

The four uses of the high-Alpine zone – hunting, mining, high-altitude grazing and transit – have all, of course, left archaeological traces. These are labelled, according to the specific local conditions of their discovery, as lost objects, deposited objects, pass objects, sacrificial offerings, or simply high-altitude finds. They prove, above all else, that the high mountain region served an economic purpose and that the main ridge of the Alps represented no insuperable barrier.

Similarly, comparable shapes of finds north and south of the main ridge leave no doubt that a lively traffic of persons and goods took place across the mountains in both directions at all times. However, Italian imports and influences are more readily perceptible in the zone north of the Alps, which may be due to the fact that the typical export articles from the north are virtually impossible to distinguish archaeologically. From the ancient Roman geographer Strabo we learn that these were hides, furs, honey, amber and slaves. In the eastern Alps raw materials such as salt, iron and gold may also have played a part, though these were either consumed or reworked into articles of local shape and hence lost to archaeological identification. In addition to the exchange of foodstuffs, archaeologists have been able to prove a Neolithic trade in raw materials, such as flint, jade, Mediterranean shells and copper, often over many hundreds of kilometres. Occasional finds of imported pottery do not, of course, indicate a trade in vessels but in their contents.

The only materials not originating in the narrow habitat of the Iceman are the flints, of which six of his implements consist, and the copper blade of his axe. The flint deposits in the Alps have so far not been thoroughly researched. The Central Alps, and hence also the Val Venosta, have to be ruled out on geological grounds. Natural occurrence is linked to limestone formations, which means that the Kalkalpen in the north and the Dolomites in the south, along with their foothills, are possible candidates. The nearest known deposits to the Hauslabjoch are down the Adige in the Trentino and east of Lake Garda in the Monti Lessini. At present, however, nothing definite can be stated about the origin of the Hauslabjoch flints, any more than about that of the copper of the axe. It is certainly not impossible that these implements, or at least their raw materials, reached the Iceman by way of barter. But we must reject the idea that his final journey was a trade mission. Among his equipment there was not one article not needed for personal use, which he could have offered for sale.

There is much more justification for describing the Iceman as a hunter; that he was equipped with bow and arrows removes any doubt. His anxiety to replace the lost bow and the spent arrows as quickly as possible also testifies to the importance he attached to that part of his equipment. Today we are amazed at the precision with which he manufactured his weapons. But such skills may then have been acquired as a matter of course by every male.

Tests with modern yew-wood bows modelled on prehistoric originals exhibit an enormous range with a proportion of hits. Capable of being drawn back 71.5 centimetres, these replica bows develop a force of between 20 and 40 kilograms. With impeccable sureness of aim lethal hunting distances of between 30 and 50 metres are achieved, and skilled archers may expect a high proportion of direct hits even at 90 metres. Maximum range without appropriate precision is around 180 metres.

Obviously the bow was not only used for hunting, but also to fight hostile humans. This is proved by (not particularly rare) finds from Neolithic graves of human bones in which flint arrowheads are embedded. Most of these arrow wounds show no trace of callus formation at the bone, which would indicate healing – meaning that the victim did not survive the shot. But there are also a few instances where the arrowhead lodged in the bone is enveloped in callus tissue. In these cases the arrow wound had healed – testimony to effective treatment.

Further support for the hunter hypothesis is to be found in the net, which could have been used for trapping birds. The end of the axe helve, which is thickened like a cudgel, could have been used for killing a wounded quarry. The dagger, like the flint scraper and the long stag-antler spike, is suitable for skinning and dismembering the kill, cleaning the hide and cutting the meat into strips for drying.

The splinters of ibex bone are vivid testimony to the Iceman's successful hunting, and there can be no doubt that hunting played an important part in his life. At the same time, the dangerous bow would provide adequate safety against adversaries of all kinds.

Nevertheless, we know that while hunting may have been a more or less major factor in the Late Neolithic population's food supplies, the main weight of the economic system rested upon arable and stock farming. It is not therefore surprising that some people have seen the Iceman as a peasant, indeed a 'rich' peasant. It was not explained whether this meant a crop-farmer or livestock-breeder, or both. As for the adjective 'rich', this must have related solely to the copper axe. But I have already emphasized that 'rich' or 'valuable' are very subjective concepts, even from our present-day viewpoint, and that they cannot be simply transferred to Neolithic circumstances. In view of our ignorance about the distribution of economic assets in Neolithic times, any attempt to assess the value or otherwise of an object is simply beyond the methodological facilities of our discipline. For that reason we shall not pursue this question any further, and it should not be overemphasized.

Enough has been said to prove that the Iceman had contacts with a farming community. This is proved by the grains in his clothing and the threshing remains in his ember-carrier. They even indicate the time when he was last in his village: after harvest, during threshing, when every able-bodied man was needed. This, of course, is an interesting detail concerning the Iceman's life, but it is also all that we can say about his participation in the agricultural routine of his native village.

A more important aspect is the fact that the Iceman was equipped for more prolonged absences from his settlement community. A peasant who has to tend his crops and look after his lifestock cannot afford to absent himself even for a few days at a time, a fact probably just as true in the Neolithic as it is today. I live with my family in a small Tyrolean farming village. My neighbours, who are full-time farmers, do not know the meaning of holidays. For a farmer's wife to this day giving birth is about her only 'recreational leave'. The unmarried son, on the other hand, or the farmhand, or an unmarried brother, moves to the highland pastures as a shepherd for several months in the summer. Now we are getting close to an agricultural model whose peculiar nature helps us explain satisfactorily everything we have learnt so far about the Iceman.

With this model we proceed on the assumption that the nearest settlement area to the Hauslabjoch in the Neolithic, as today, was the Val Venosta to the south. No one could reasonably question this. It is also indisputable that the natural mountain grasslands above the treeline in the Upper Ötztal have been used by man for grazing since at least 4000 BC. Such high-lying pastures are ideally suited for grazing sheep and goats. As we have mentioned before, research has shown that in mountainous locations in the Neolithic the relative importance of sheep and goats was greater than that of cattle and pigs, so we may conclude that in the Val Venosta, too, the keeping of sheep and goats took priority. The meadow woodlands of the valley floor were insufficient to provide adequate summer grazing; the high-altitude grasslands were available. Our model postulates a limited transhumance from the Val Venosta across the main ridge of the Alps into the Niedertal and Rofental, and perhaps even further north.

In this connection we might ask, with the same justification, whether for the even longer winter season a major transhumance should not also be assumed to have taken place towards the south, in the direction of the Po plain. This would lead through an area where the cornelian cherry grows and where, further south at the mouth of the Adige valley in the Monti Lessini east of Lake Garda, there are good flint deposits. I will not pursue this idea any further, as it would involve a good deal of speculation, but it should not be forgotten.

The economic system outlined here for the Neolithic period in the Val Venosta naturally also includes the people. These are, first of all, the settled peasants with their families who remain at home, or have to remain at home, all the year round.

In addition, there are other people who are dispensable and who practise a roaming grazing economy. We are interested principally in the brief summer transhumance from the Val Venosta over the saddles of the main ridge to the high-altitude grazings of the Upper Ötztal. How are we to visualize the shepherds who drove the flocks up into the mountains in June and drove them back again in September? The Iceman had with him everything that such an activity would require:

At his age he was experienced enough to do the responsible job of looking after a herd of goats and sheep over a period of several months.

He had a strong physique, even though, as a result of his work in wind and weather, the first signs of wear and tear had begun to appear in his joints.

His equipment would have enabled him to survive for several months away from regular supplies.

His hunting weapons would have enabled him to keep himself supplied with additional meat.

At the same time his weapons – and his axe would have to be included here – enabled him to defend himself and his animals against adversaries, either animal or human predators.

The roaming area through which he moved extended from the valley floor to the zone above the treeline.

His clothing was perfectly suited to the climatic conditions of the high mountains, where night frosts must be expected even in the summer months.

Finally, the site of his death marks the route by which the herds would cross the main ridge of the Alps. This also fits in with the time of his death. The summer grazing season in the Upper Ötztal was over.

There is in fact no detail which does not fit this interpretation. We might mention, in conclusion, that straw or grass cloaks continued to be a typical characteristic of shepherds down to our own century.

We can therefore state that the main occupation of the Iceman was itinerant flock herding, and that he was firmly integrated into the agricultural system of his Neolithic community. That system became the foundation of the economic development of the Alpine mountain regions, which survived in some aspects down to our century.

Studies of animal bones in Neolithic settlements have shown that as a rule, the number of sheep greatly exceeded that of goats – as is the case today. It may therefore be assumed that the Iceman's flock consisted mainly of sheep with just a few goats. Nothing definite can be said about the size of his flock; it could have numbered a few dozen or a few hundred animals.

Equally there is nothing to indicate whether the animals were tended by just one person or by several shepherds. This would depend chiefly on the size of the flock. Besides, predators then represented a substantially greater danger than they do now, when, indeed, there are none left in the Alps. The most unpleasant enemy of the shepherd was the wolf. We may assume, therefore, that the shepherds had their sheep and goats guarded by trained dogs, whose tasks also included finding and rounding up stragglers.

In the course of archaeological work in the Spanish Estremadura west of Madrid towards the Portuguese frontier I frequently saw huge forest fires. These had been started deliberately by the shepherds whose sheep and goats were grazing in those areas. The purpose, I was told, was to bring the hated wolves out of their lairs to face the shepherds' rifles. At the same time these fires created further grazings by clearing the land.

Charcoal layers from the pollen profiles of the Upper Ötztal likewise indicate extensive fire-clearance by Neolithic man in the high-level forests of the Alps. I am convinced that these measures were also intended to decimate predators and that the creation of new grazing areas was simply an added bonus. This explains why the economic development of the Alpine valleys proceeded from the top downwards, from the natural grasslands above the treeline down towards the valleys.

While we have now decided on the occupation of the Iceman, we do not wish to disregard entirely the other possibilities discussed above.

Of course he was no outcast, but he inevitably and regularly lived away from his kith and kin for months on end.

Nor was he a shaman or a priest, although no doubt he was influenced by religious ideas and sentiments. He would pray or sacrifice at the sacred picture stones in the valley for the success of his task before he set out to drive his herd into the mountains, and he would give thanks for his successful return. His amulet, the white marble bead, was to protect him against misfortune.

When he was on the move with his herd he would gather anything he came across that might come in useful. He would not have deliberately searched for those green or blue stones from which shining copper could be smelted. But if he found any he would not have left them carelessly lying by the wayside.

If on his journeys he encountered a trader he might have acquired a few flint implements in exchange for a fat lamb, or a copper axe in exchange for a virile ram.

Of course he also hunted – and not only for his own meat supply. The mountain animals were fodder rivals for his herd, and he also had to fend them off to avoid cross-breeding, for instance, by ibex. Predators such as wolves, bears, lynx and foxes, and aggressive birds like vultures and eagles, had to be driven off.

Return to his people and his family in spring and autumn would not be for the sole purpose of shearing the sheep or slaughtering any superfluous animals. If a hand had to be lent with sowing or harvesting he would certainly be available.

But his life revolved around the herd which he had to take to the grazing and guard together with his dogs. It assured him and his family of a livelihood and survival in a still largely untouched wilderness, which Neolithic man was just then beginning to bring under his control.

10 The 'Disaster'

We have now come to the final chapter in the Iceman's life, which ended at the Hauslabjoch. It appears that he probably spent the summer on the mountain pastures of the Ötztal with his flock of sheep and goats. Before the onset of the first winter snows he gathered up his flock and led it down to his settlement in the Val Venosta. This could have been in company with other shepherds or with helpers who had come up specially from the valley. There is much to suggest that this operation took place according to plan and that the Iceman reached his village safely with his sheep and goats some time during the second half of September.

There the harvest was already in full swing. All available hands were needed to store up supplies for the long winter. The ears were threshed, the grain was poured into large clay containers and placed in the stores. The sheep were shorn and any superfluous animals, now well fed, were slaughtered unless they were needed for breeding.

At some point about then a disaster must have occurred, whose nature, of course, as the Iceman is our only source of information, we cannot reconstruct. There was certainly a violent conflict, perhaps more than one, as a result of which the Iceman had to flee. In the process he lost some of his equipment and other items were damaged. He himself suffered a serial rib fracture. For his escape route he chose the one with which he was familiar from the annual transhumance, until death caught up with him at the Hauslabjoch.

Naturally, one is tempted to ask what form such a disaster might have taken. Any answer, however, must be speculative, as too many possibilities exist. Basically, we may proceed from the fact that the man, in the year of his death, spent the summer with a herd away from his native settlement. One possibility is that, having been away for several months, the family structures in his village may have changed. This need not necessarily have involved a drama of jealousy. There could equally well have been hierarchical conflicts (power struggles). We cannot even rule out some personal culpability, such as the loss of sheep or goats or an offence against his community's norms, as the cause of an entirely personal disaster.

At the same time we might be dealing with an event affecting his entire community. If our conclusions about the location of the Iceman's village and the extent of his wanderings are correct, then such a disaster would have occurred in the Val Venosta. Again, the time of the event can be narrowed down to the late summer or early autumn, i.e. the months of September and October. This suggests conclusively – and the threshing remains on the Iceman confirm it – that the event culminated during the storing up of supplies for the winter. The barns were filling and the village was in particular jeopardy.

It can be no accident that so many Neolithic settlements are in locations that favour defence; it suggests that threats were expected. Moreover, it may confidently be assumed that these were directed not only against the villagers' lives but mainly against their possessions. Needless to say, the richest booty would be found just after the harvest had been brought home. The theory I am putting forward here would be supported if it could be proved that rival groups existed in Neolithic times with disparate material wealth.

However, only in the rarest instances can archaeology throw any light on this problem. Conflagration levels have indeed been found during the excavation of several Neolithic settlements, suggesting a catastrophe. As a rule, however, it is impossible to decide whether the village was burnt down in the course of hostilities or as a result of negligence or, indeed, whether decrepit houses were deliberately burnt down to make room for new ones.

But there is one archaeological find that testifies beyond a shadow of doubt to a warlike conflict in the Neolithic period. Here I am not trying to make a connection with the events immediately preceding the Iceman's death; I merely wish to illustrate one of many possible scenarios, even if in this shocking example all the evidence comes from the losing side.

A short distance to the west of the centre of Talheim, in the district of Heilbronn in southwest Germany, are the buried ruins of a Neolithic village. Excavation has not yet begun there. We only know of the settlement from a few chance finds which indicate that it existed about the middle of the sixth millennium BC and belonged to the early Neolithic Bandkeramik culture. On the edge of this Neolithic village a mass grave was discovered in 1983. In a pit barely 3 metres long, up to 1.5 metres wide and originally about 1.5 metres deep thirty-four skeletons were found. The scientists – the anthropologist Dr Joachim Wahl and the forensic expert Dr Hans Günter König, both from Tübingen – described the find in poignant terms:

> More than half of the individuals were lying on their stomachs, their arms and legs sticking out in all directions, sometimes almost in defiance of anatomy and at absurd angles. Others, while their upper bodies lay on their backs, had their extremities unnaturally dislocated. The arms and legs of one individual were lying above and below a second or third one, the leg of a fourth sticking out above them, etc. The individual bodies were pushed into each other and wedged into one another...

They look as though they had been flung on top of each other.

The first archaeological observation was that the dead had no grave goods of any kind or any equipment with them. This is extremely unusual for that time: as a rule Bandkeramik graves are very well furnished with ornaments, implements, weapons and vessels. There is no doubt that the corpses were stripped of their belongings before their disposal: by the time they were flung into the ground they had no personal belongings with them whatsoever.

Anthropological evaluation showed that there were sixteen children or juveniles, seven of whom were between one and six years, and eighteen adults. Four of the adults were over fifty, one an old man between sixty and seventy. Of the adults, nine were men and seven women. The sex of two skeletons could not be determined; moreover, with children and juveniles any determination of sex can only be tentative. But indications were that the sex ratio was about even. Six women were of childbearing age. The Talheim community of dead thus represents the complete population range of a Neolithic village.

Forensic examination revealed the full horror of the drama. Everyone had been murdered. Marks on the bones proved that stone axes and cudgels ('blunt instruments') were used and two people had been killed by stone arrowheads. Some of the blows had been administered while the bodies were upright, others while the dying were lying on the ground. An interesting feature is that there are no defensive fractures – injuries to the bones sustained, for instance, when the attacked person raises his arm to protect his head (a 'parrying fracture'). Thus, in some cases at least, one might assume regular execution. Afterwards the bodies were carelessly tossed into a pit.

Analysis of the age structure and sex ratio of the Talheim mass grave suggests the – purely theoretical – possibility that some individuals may have survived the massacre. There is an absence of neonate bodies, i.e. children of up to twelve months, whom one would expect to find among such a population. Unless this is due to the poor preservation of the delicate infant bones – though the workers at the dig kept a particular eye out for them - one might think of child abduction, practised by some primitive peoples to add new blood to their own population. Certain age limits were set in order to facilitate the linguistic and cultural integration of these young children into their new community.

At least it is certain that, after the massacre, there was no cannibalism, as has not infrequently been observed at other Neolithic sites. The Talheim bodies lack the tell-tale cutting and slaughtering marks on their bones.

The Talheim mass grave demonstrates that in Neolithic times, too, warlike conflicts could mean the extermination of entire village populations, and that the victor treated the vanquished mercilessly – which rules out any ideologically motivated 'ethnic cleansing'. The total lack of grave goods strongly suggests that the principal aim was plunder, the unlawful acquisition of material assets, and possibly also child abduction. One merely asks oneself why, after such a seemingly total annihilation of this community consisting of three or four families, the victims were buried at all, however contemptuously.

One might therefore suppose that the attackers were not out for the quick seizure of movable property alone. If they took possession of the entire village, complete with houses, livestock, fields and stores, then it would, of course, have been unpleasant to have bodies lying about.

The Talheim disaster casts a striking light on the behaviour of 'primitive' populations. In view of the often astonishing achievements of our ancestors I do not normally use this derogatory term to characterize human behaviour; I have chosen it here advisedly. For despite its history over many thousands of years mankind to this day has not shed this macabre legacy, which we can trace back, beyond the Neolithic period, to much more ancient stages of our evolution.

It is conceivable that a fate similar to that of the Talheim settlement also befell the Iceman's native village. In a hopeless situation he succeeded in fleeing from the enemy. Evidently his escape was noticed and every effort was made to capture him. If, as we know from Talheim, even women and children were massacred, how much more dangerous would a grown man's escape from the pogrom seem to the victors. His only advantage, apart from saving the remains of his equipment, was his superior local knowledge, so that flight and pursuit extended over several days and weeks. There was no hope of return, at whatever time and for whatever purpose. So the man set out in the direction of the Hauslabjoch, hoping that, beyond the main ridge of the Alps, he might escape his pursuers.

VI
Reactions to the Find

1 An Extraordinary Event

Spectacular prehistoric and early historical finds, those great moments in archaeology, usually arouse wide public interest. Examples are Heinrich Schliemann's discovery of the putative golden treasure of King Priam in Troy, Howard Carter's opening of the tomb of the pharaoh Tutankhamun, or Jörg Biel's excavation of the burial of the Celtic ruler at Hochdorf. To this day, none of these sensational finds has lost its fascination. And yet there is a striking difference between them and the Iceman. They - and others like them that could be listed here – are all marked by an immeasurable wealth of gold. Public interest in an archaeological event is evidently linked to the discovery of a pile of gold, of which everyone dreams but which most are denied.

On a smaller scale we experienced this recently at our own institute. The period of the Carolingian, Ottonian and Salic emperors (eighth to eleventh century AD) is regarded as the 'dark ages' of Tyrolean history. Written sources are scanty and there are virtually no archaeological finds. There was, therefore, considerable scholarly interest when an eleventh-century site was discovered at the ancient church of St Justina at Assling in eastern Tyrol, and when we started excavations there. For a time, however, no one outside the academic community took the slightest notice.

But on 29 April 1993, when the director of the dig, Dr Harald Stadler, while sieving the sediments of a cultural level, unexpectedly found parts of a gem-studded gold ornament, the situation was instantly transformed.

Every archaeologist knows that finds of gold are not normally made at planned digs, let alone during the examination of settlements. To that extent the discovery of gold objects in an eleventh-century level was in itself a sensation. But there was also something special, not to say unique. The jewels belonged to a garment of the kind originally worn by the Byzantine rulers of the Eastern Roman Empire. A jewelled collar (*maniakion*) and the ruler's sash (*loros*) were embroidered with a fine network of delicate gold chains, at the crossing-points of which gold-mounted precious stones, pearls and antique gems were set. The European high nobility of the Middle Ages took over these sovereign insignia, almost as a kind of international fashion. With a single exception none of these raiments survived, though we have illustrations in medieval manuscripts and gospels of crowned heads thus attired. Only in the Mainz treasure, now attributed to the German empress Agnes of Poitou (1043–62), the wife of Henry III (1046–56), had the jewels of a *maniakion* and a *loros* been preserved. With that treasure alone could the new finds in eastern Tyrol be compared. We realized that we were dealing with the highest level of the nobility of the High Middle Ages.

Although we always try not to reveal details of finds and sites, so as not to attract robbers, the news leaked out. Instantly – just as in the days when the Iceman was the world's hottest news – all hell broke loose at the institute. All the telephone lines were swamped; television, radio and press journalists arrived in droves. Probably this is what always happens when gold is found.

The Iceman was a find of a very different order. A pitiful bundle of humanity, wrapped in furs and grass, with implements made of wood, bones and stone. Even his axe was only made of copper. How different from the gleaming gold and the sparkling jewels which usually capture the public's imagination, though we are not exactly short of genuine and man-made sensations. Why then did the public show such an astonishing level of interest in the Iceman? Was it the loneliness of his place of death high up among the eternal ice of the Alps? Was it the mummified remains of his body, when normally one finds only bleached bones? Was it the over 5,000-year-old face that, marked by death, gazed upon us? Or was it the exciting circumstances of the body's discovery and recovery? No, it must be something more.

Within hours modern communications catapulted the news of the discovery via press, radio and television to all corners of the globe and back again to the Innsbruck institute. Never before had an archaeological event so completely dominated the headlines. Comparisons with other major finds are impossible as the Iceman clearly represents an absolutely unique case. For days it pushed everything else off the front pages. Even now, two years after the original discovery, its attraction has not diminished, although minor aspects of the story have given rise to fatter headlines than the true scientific significance of the find.

Naturally journalists like a story they can communicate easily to their readers, listeners or viewers. This was entirely possible with the Iceman. The public always wants to hear the latest results of scientific research, and as a rule they are content with the information provided by the media. The Iceman, however, developed a dynamic of his own. Interest in everything that was happening in connection with him set entirely new standards, to which some of the professional journalists were evidently not quite equal. Thus two closely interlinked groups evolved, one to serve the marketing, the other to satisfy the demand for information. Scientists, dealers, reporters and even private citizens were all involved, each group with its own entourage of lawyers, management experts, politicians and sponsors.

The scientists fall into two categories – those actively involved in the Iceman project and those not involved in research but nevertheless anxious to make themselves heard. The scientist finds himself on the horns of a dilemma. Basically he would like to pursue his studies in peace, as well as fulfilling his other obligations, such as teaching, administration, giving expert opinions and research. These more than fill up his day. As for his findings, he prepares his scientific publications. Informing the public about his results is not really his job, but he is expected to do it. Often he will meet these demands, provided he feels willing and has the time. Such publicity work may take the form of lectures to a variety of audiences, or go through the press, radio and television. In many cases journalists and reporters play the part of mediator.

Interest in the Iceman project, however, far exceeded our normal capacity to deal with such events. For reasons of time, if for no other, information had to be channelled. Some thirty publishing houses, eager to bring out popular-scientific 'Ötzi books' sent us offers. Over a hundred newspapers, magazines and journals, some internationally renowned, also wanted to print articles by us.

Invitations to lecture ran into hundreds, and came not only from Europe but also from overseas, including America and Australia. Interviews for the daily papers, radio and television became so numerous we lost count. They probably amounted to a few thousand, despite the fact that we were forever giving press conferences, sending out press releases and statements by telephone, fax and mail. I shall return to these briefly later. They also ran into thousands. Never before had there been such an all-consuming reaction by the public to an archaeological find. It is therefore worth dealing with this aspect of Iceman research as well and perhaps drawing some tentative conclusions which throw light on our own society.

2 Books, Books and More Books

Within a few days the Iceman had become a cult figure and the public's demand for information rose correspondingly. There was a new occupation – that of self-appointed Iceman expert, who entered into commercial competition with the scientists engaged on the project. Although from the very beginning we had pointed out that, beyond the publication of our findings in learned periodicals, we were also preparing a popular version, various authors made the most of the interim period.

A serious academic cannot simply, on the strength of first impressions, theories and speculation, write a popular scientific book. To do that he needs a certain minimum of firm results which he will then try to translate from his dry, academic technical terminology into language accessible to the general public. Which is why this book is being published more than two years after the find at the Hauslabjoch.

Nevertheless, within a few weeks of the recovery of the corpse two books appeared almost simultaneously, one in German and the other in Italian. Both books are a string of annotated newspaper articles, bolstered by a few notes on cultural history. The authors, journalists on the Austrian daily *Kurier* and the Italian-language South Tyrolean daily *Alto Adige*, beefed up their publications with various criticisms of the Innsbruck scientists involved in the recovery and examination of the corpse. Even the actual course of events at the Hauslabjoch had not, at the time, been clearly established, and these two books in particular did much to promote the spread of false information.

They were followed, in the autumn of 1992, by two further, substantially more voluminous, publications – this time written by archaeologists. The authors used underhand means to obtain certain information from the international symposium on 'The Man in the Ice', held by us in Innsbruck in June 1992, and hastily worked our findings into their manuscripts. The publishers were in a hurry to catch the Christmas market. These two books, compiled in great haste, did a lot of harm to the publishing activity of the University of Innsbruck by more or less inferring that they were semi-official, authoritative accounts.

Nevertheless, our symposium volume appeared even before the publication of those two hasty accounts, punctually on the first anniversary of the discovery, 19 September 1992. It was entitled *The Man in the Ice: Volume 1. Report on the International Symposium 1992 in Innsbruck*. For a purely scientific book its success was overwhelming. The first edition sold out in six weeks, and a reprint had to be ordered as fast as possible.

One would have thought that these publications would, for the time being, have satisfied the market demand. But publishing houses soon discovered new angles. A children's novel appeared. This could have provided an opportunity to fill gaps in their education and, in language comprehensible to young readers but full of drama and excitement, to present an accurate account of the Iceman's life in the Neolithic. Instead, a scenario of murder and killing was created, which someone evidently thought appropriate to convey to our children an understanding of prehistory. The intellectual was also catered for. In Switzerland a little book was published, conjuring up a satirical version of future 'Ötzi' research.

3 A Fraud?

The nadir of unauthorized publishing was a book released by the normally reputable Rowohlt Verlag under the significant title *The Ötztal Fraud: Anatomy of an Archaeological Grotesque*. The subtitle straight away refers to the three principal targets of the author's attack as a 'trio', a phrase reminiscent of such disparaging terms as a 'trio of crooks' or 'gang of three'. These are Professor of Anatomy Werner Platzer, the professor of German studies Dr Hans Moser (in his capacity of Rector or Vice-Chancellor of Innsbruck University and Director of the Research Institute for Alpine Prehistory), whose duties include coordination of research into the Iceman, and myself as an archaeologist.

The author of *The Ötztal Fraud* was a Munich television journalist, and the pictures were provided by an Innsbruck press photographer. What is hushed up is the fact that this photographer did not himself take the crucial pictures of the recovery of the glacier corpse, although they are attributed to him in the list of illustrations. The photos were in fact taken by Anton Prodinger, the helicopter pilot to whom the photographer gave his camera at the intermediate landing in Vent.

Quite despicable, and a disgrace as much for serious scientific journalism as for a reputable publishing house, is the aggressive way in which the academics involved in the Iceman project are disparaged and ridiculed as 'scholarly cabaret artists'. The author falsifies information, deliberately highlighting things which cannot – certainly not easily – be verified by the reader. I don't hesitate to quote a few examples:

> ... then there is Spindler ... a quick learner and not averse to courting publicity. During the first few days, at the table in Pathology, he still dances around the body in jeans. Subsequently, when a sloe was found at the Hauslabjoch ... the humanities scholar Spindler puts on a white coat in order, before the eyes of the press, to examine the berry under the microscope.

The implication is that I presumptuously slipped into the role of a natural scientist by donning a white coat. Except during my original studies as a medical student, in the dissection room of the Anatomical Institute of the University of Münster in Westphalia, and for a holiday job more than thirty years ago, I have never in my life put on a white coat. But the author lies shamelessly to support his campaign of defamation.

Besides, the above brief quotation, like everything else, is full of errors. The corpse was never in 'Pathology', but – only temporarily – in Forensic Medicine. Nor was it lying there 'during the first few days', but only for a single day, before being transferred to Anatomy – so that I could have 'danced around' it for half a day at most. The blackthorn does not produce 'berries' but fruits – but that difference is probably lost on the authors, one of whom, on his own admission, is a history graduate and so not an expert in our field. The instrument is not a 'microscope' but a reflected-light binocular.

What I find particularly outrageous is the way in which the author denies the participating scientists' qualifications for working on the Iceman project. Thus he claims that my doctoral thesis was not in prehistory, as would have been required for the case in hand, but in 'medieval archaeology'. This is incorrect: my thesis was in prehistory. With unparalleled insolence he implies a lack of specialized knowledge in order to persuade his readers that, because of my lack of understanding, I have been the victim of a fraud.

He states:

> And the professor is eloquent in earnestly telling gullible journalists about a glacier which is supposed to have encapsulated the dead man at the Hauslabjoch – a glacier which Konrad Spindler has never seen and never will see, because some decades ago the ice retreated from the site of the find in the direction of the Niederjoch and the Similaun.

Of course I saw the glacier at the site of the find, and I am not alone. Literally dozens of other people are willing to testify on oath to the existence of that glacier. The reports of the different recovery parties, who tried hard to release the body from the ice, provide sufficient evidence of the existence of a residual glacier between 0.6 and 0.8 metres deep, in the lower portion of the gully at the Hauslabjoch. As 'proof' of his allegations the author shows a photograph in which the gully is indeed free of ice. This picture was taken in August 1992, after our digging party had manually removed the ice. The provocative caption reads: 'Where is the glacier which is supposed to have released the body? For years ... there has been no glacier at the site.' The caption should have read: 'As from yesterday, there is no glacier at the site.'

When nothing else will do, the author twists the facts. He claims to be intimately familiar with the local topography. Proudly he has himself photographed at the end of the gully, sitting in the snow and ice of the Ötztal Alps. Despite his alleged local knowledge he describes the gully as running north-to-south, 'in the direction of the fall line'. If this were so, the descending glacier would have had to pass through the gully and would have destroyed the corpse. In fact, the gully runs approximately at right angles to the fall line, which is almost exactly east-west, so that, at times of high glacier levels, the ice passed over it. The author wilfully turned the fall line through 90 degrees.

Next he asserts that the Iceman is not, in fact, a Late Neolithic inhabitant of the Alps who met his death in the high mountains. Instead he concludes: 'The dead man comes from a distant, strange country. Not from the Val di Senales or from the Ötztal. Hence he does not come from a glacier. He was deposited there.' A 'joker' had dragged the corpse up to the Hauslabjoch in order to fool the scientists! At the same time, this unknown person had acquired the various implements and draped them around the corpse in order to complete the fake.

The author describes at some length the ins and outs of modern mummy smuggling. He cites the German mountaineer Ernst-Eugen Stiebitz, who successfully smuggled Peruvian mummies through customs in Lima and Munich. He points to the centuries-old trade in Egyptian mummies which were pounded into miracle drugs in European pharmacies. Nor does he flinch from involving Heinrich Himmler, the Reichsführer SS, who mounted an expedition to Tibet in 1938, and suggests that the leader of that expedition, the ornithologist Dr Ernst Schäfer, known as Tibet-Schäfer, could have brought some Asian mummies home with him, one of which was possibly, fifty years later, secretly taken up to the Hauslabjoch.

While indulging in these, and even more absurd, ideas about where the putative 'hoaxer' might have obtained the body, the author persistently keeps silent about the fact that the smuggled mummies he refers to came from graves. Buried corpses all lie in a typical position, in line with the interment ritual of their particular cultures. South American mummies mostly lie with their knees drawn up to their chins and tied into bundles; Egyptian and all other known Asian mummies normally rest extended on their backs, with arms lying along the sides of their bodies or folded across the chest. The Iceman's body position, with his arms pointing to the right, deviates completely from such patterns, simply because he was not in a burial posture but in an actively chosen sleeping position, modified somewhat by slight movement of the ice. The hoaxer postulated by the author, therefore, would have had to find not only a corpse, but one for whose particular body posture there is not, and cannot be, a parallel in the whole world.

The author goes on to claim that the absence of grave-wax formation in the corpse proves that sleight-of-hand was performed at the site of the discovery. Yet elsewhere in his book he specifically refers to permafrost corpses which likewise exhibit no grave-wax formation. He even illustrates a grave-wax body with the caption: 'This is what bodies from the ice look like.' (Very wisely, he does not illustrate a permafrost corpse.) The naivety which the author presupposes on the part of his readers is demonstrated by the very picture chosen by him as an example. In that corpse head and body have been converted into grave wax, while forearms and hands are dry-mummified. Its hand, perfectly preserved to its fingertips and its spread-out thumb, contrasts clearly with the white shroud. Yet the picture of the identically mummified hand of the Iceman is commented on quite differently: 'The glacier did not even break the little finger of the dead man's hand. Such pictures have never been seen in the history of glaciology.' He contradicts himself with his clumsy choice of illustrations, so that one must ask, who is the real fool?

With the viciousness of the intriguer the author attempts to pillory the story of the discovery, so carefully investigated by us, as an 'absurdity'. We published this 'Reconstruction account' as an abridged summary in our scientific symposium volume. Here the author goes to town:

> We can now read in the same documentation that the [Messner] group saw the face of the body as it lay on its stomach on a boulder; it is described as 'compressed'. Kammerlander records that even the eyes were still preserved. Kammerlander thus looked into the eyes of a dead man whose lower body is held in the ice and whose right arm is held under a stone slab. How does one get to see them without breaking the spine of the dead man, who would have to be lifted up, at hip level, and without having to amputate the trapped right arm? According to his account, Kammerlander, lying on his back, would have had to wriggle under the face of the dead man, which may have been slightly raised by his comrades, in order to look into the eyes of the corpse. This surpasses our powers of imagination. But Kammerlander looked into the dead man's eyes – that is what it says in the university's official account.

That it was entirely feasible to look into the dead man's face at the spot where he was found, without any amputation and without breaking his back, is proved by the photos taken at the time –

photos about whose existence the author is careful to keep quiet.

One could go on and on refuting the author's 'chain of arguments'. His methods are perfectly clear: denigration of the scientists engaged in the project; falsification of actual events; factual distortion; lies; deliberate omission of material evidence; exaggeration of ordinary events; and finally, untruthful answers to his own questions.

Thanks to their connections the two authors were able to place their book in the hands of an eagerly receptive media. With all this free publicity the success of their concoction appeared to be a foregone conclusion. Indeed, in line with the ancient Latin proverb *Mundus vult decipi* ('the world wants to be deceived'), at first the book didn't sell at all badly. Then, in their euphoria and a few weeks after publication, the authors rashly agreed to take part in a discussion programme. In a ninety-minute *Club 2* transmission for Austrian television, also relayed to Germany, they had to stand up to questioning by two Innsbruck scientists. One almost had to pity them; they were taken to pieces and made to look complete fools. Since then a deafening silence has descended on the subject of the 'Ötztal fraud'.

4 Ötzi and the Cartoonists

Apart from such absurdities, which hardly anybody took seriously anyway, there were other groups who laid claim to our Iceman. 'Ötzi' was promoted – if that is the right word – to the role of cartoonists' hero, or victim. Prominent and not so prominent cartoonists adopted him, so helping to popularize the Hauslabjoch find. Their mixed offerings consist of humour as well as bad taste, of satire and ridicule.

Messner and Kammerlander are represented at the moment of discovering the corpse: 'Look, the yeti, and right at our door!' Another cartoon: the Iceman's angled arm is raised as though in a Nazi salute while two spectators remark: 'According to the artefacts this is a thousand-year corpse,' so joining in the initial arguments about his age. Along the same lines, an Ötzi drawn by a childish hand objects: 'I don't want to play doctors any more: you make me older each time!' Not surprisingly, a lot of space was taken up by the frontier issue: two Tyroleans pull the miserable-looking Iceman in opposite directions, one to the north, the other to the south. Or two quarrelling individuals are seen poised over a map: 'Well, there, he's Austrian.' 'No, no, Italiano.' A southerner, clearly identifiable as a mafioso with a submachine-gun, generously waves his hand: 'No, you keep him. We make our own corpses.' Even the lacerations in the genital region were seized upon. Among a cosy group of beer drinkers someone frowns: 'What I don't understand is his whole body is there, but he's got no cock!' The man across the table raises his forefinger: 'Must have peed into the snow and his cock froze up like an icicle and broke off!'

Allusions to marketing efforts soon followed. One heading ran 'Advertising agencies are already cashing in on the Iceman's popularity.' The cartoon shows a naked old man in the photographer's studio holding a quiver and an ice lolly.

Some cartoonists even combined topical news items with the Hauslabjoch find. Franz Viehböck, the first Austrian astronaut, was then in space, but even this sensational event could not displace the Iceman from the headlines. Thus we see him crouching on the bare surface of the moon. The spaceship sails past in the darkness of deep space. A balloon rises from the Iceman's head: 'It's all very well for Viehböck, he'll get down all right ... But how do I get back to the Bronze Age?'

Another example. The arrest of the 'prison poet' and presumed murderer of prostitutes Jack Unterweger in Miami in June 1992 inspired the cartoonists: 'Ötzi – Unterweger's latest victim'. Two amateur sleuths self-importantly present a shoe: 'Ötzi is a woman because the body has no prick. We looked for it for months in the snow . . . without success. Instead we found this lady's shoe: exactly Ötzi's size!' His partner chips in: 'And Ötzi wasn't five thousand years old either, but a dashing young thing. And when they found her she was naked ... just like the murderer's other victims! Besides, we know local witnesses who can remember a stranger with a very unusual toboggan . . .' This is shown in the next picture. Two Tyroleans suspiciously eye a toboggan with the tag 'Jack 1', on which sits a massively built chap with a fur cap. Finally we see a broad-hipped lady in Tyrolean dress, a cigarette between her fingers: 'And another thing: the lady known as Glacier Gertie, a well-known Austrian prostitute, suddenly vanished from her beat after the stranger's arrival.' In front of a beautiful mountain background the manned toboggan runs straight at her. From her mouth

Cartoons by Manfred Deix in *Krone Bunt*.

comes a heart-shaped speech balloon: 'How about it, dearie?'

The cartoonists also followed the Iceman through the seasons. He tears off the calendar page for 31 December 1991 with the comment: 'Another year gone – well, none of us is getting any younger.' For carnival some suggestions are made for costumes – with only one shoe: 'Why not go as an archaeological sensation?

A swimming cap, several bottles of Tyrolean nut-oil (self-tanning) - and you have the genuine Similaun man.'

More than fifty cartoons on the Iceman theme have come our way; probably there are many more. These are just cuttings sent to us or brought by friends. At this point we should like to say 'Thank you' to all our Ötzi friends. No doubt the number of Ötzi cartoons makes many an international politician green with envy – at least if he gauges his popularity by the number of caricatures of himself. The cartoons quoted here come from the following papers: *Dolomiten* (Josef 'Peppi' Tischler), *Kurier* (Dieter Zehentmayr), *Krone Bunt* (Manfred Deix) and *Profil* (Gerhard Haderer).

5 Souvenirs

It was inevitable that the Iceman should be turned into kitsch. As a lifesize papier maché figure, dressed in jeans, he lies in the window of a Tyrolean fashion shop. His death-marked features in brilliant colours gaze out from T-shirts. As a 2-centimetre-high miniature cast in solid silver he can be obtained from a jeweller in Merano, South Tyrol. He is available in two versions – either with a clip or as a pendant, with a small ring at his head. On the *People Plus* audio cassette produced by *Time* magazine in the USA 'The 4,600 Year-Old Man' appears alongside Elizabeth Taylor, Princess Diana, George Bush, and Magic Johnson among the top twenty-five personalities of 1991.

His fate was even sung about on gramophone records. The Japanese motor manufacturer Suzuki brought out a special model, the Vitara 'Ötzi'. As the car is being launched in September, this can be neither an April fool's joke nor a carnival prank. In its publicity the firm states: '... the Vitara Ötzi for all those valuing independence ... as weather-resistant, as unaffected by sun, rain, snow and ice as Ötzi ... with numerous extras, such as ... unique Ötzi décor on both sides, bonnet and spare-wheel cover ...' Price 28,900 Swiss francs. A Vienna leather company developed 'ICEMAN The Original Boot' and immediately distributed appropriate stickers. Their publicity leaflet explains: 'it was the European proto-shoe which inspired us ... spanning five thousand years of evolution ... the proto-design ... ICEMAN completes the evolution of shoe history ...'

Actors and private individuals have themselves fantastically made up and costumed as the Iceman. But anyone thinking that the actors would confine themselves to carnival processions and balls would be mistaken. The Iceman was even brought into the last election campaign. One advertisement shows a politician from the Austrian Liberal Party (FPÖ) with an actor disguised as the Iceman: 'Ötzi too has decided to vote for W.T. for mayor.' A thousand bottles of Blue Burgundy were specially bottled at the Köflgut vineyard in Kastellbell for a special exhibition. The Iceman winks from the label: '*Vinum tirolensis*'. Even a health club uses a picture of him for publicity: '... found 1991 in the Ötztal in modified ATNR position: the oldest Bobath-Therapist in Europe!'

Nor did the postcard industry miss the opportunity: on sale are not only original photographs of the body at the discovery site or against the background of the Ötztal Alps, but also the Iceman as a cartoon character. And anyone who still doesn't have an Iceman souvenir might just manage to get hold of an Ötzi key ring, 8.5 centimetres high, moulded in plastic, brown to black, with a safety catch, and the word 'China' across his bottom.

Appendix on Mummies, in particular Permafrost Mummies

The unique value of the Iceman becomes clear if we consider a few especially important mummy finds. The word mummy is derived from the Persian word *mummeia* or *mum*, meaning asphalt or bitumen, a substance which was occasionally used for embalming.

Egypt's extremely hot and dry desert climate naturally favoured the custom of artificial mummification, which began there in about 2600 BC and lapsed in the sixth century AD. It was applied, on an almost unimaginable scale, not only to humans but also to animals. There are mummy cemeteries for embalmed cats, monkeys, crocodiles and ibises which were regarded as sacred in the cult of deities. A lively trade in Egyptian mummies and parts of mummies began as early as the Renaissance, not only for equipping princely cabinets of rarities and miracles, but also for sale, in powdered form, as an aphrodisiac or medication for all kinds of ailments. The virtually inexhaustible stock was even used there for firing the steam boilers of nineteenth-century railway engines. For the Egyptian railway between Cairo and Khartoum they represented the cheapest fuel in the country.

In the fifth century BC the Greek geographer and historian Herodotus described Egyptian embalming techniques in detail. His report has been confirmed and supplemented by modern research. As the Egyptians believed that the soul re-entered the body after death, it was necessary to conserve the body. The principle of mummification was based on the removal of the brain and the intestines, which were interred separately in vessels known as canopic jars. The brain was drawn out with a hook-like implement through the nostril or the occipital *foramen magnum*. Herodotus states that the cavity of the body was first rinsed out with palm wine and then rubbed with pounded incense. The next step was the dehydration of muscular tissue with soda, which withdraws the body fluids. According to Herodotus this procedure took seventy days. To preserve the shape of the body, its cavity was again filled – with sawdust, linen, Nile mud, resinous oils, fragrant lichens, etc. The body was then swathed like a cocoon in linen strips, with various substances being applied at this stage, such as beeswax, conifer resin, bitumen and gum arabic. On these bandages a face was sometimes painted, or breasts and genitals were shaped from linen strips in order to lend the mummy as lifelike an appearance as possible. Occasionally an excessive application of embalming substances, especially for persons of high rank, was, if anything, detrimental to the preservation of the body. The body of Tutankhamun suffered more than others.

More or less effective measures for the conservation of corpses by mummification and embalming techniques are found among many ancient societies. They are all based on principles similar to those used in Egypt. Intestines and brain are removed as a matter of course, as decay starts in these tissues. Then follows the desiccation of the body. Alongside the use of hygroscopic salts there is smoke-drying, as with the shrunken bodies of Ecuador, or sun-drying, as with the Guanchos of the Canary Islands. Particularly interesting in this connection are the human mummies from the Siberian burial mounds of Pazyryk in the Altai Mountains. These reveal embalming measures, but in addition owe their superb preservation to their storage in permafrost. It must be pointed out that this preservation is due not to their artificial mummification but solely to the permafrost. Their excavation was directed by the Russian archaeologist Sergey I. Rudenko and was carried out, with interruptions, between 1929 and 1949. The necropolis consisted of forty kurgans, five of which stood out because of their conspicuous size; only underneath these had preserving ice layers formed. The bodies rested in wooden, ice-filled grave chambers. Bringing water from several kilometres away to thaw the ice posed a major challenge to the archaeologists. The finds included precious wall-hangings, a wagon made of birchwood with a felt hood, implements for hemp-vapour drug inhalation, and other objects. Even horses were included, complete with saddles and saddlecloths. The bodies belonged to a Scythian nomadic nobility living, according to dendrological and carbon-14 tests, in the second half of the fifth and the first half of the fourth century BC in the south Siberian steppes.

These bodies, too, had had their brains and intestines removed and their body cavities filled with vegetable materials before being sewn up again. Because they were discovered during the Second World War scientific investigation was somewhat limited. In view of the fact that the preservation of the mummies at the St Petersburg Hermitage is evidently causing problems, it is doubtful whether analyses of the stuffing or embalming substances are still possible at this stage. The impressive tattoos on the body of the man from the second Pazyryk kurgan show intertwined fabulous creatures in the Scythian animal style. In addition there were tattoos in the form of rows of dots for therapeutic purposes. Of special importance is the finding that, owing to their storage in the ice, there was no grave-wax formation in the Siberian bodies, any more than in the Hauslabjoch man.

Like the Pazyryk mummies, the Eskimo mummies of Qilakitsoq, discovered on 9 October 1972 on the central west coast of Greenland, gained international fame. They were recovered in 1978 and scientifically examined at the National Museum in Copenhagen under the direction of Professor J. P. Hart Hansen. The bodies rested in two graves constructed of stone slabs close to one another and protected against sun and precipitation by an overhanging rockface. Tomb 1 contained five bodies, Tomb 2 contained three. Carbon-14 dating placed them in the second half of the fifteenth century AD. They were probably all members of the same family. There were six adult women aged between eighteen and fifty, a four-year-old boy and a male infant of about six months. The bodies were naturally mummified by a combination of low temperatures and dry air. Temperatures at Qilakitsoq drop to minus 40° Centigrade in winter and scarcely rise above 5° Centigrade in summer. In addition, one may assume that there was some degree of air circulation between the stone slabs of the tombs, favouring the mummification process. In this way total dehydration occurred through freeze-drying, without any embalming measures. The extremely dry Arctic air also prevented grave-wax formation. While there were no actual grave goods, the clothes of seal and reindeer skin, as well as of birdskin, were in an excellent state of preservation. In addition, the bodies were wrapped in furs. Examination under ultraviolet light revealed the ornamental facial tattoos typical of the Eskimo culture.

A marked contrast is provided by the human remains, from much the same period and locality, recovered by Dr Paul Nordlund in 1921 from the graveyard of Jerjolfs-nes. This settlement is also on the west coast of Greenland, but very near the island's southern tip. In the Middle Ages it was inhabited by Europeans descended from the Norsemen. These descendants of Eric the Red, who settled Greenland with several shiploads of likeminded people about AD 1000, became extinct towards the end of the fifteenth century. They had not adapted to the Eskimo lifestyle but clung to their European customs to the bitter (and as yet unexplained) end. Their graveyard lay within range of the coastal waves. The graves had been sunk into the permafrost layer, and as a result the wooden coffins and crosses have survived perfectly. The discovery, moreover, of numerous garments, hoods, caps and stockings, all of them made of woollen twill cloth, was a sensational contribution to the history of medieval costume and textiles. They reflect the continental fourteenth-century costume of Burgundian hat and ribboned hood, until then known only from contemporary illuminations. The human remains proper, on the other hand, are less well preserved. Although bones and skulls were recovered, no mention is made of mummified soft tissues, which, strictly speaking, should have been present. Only one old report speaks of a skull with 'blond hair', i.e. that of a European and not an Eskimo. Moreover, as a result of storage next to the ship's steam boiler, the bones had shrunk and become twisted from the heat during the passage to Denmark; this caused the archaeologists to reconstruct a degenerate race, with small skulls, hunchbacked and unhealthy – a society that was dying out. These findings are now open to serious doubt.

Arctic archaeology of human permafrost mummies reached its unchallenged peak when the graves of the Franklin sailors were exhumed. On Monday, 18 May 1845, an expedition commanded by Sir John Franklin set out from London with one hundred and thirty-four men aboard the ships *Erebus* and *Terror*. The purpose of the expedition was the exploration of a northwest passage, a sea route from the Atlantic to the Pacific along the North American coast. An accompanying vessel, which left the expedition on 12 July, took aboard five men of Franklin's crew. There was a last sighting at the beginning of August 1845 by the two whalers *Prince of Wales* and *Enterprise*. From that time onward the remaining one hundred and twenty-nine sailors were neither seen nor heard of again. Numerous rescue parties were dispatched until, on 27 August 1850, more than five years after the expedition's launch, three graves were discovered in which members of Franklin's expedition had been buried. It was a moving moment for the search party when it landed on small Beechey Island in the Arctic archipelago between Canada and Greenland. Three headboards stood on the beach: John Torrington, died 1 January; John Hartnell, died 4 January; and William Braine, died 3 April 1846. It was hoped that an autopsy would provide clues to the expedition's failure. The Canadian forensic expert Professor Owen Beattie exhumed the bodies in 1984 and 1986. The workers were faced with an astonishing sight. The bodies, resting in their coffins frozen into blocks of ice, were so well preserved that they looked like men sleeping. There was only slight dehydration, which had reduced the body weight to between 40 and 50 kilograms. After thawing the limbs could again be moved. There had been no grave-wax formation whatsoever. Dissection and subsequent histological examination revealed a high level of lead, due undoubtedly to the fact that the expedition's tins of food had been inexpertly soldered. Thus the riddle of the Franklin expedition was solved. The crew died of progressive lead poisoning – the weakest of them, Torrington, Hartnell and Braine, who were also suffering from other diseases, died as early as the first winter. The fittest, who succeeded in battling their way south for another 500 kilometres, may even have reached the American coast to die there in the winter of 1848/9.

Glossary of Technical Terms

Ablation The melting of glaciers.

Abrasion Wearing down of teeth through chewing. High levels of abrasion are caused by chewing tough dried meat or cereal foods contaminated with stone dust, also by the unconscious grinding of teeth during sleep.

Adipocere See **Grave-wax**

Altheim culture A Neolithic culture (*c.* 3900–3400 BC) prevalent mainly in Bavaria. Its typical feature is virtually undecorated pots with flat bottoms and thickened rims.

Anaerobic Taking place in the absence of free oxygen.

Annual-ring radius Half-diameter of the growth rings of timbers, usually arranged in circles in the cross-section of the trunk.

Anobium Genus of beetle whose larvae destroy wood, known as 'wood beetles'.

Arteriosclerosis Hardening of the arteries caused by deposits of fatty tissue in the lining of the arteries.

Autolysis Destruction of cells or tissues by their own enzymes without the help of bacteria.

Bacteriocidal Germ-killing. Poisonous metal salts, for instance, often kill bacteria.

Baden culture Neolithic culture (*c.* 3400–2900 BC) prevalent mainly in eastern Austria and western Hungary. A typical feature is the bellied shape of pots with channelled and burnished decoration, often with handles of *ansa lunata* type.

Bast Cord made from the fibrous matter in tree barks of various species, especially e.g. lime.

Bell Beaker culture Late Neolithic culture (*c.* 2800–2200 BC) extensive over western Europe. A typical feature is inverted bell-shaped beakers decorated with horizontal bands.

Biogenic Produced or brought about by living organisms.

Bulb In archaeology the thickening on a flint blade, caused at the point of impact of the blow which separates it from the flint core.

Calcification The deposit of calcium carbonate or other calcium compounds.

Callus The bony healing tissue which forms around the ends of broken bones.

Capercaillie A large mountain-dwelling bird related to the grouse.

Carbon-14 dating See **Radiocarbon**

Cham culture Neolithic culture (*c.* 3200–2500 BC) mainly prevalent in Bavaria. A typical feature is massive bellied pots with appliqué cordons and often a not very smooth surface.

Chromatography Various methods of chemical separation or analysis, in which the sample is first dissolved in a liquid or evaporated into a gas, then passed through absorbent material. The differing rates of passage of the components of the mixture are analysed, and the resulting chromatogram gives a band picture peculiar to that substance.

Chronometer An instrument for measuring time designed to be accurate at all temperatures.

Colloid A jelly-like substance found in the body.

Computer tomography A technique in which X-rays or ultrasound are used to provide images of successive plane sections of the human body or other solid objects, which are then processed by computer to give a three-dimensional image. *See also* **Spiral computer tomography.**

Corded Ware culture Neolithic culture (*c.* 2900–2500 BC) prevalent in Central Europe. Typical are amphorae and beaker-shaped pots decorated with an imprint of twisted cords.

Dehydration Loss or removal of water (H_2O). Not to be confused with dehydrogenation, meaning the removal of hydrogen (H).

Dendrochronology A scientific method for determining the age of timber, based on the variation in the width of the annual growth rings in tree trunks.

Einkorn An early kind of wheat, *Triticum monococcum*, with one-grained spikelets, which was eaten in prehistoric times but is now grown only as animal fodder.

Ergonomics The study of the suitability and organization of different materials for various purposes.

Fertile Crescent An arc-shaped region of the Middle East, taking in Jordan, Israel, Lebanon, Syria, southeastern Turkey, northern Iraq and western Iran. It is thought that arable farming and stockbreeding originated here.

Follicle pattern The pattern of the points where hair emerges from the skin, which is used in determining the provenance of fur.

Frenulum A small ligament or membranous fold serving to check the motion of the part to which it is attached, such as those underneath the tongue or foreskin.

Geodesy Land surveying, and the branch of mathematics which deals with the shape and area of the earth or of large parts of it.

Glume Any of certain membranous scales in the flowering part of grasses or related plants.

Grain side A furrier's term for the outside of leather or fur-skin.

Granular impressions Small dents, e.g. on the human skin, caused by storage on a rough surface.

Grave-wax formation Also, lipocere or adipocere formation. The conversion of fat-containing organs of a dead body into an unstructured white mass, common in glacier corpses and drowned bodies.

Hallstatt culture Early Iron Age culture (*c.* 800–400 BC) in southern Central Europe. Named after a burial ground at Hallstatt, Upper Austria.

Hauslabjoch A crossing of the main ridge of the Alps which, along with other passes, links the valleys of the Ötztal and Val Venosta. The Iceman was discovered 300 metres south of the Hauslabjoch on 19 September 1991.

Histological examination Microscopic examination of organic tissue.

Horgen culture Neolithic culture (*c.* 3500–2800 BC) prevalent in western Switzerland and around Lake Constance. A typical feature is thick-walled bucket-shaped pots with sparse decoration.

Humerus Bone of the upper arm.

Hygroscopic salts Salts which absorb moisture, e.g. soda, used in artificial mummification in order to dehydrate the soft parts of the body. Hygroscopic salts are also used for reducing air humidity.

Integument The natural outside covering of part of an animal or a plant.

Kurgan The local name for a grave mound in the Eurasian steppe regions, especially in the Ukraine and southern Siberia.

Lagozza culture Neolithic culture (*c.* 3800–3400 BC) prevalent in northern Italy. The typical pot shape is carinated, sometimes with a round bottom, and with sparse decoration.

Lead solder Lead alloy used because of its low melting point for soldering thin metal sheets.

Leaf-point A prehistoric projectile point shaped like a leaf, made of flint or similar rock.

Lipocere See **Grave-Wax**

Loess A fine, yellowish-grey loam composed of peri-glacial material transported by the wind during and after the Ice Age.

Main ridge of the Alps The chain of the highest Alpine peaks, running roughly east–west.

Mesentery A fold of tissue which attaches an organ, especially the small intestine, to the posterior wall of the abdomen.

Michelsberg culture Early Neolithic culture (*c.* 4400–3800 BC) prevalent mainly in western and southern Germany. Typical pot shapes are bellied and conical, almost invariably without decoration. Typical forms are tulip-shaped beakers and loop-rimmed flasks with round bases.

Mondsee culture Neolithic culture (*c.* 3700–2900 BC) prevalent mainly in Upper Austria. Typical is a richly decorated pottery with deep-engraved designs, into which a white paste was brushed ('incrustation').

Morphology The branch of biology that deals with the form of organisms.

Mycology The scientific study of fungi.

Necropolis An ancient burying-place.

Ore outcrop The place where otherwise underground metal-bearing rock appears on the surface.

Oversewing A sewing technique for joining two pieces of material or leather. The thread is taken from one hole over the edge to the next hole. Because of the constriction of the thread, the edge looks toothed.

Pfyn culture Early Neolithic culture (*c.* 4200–3500 BC) prevalent around Lake Constance. Typical pot shapes are bellied and conical, with flat bottoms and sparse decoration. Jugs with handles have also been found.

Photogrammetry The technique of using photographs to obtain and plot measurement of what is photographed.

Pneumatic chisel A pistol-shaped tool driven by compressed air, with a chisel fitted into the 'barrel'. Used by the mountain rescue service for freeing glacier corpses.

Precipitation Water from the atmosphere that falls to or condenses on the ground, such as rain, snow, dew, hail, etc.

Pyrogallic ester Organic salt of pyrogallol, a phenolic compound of tannin, produced in small quantities during such processes as the smoke-tanning of leather.

Radial fletching Flight-stabilizing elements, usually feathers, fitted to the end of the arrow-shaft, usually in threes, with the feather spine running parallel to the shaft and the feathers perpendicular to it. In tangential fletching, the underside of the feather spine is cemented to the shaft, so that the feather lies sideways against the shaft.

Radiocarbon or carbon-14 dating A scientific method for determining the age of organic materials, based on decay over time in the natural radioactivity of carbon present in all organic matter.

Reflected-light microscope An optical instrument in which the object is seen under incident light, in contrast to the transilluminating microscope, in which wafer-thin samples are observed in the light passing through them.

Remedello culture Neolithic culture (*c.* 3300–2300 BC) prevalent in northern Italy. Typical are bi-conical pots with bands of hatching scratched into the clay, as well as daggers and axes made of copper.

Repoussé work Metalwork in which an ornamental relief pattern is made by hammering from the reverse side.

Roman-Germanic Central Museum, Mainz, Germany One of the most important museums of European prehistory and early history. It has a collection of facsimiles from all major archaeological sites, and an extensive collection of original pieces. Its workshops are at the forefront of modern conservation techniques.

Ruminant An animal that chews the cud.

Similaun glacier There is no Similaun glacier. The phrase is an invention of the media, based on the Similaunhütte, the refuge nearest the spot where the Iceman was found. The body was lying in a small unnamed secondary glacier which originally flowed into the Niederferner but, due to the marked recession of glaciers in recent decades, has long been totally separated from it.

Spectroscopy A technique which analyses the colour spectrum of a given substance or substances.

Spiral computer tomography A specialized method of computer tomography, in which the data are registered not two-dimensionally but 'spirally', i.e. three-dimensionally. The recorded data can be converted by the computer into images and can also be used to make three-dimensional models.

Square-Mouthed Pot culture (Bocca Quadrata) Early Neolithic culture (*c.* 4300–3500 BC) prevalent in northern Italy. Typical are pot-shaped and bowl-shaped vessels with a square-pinched mouth, almost invariably with round bottoms.

Stereolithography A technique using laser beams, by which an exact three-dimensional model of an object – such as a skull – can be produced.

Stereomicroscope A binocular microscope that gives a three-dimensional view of the object under examination.

Sudorific Producing perspiration.

Syncretism The merging of ideas of different religions.

Terrestrial-tachymetric survey Fast surface survey method by means of the simultaneous measurement of distance and altitude with a special instrument, the tachymeter.

Theodolite An optical instrument for measuring horizontal and altitudinal angles, used in surveying land.

Tisenjoch A crossing in the main ridge of the Alps which, along with other passes, links the valleys of the Ötztal and Val Venosta. The Iceman was discovered 80 metres northwest of the Tisenjoch on 19 September 1991. The name exists only in oral tradition and is not found on official maps.

Transhumance The seasonal transfer of grazing animals to different pastures, often over long distances.

Twelve as a sacred number The number 12 has always played an important role in cult and myth: 12 apostles, 12 months, 12 nights, 12 prophets, 12 cities, 12 tablets.

Zygomatic arch A bony arch on each side of the skull, formed by the cheekbone and its linkage with the temporal bone.

Experts Involved in the Research Project

Dr. Peter Adolff, Kriminaltechnisches Institut, Bundeskriminalamt Wiesbaden
Prof. Otto Appenzeller, M. D., Ph. D., NM EMC Research Foundation, Albuquerque
Dr. Walter Asci, Dipartimento di Biologia, Moleculare Cellulare e Animale, Università di Camerino
Dott. ssa Alessandra Aspes, Sezione Preistoria, Museo Civico di Storia Naturale, Verona
Prof. Dr. Horst Aspöck, Abteilung für Medizinische Parasitologie, Klinisches Institut für Hygiene, Universität Wien
Dr. Herbert Auer, Abteilung für Medizinische Parasitologie, Klinisches Institut für Hygiene, Universität Wien
Dipl.-Ing. Dr. Gerd Augustin, Institut für Geodäsie, Universität Innsbruck
Prof. Bernadino Bagolini, Facoltà di Lettere e Filosofia, Dipartimento Storia e Civiltà Europea, Università di Trento
Priv.-Doz. Dr. Svetla Balabanova, Abteilung Rechtsmedizin, Universität Ulm
Antonio Belli, FO. A. R. T. srl, Fotogrammetria Architettonica-Rilievo Terreste, Parma
Prof. Dr. Wolfram Bernhard, Institut für Anthropologie, Universität Mainz
Dipl.-Ged. Alex Binsteiner, Regensburg
Dr. Georges Bonani, Institut für Mittelenergiephysik, Eidgenössische Technische Hochschule, Zürich
Prof. Dr. Sigmar Bortenschlager, Institut für Botanik, Universität Innsbruck
Dr. Christopher R. Bronk, Radiocarbon Accelerator Unit, Research Laboratory for Archaeology and the History of Art, Oxford
Dr. D. R. Brothwell, Institute of Archaeology, University College London
Dott. Luigi Capasso, Soprintendenza Archeologica dell' Abruzzo, Chieti
Prof. Dott. Arnaldo Capelli, Istituto di Anatomia Patologica, Università Cattolica del Sacro Curoe, Roma
Dir Dr. Lanfredo Castelletti, Museo Civico Archeologico di Como
Prof. Dipl.-Ing. Dr. Günter Chesi, Institut für Geodäsie, Universität Innsbruck
Dr. Lorenzo Dal Ri, Landesdenkmalamt, Bozen
Prof. Raffaele C. De Marinis, Istituto di Archeologia, Università degli Studi di Milano
Dr. Koert P. Dingemans, Department of Pathology, University of Amsterdam
Prof. Dr. Ekkehard Dreiseitl, Institut für Meteorologie und Geophysik, Universität Innsbruck

Experts Involved in the Research Project

Prof. Dr. Angela von den Driesch, Institut für Palaeoanatomie, Domestikationsforschung und Geschichte der Tiermedizin, Universität München
Dr. Leo D. Dulk, Department of Dermato-Venereology, Academic Hospital Rotterdam
Dr. Michael Ebner, Mailand
Dr. Markus Egg, Römisch-Germanisches-Zentralmuseum, Mainz
Dr. Brigitta Erschbamer, Institut für Botanik, Universität Innsbruck
Ingrid Faé, Institut für Blutgruppenserologie, Universität Wien
Prof. Dott. Leone Fasani, Dipartimento di Scienze della Terra, Università degli Studi di Milano
Prof. Walter Ferri, Istituto di Geotesia, Topologia e Fotogrammetria, Facoltà Ingegneria, Università di Pisa
Dr. Gottfried Fischer, Institut für Blutgruppenserologie, Universität Wien
Laurent Flutsch, Kulturgeschichte 1 des Schweizerischen Landesmuseums, Zürich
Dr. Romana Fornwagner, Forschungsinstitut für Alpine Vorzeit, Universität Innsbruck
Prof. Dott. Luigi Frati, Dipartimento di Medicina Sperimentale, Università 'La Sapienza', Roma
Prof. Dr. Othmar Gaber, Institut für Anatomie, Universität Innsbruck
Dr. Stefan Galler, Institut für Zoologie, Universität Salzburg
Prof. Dr. Kurt Gausch, Univ.-Klinik für Zahn-, Mund- und Kieferheilkunde, Universität Innsbruck
Prof. Dr. Franz Gerstenbrand, Univ.-Klink für Neurologie, Universität Innsbruck
Stefano Giuliani, Institut für Strafrecht und sonstige Kriminalwissenschaften/Gemeinsame Einrichtung für Italienisches Recht, Universität Innsbruck
Roswitha Goedecker-Ciolek, Römisch-Germanisches-Zentralmuseum, Mainz
Prof. Dr. Rainer Gothe, Institut für Vergleichende Tropenmedizin und Parasitologie, Universität München
Prof. Dott. Giorgio Graziosi, Istituto di Biologia, Università degli Studi di Trieste
Prof. Dr. Willy Groenman-van Waateringe, Instituut voor Pre- en Protohistorische Archeologie Albert Egges van Griffen, Universiteit van Amsterdam
Prof. Dr. Gisela Grupe, Institut für Anthropologie und Humangenetik, Universität München
Dr. Verena Günther, Univ.-Klinik für Psychiatrie, Universität Innsbruck
Prof. Dr. Jens Peder Hart Hansen, Department of Pathology, University of Copenhgaen
Prof. Dr. Kurt Haselwandter, Institut für Mikrobiologie, Universität Innsbruck
Dr. Anreas Haßl, Abteilung für Medizinische Parasitologie, Klinisches Institut für Hygiene, Universität Wien
Dr. Gertrude Hauser, Histologisch-Embryologisches Institut, Universität Wien
Dr. Dipl. Ing Erich W. H. Hayek, Institut für Allgemeine Chemie der Technischen Universität Wien
Dr. Robert E. M. Hedges, Radiocarbon Accelerator Unit, Research Laboratory for Archaeology and the History of Art, Oxford
Prof. Dr. Rainer Henn (†) Institut für Gerichtliche Medizin, Universität Innsbruck
Prof. Dr. Bernd Herrmann, Anthropologisches Institut, Universität Göttingen

Experts Involved in the Research Project

Prof. Dr. Hartmann Hinterhuber, Univ.-Klinik für Psychiatrie, Universität Innsbruck
Dr. T. C. Holden, Museum of London
Dr. Rupert A. Housley, Radiocarbon Accelerator Unit, Research Laboratory for Archaeology and the History of Art, Oxford
Prof. Dr. Kurt Johann Irgolic, Institut für Analytische Chemie, Universität Graz
Mag. Susan D. Ivy, Institut für Mittelenergiephysik, Eidgenössische-Technische-Hochschule, Zürich
Dr. Dipl. Ing Ulrich Jordis, Institut für Organische Chemie der Technischen Universität Wien
Dr. Georg Kaser, Institut für Geographie, Universität Innsbruck
Dr. Gert Jaap van Klinken, Radiocarbon Accelerator Unit, Research Laboratory for Archaeology and the History of Art, Oxford
Werner Kofler, institut für Botanik, Universität Innsbruck
Albert Kremer, Römisch-Germanisches-Zentralmuseum, Mainz
Dr. Karl-Heinz Künzel, Institut für Anatomie, Universität Innsbruck
Dipl.-Chem. Joachim Lange, Lehr-, Prüf- und Forschungsinstitut, Westdeutsche Gerberschule Reutlingen
Dipl.-Geol. Dr. phil Gabriele Lass, Seminar für Ur- und Frühgeschichte, Universität Münster
Ing Henk Leenders, Laboratory of Otobiology and biocompatibility, Leiden
Prof. Univ.-Doz. Dr. Walter Leitner, Institut für Ur- und Frühgeschichte, Universität Innsbruck
Prof. Dr. Andreas Lippert, Institut für Ur- und Frühgeschichte, Universität Wien
Dr. Thomas H. Loy, Department of Prehistory, Australian National University, Canberra
Prof. Dr. Gert Lubec, Abteilung für Pädiatrie, Universität Wien
Prof. Dott. Renato Mariani-Costantini, Facolta di Odontologia, Universität 'Gabriele D'Annunzio', Chieti
Mag. Michael Marius, Bundesdenkmalamt, Wien
Gerhard Markl, Institut für Meteorologie und Geophysik, Universität Innsbruck
Dipl. Ing Franz Markowski, Gesellschaft der AVT, Imst
Dr. Isolina Marota, Dipartimento di Biologia, Moleculare Cellulare e Animale, Università di Camerino
Dr. Jochen Martens, Institut für Zoologie, Universität Mainz
Prof. Dr. Herbert Maurer, Institut für Anatomie, Universität Innsbruck
Dr. Ullrich Meise, Univ.-Klinik für Psychiatrie, Universität Innsbruck
Ing. Isabelle Mosberger, Institut für Klinische Pathologie, Universität Wien
C. Mühlenhof, Klinik für Dermatologie, Universität Wien
Andreas Müllner, Institut für Humanbiologie, Universität Wien
Prof. Dr. William A. Murphy. St Louis, Mo.
Dr. Bernard Naafs, Department of Dermatology, Academisch Ziekenhuis Rotterdam
Prof. Dr. Karl Narr, Seminar für Ur- und Frühgeschichte, Universität Münster
Prof. Dr. Dieter zur Nedden, Radiologische Abteilung der Univ.-Klinik für Innere Medizin, Universität Innsbruck
Dr. Dr. Kurt Nicolussi, Forschungsinstitut für Hochgebirgsforschung, Universität Innsbruck

Experts Involved in the Research Project

Dipl. phys. Thomas R. Niklaus, Institut für Mittelenergiephysik, Eidgenössische-Technische-Hochschule, Zürich
Dr. Hans Nothdurfter, Landesdenkmalamt Bozen
Mag. Jörg Obereder, Naturhistorisches Museum, Wien
Dr. Klaus Oeggl, Institut für Botanik, Universität Innsbruck
Dr. Nikolaus Oswald, Sektion Forschung und Entwicklung des Schweizerischen Landesmuseums, Zürich
Prof. Dr. Svante Pääbo, Zoologisches Institut, Universität München
Prof. Dr. Gernot Patzelt, Forschungsinstitut für Hochgebirgsforschung, Universität Innsbruck
Prof. Dr. Margit Pavelka, Institut für Histologie und Embryologie, Universität Innsbruck
Dott. ssa Annaluisa Pedrotti, Ufficio Beni Archeologici, Provincia Autonoma di Trento
Mag. Ursula Peintner, Institut für Mikrobiologie, Universität Innsbruck
Dr. Josef Penninger, Ontario Cancer Institute, Dep. of Biophysics and Summonology, University of Toronto
Prof. Dr. Joris Peters, Institut für Paläoanatomie, Domestikationsforschung und Geschichte der Tiermedizin, Universität München
Dr. Otto Picher, Abteilung für Medizinische Parasitologie, Klinisches Institut für Hygiene, Universität Wien
Prof. Dr. Sandro Pignatti, Istituto ed Orto Botanico, Università degli Studi di Trieste
Prof. Dr. Werner Platzer, Institut für Anatomie, Universität Innsbruck
Dr. Reinhold Pöder, Institut für Mikrobiologie, Universität Innsbruck
Dr. Thomas Pümpel, Institut für Mikrobiologie, Universität Innsbruck
Dr. Sebastiaan C. J. van der Putte, Department of Pathology, Academic Hospital Utrecht
Prof. Dr. Reinhard Putz, Anatomisches Institut, Universität München
Dr. Oskar Reinwarth, Abteilung Glaziologie, Bayerische Akademie der Wissenschaften, München
Mag. Alarich Riss, Umweltbundesamt, Wien
Dott. Franco Rollo, Dipartimento di Biologia, Molecolare Cellulare e Animale, Università di Camerino
Dr. Elsebet Sander-Joergensen, Silkeborg Museum, Silkeborg
Prof. Dr. Friedrich Sauter, Institut für Organische Chemie, Technische Universität Wien
Michael Schäfer, Institut für Ur- und Frühgeschichte, Universität Wien
Dr. Heralt Schneider, Institut für Mathematik, Universität Innsbruck
Dr. Werner Schoch, Labor für Quartäre Hölzer, Adliswil
Heidrun Schöl, Institut für Vergleichende Tropenmedizin und Parasitologie, Universität München
Prof. Dr. Rudolf Schweyen, Institut für Mikrobiologie und Genetik, Universität Wien
Prof. Dr. Horst Seidler, Institut für Humanbiologie, Universität Wien
Dr. Ingrid Simonitsch, Institut für Klinische Pathologie, Universität Wien
Prof. Dr. Torstein Sjøvold, Osteological Research Laboratory, Stockholm University

Experts Involved in the Research Project

Gerhard Sommer, Institut für Ur- und Frühgeschichte, Universität Innsbruck
Dipl. Ing Dr. Dr. Gerhard Sperl, Erich Schmid-Institut für Festkörperphysik, Österreichische Akademie der Wissenschaften, Leoben
Prof. Dr. Konrad Spindler, Institut für Ur- und Frühgeschichte, Universität Innsbruck
Prof. Dipl. Ing. Dr. tech. Herbert Stachelberger, Institut für Angewandte Botanik, Technische Mikroskopie und organische Rohstofflehre der Technischen Universität Wien
Martin Steinlechner, Institut für Anatomie, Universität Innsbruck
Dr. Martin Suter, Institut für Mittelenergiephysik, Eidgenössische-Technische-Hochschule, Zürich
Dr. Maria Teschler-Nicola, Anthropologische Abteilung, Naturhistorisches Museum, Wien
Prof. Dr. Friedrich Tiefenbrunner, Institut für Hygiene, Universität Innsbruck
Dr. Rupert Timpl, Max-Planck-Institut für Biochemie, München
Dr. Martina Traindl-Prohazka, Institut für Humanbiologie, Universität Wien
Dr. Erwin Tschachler, Klinik für Dermatologie, Universität Wien
Prof. Dr. Gerd Utermann, Institut für Medizinische Biologie und Humangenetik, Universität Innsbruck
Dr. med. vet. Aumaid Utman, Klinik für Dermatologie, Universität Wien
Edmund M. van der Velden, Department of Dermato-Venerelogy, Academic Hospital Rotterdam
Dr. Beatrix Volc-Platzer, Klinik für Dermatologie, Universität Wien
Dr. Vojislav D. Vuzevski, Department of Pathology, Erasmus University, Rotterdam
Dr. Gerhard Weber, Institut für Humanbiologie, Universität Wien
Prof. Dr. Karl Weber, Institut für Öffentliches Recht, Universität Innsbruck
Prof. Dr. Georg Wick, Institut für Allgemeine und Experimentelle Pathologie, Universität Innsbruck
Dr. Klaus Wicke, Abteilung für Röntgendiagnostik und Computertomographie, Univ. Kliniken Innsbruck
Dr. Harald Wilfing, Institut für Humanbiologie, Universität Wien
Dr. Josef Winiger, Vinelz
Dr. Manfred Wittig, Kriminaltechnisches Institut, Bundeskriminalamt Wiesbaden
Franz Josef Wortmann, Deutsches Wollforschungsinstitut, Aachen
Dr. Gabriele Wortmann, Deutsches Wollforschungsinstitut, Aachen
Mag. Elisabeth Zissernig, Forschungsinstitut für Alpine Vorzeit, Universität Innsbruck

Bibliography

compiled by Christiane Ganner

Acanfora, Maria Ornella: *Fontanella Mantovana e la cultura di Remedello*. In: *Bullettino di Paletnologia Italiana*, N.S. 10, 65, 1956, Fasc. 1, pp. 321–385.

Allely, Steve / Baker, Tim / Comstock, Paul / Hamm, Jim / Hardcastle, Ron / Massey, Jay / Strunk, John: *The Traditional Bowyer's Bible*, Vol. 1, Fort Worth 1992.

Ambach, Edda / Tributsch, Wolfgang / Puffer, Peter / Rabl, Walter: *Ungewöhnliche Auffindung zweier Leichen im Gletscher – Forensische und glaziologische Aspekte*. In: *Beiträge zur Gerichtlichen Medizin* 49, 1991, pp. 285–288.

Ambach, Walter / Schneider, Heralt / Ambach, Edda / Tributsch, Wolfgang: *Nach 25 Jahren vom Gletscher freigegeben – Glazialmeteorologische Aspekte*. In: *Wetter und Leben* 42, 1990, No. 3/4, pp. 183–188.

Arch, Harwick W.: *Noch ein Mann vom Hauslabjoch*. In: *Appell – Zeitschrift der Offiziers-Gesellschaft Tirol* 1992, No. 13, p. 14.

Arch, Harwick W.: *Noch ein Mann vom Hauslabjoch*. In: *Der Schlern* 65, 1991, pp. 635–636.

Artamanonov, Michail I.: *Frozen Tombs of the Scythians*. In: *Scientific America* 212, 1965, pp. 101–109.

Aspes, Alessandra / Barfield, Lawrence / Bermond Montanari, Giovanna / Burroni, Daniela / Fasani, Leone / Mezzena, Franco / Poggiani Keller, Raffaela: *L'Età del Rame nell'Italia Settentrionale*. In: *Rassegna di Archeologia* 7, 1988, pp. 401–440.

Aspes, Alessandra / Fasani, Leone: *Tentativo di classificazione delle asce piatte della regione sudalpina centrale e padana*. In: Höpfel, Frank / Platzer, Werner / Spindler, Konrad: *Der Mann im Eis*, Vol. 1, *Veröffentlichungen der Universität Innsbruck* 187, Innsbruck 1992, pp. 378–388.

Aspes, Alessandra: *Il Veneto nell'Antichità – Preistoria e protostoria*, Vol. 1, Verona 1984.

Aspöck, Horst / Auer, Herbert: *Zur parasitologischen Untersuchung des Mannes vom Hauslabjoch*. In: Höpfel, Frank / Platzer, Werner / Spindler, Konrad: *Der Mann im Eis*, Vol. 1, *Veröffentlichungen der Universität Innsbruck* 187, Innsbruck 1992, pp. 214–217.

Ausserer, Carl: *Der Alpensteinbock*, Wien 1946.

Bagolini, Bernardino / Barfield, Lawrence H: *The European Context of Northern Italy during the Third Millennium*. In: *Saarbrücker Beiträge zur Altertumskunde*, Vol. 55, Bonn 1991, pp. 287–300.

Bibliography

Bagolini, Bernardino / Pedrotti, Annaluisa: *Vorgeschichtliche Höhenfunde im Trentino-Südtirol und im Dolomitenraum vom Spätpaläolithikum bis zu den Anfängen der Metallurgie*. In: Höpfel, Frank / Platzer, Werner / Spindler, Konrad: *Der Mann im Eis*, Vol. 1, *Veröffentlichungen der Universität Innsbruck* 187, Innsbruck 1992, pp. 359–377.

Bagolini, Bernardino: *Il Trentino nella Preistoria del mondo Alpino*, Trento 1980.

Bagolini, Bernardo / Lanzinger, Michele / Pedrotti, Annaluisa: *Rinvenimento di quattro statue stele ad Arco (Valle del Sarca - Trentino Meridionale)*. In: *Atti della XXVIII Riunione Scientifica dell'Istituto Italiano di Preistoria e Protostoria*, Firenze 1992, pp. 355–370.

Bagolini, Bernardo / Pedrotti, Annaluisa: *Età del Rame – Gruppo di Remedello e Ceramica Metopale*. In: *Uomini di Pietra – Statue Stele e prima Metallurgia in Trentino – Alto Adige. Catalogo della Mostra a Castel Besena 15.7.–7.11. 1993*, in course of publication.

Bagolini, Bernardo / Pedrotti, Annaluisa: *Neolitico Medio e Recente dell'Italia Settentrionale – Cultura dei Vasi a Bocca Quadrata (V.B.Q.)*. In: *Uomini di Pietra – Statue Stele e prima Metallurgia in Trentino – Alto Adige. Catalogo della Mostra a Castel Besena 10.7.–7.11.1993*, in course of publication.

Ballinger, Erich: *Der Geltschermann – Ein Krimi aus der Steinzeit*, Wien 1992.

Barfield, Lawrence / Koller, Ebba / Lippert, Andreas: *Der Zeuge aus dem Gletscher – Das Rätsel der frühen Alpen-Europäer*, Wien 1992.

Battaglia, Raffaello: *Preistoria del Veneto e della Venezia Giulia*. In: *Bullettino di Paletnologia Italiana* 67–68, 1958/59, pp. 1–419.

Beattie, Owen / Geiger, John: *Frozen in Time: The Fate of the Franklin Expedition*, London 1987.

Becker, Bern / Billamboz, André / Egger, Heinz / Gassmann, Patrick / Orcel, Christian / Ruoff, Ulrich: *Dendrochronologie in der Ur- und Frühgeschichte*. In: *Antiqua* 11, Basel 1985.

Belli, Antonio / Capasso, Luigi / Ferri, Walter: *The Stereophotogrammetric Survey of the Val Senales Mummy – Peculiar Photographic Techniques and State of the Art*. In: Höpfel, Frank / Platzer, Werner / Spindler, Konrad: *Der Mann im Eis*, Vol. 1, *Veröffentlichungen der Universität Innsbruck* 187, Innsbruck 1992, pp. 218–226.

Bellwald, Werner: *Drei spätneolithisch / Frühbronzezeitliche Pfeilbogen aus dem Gletschereis am Lötschenpass*. In: *Archäologie der Schweiz* 15, 1992, No. 4, pp. 166–171.

Berg, Steffen / Rolle, Renate / Seemann, Henning: *Der Archäologe und der Tod – Archäologie und Gerichtsmedizin*, München und Luzern 1981.

Bernhard, Wolfram: *Multivariate statistische Untersuchungen zur Anthropologie des Mannes vom Hauslabjoch*. In: *Der Mann im Eis*, Vol. 2, *Veröffentlichungen der Universität Innsbruck*, in course of publication.

Bernhard, Wolfram: *Vergleichende Untersuchungen zur Anthropologie des Mannes vom Hauslabjoch*. In: Höpfel, Frank / Platzer, Werner / Spindler, Konrad: *Der Mann im Eis*, Vol. 1, *Veröffentlichungen der Universität Innsbruck* 187, Innsbruck 1992, pp. 163–187.

Bernhardt, Bodo: *Zur Frage der Staatsgrenze*. In: Höpfel Frank / Platzer, Werner / Spindler, Konrad: *Der Mann im Eis*, Vol. 1, *Veröffentlichungen der Universität Innsbruck* 187, Innsbruck 1992. pp. 66–80.

Bibliography

Biagi, Paolo: *La preistoria in terra bresciana*, Brescia 1978.

Biedermann, Hans: *Höhlenkunst der Eiszeit – Wege zur Sinndeutung der ältesten Kunst Europas*, Köln 1984.

Boessneck, Joachim / Jequieur, J.-P. / Stampfli, H. R.: *Seeberg Burgäschisee-Süd. Die Tierreste. Acta Bernensia* 2, Bern 1963, Part 3.

Bonani, Georges / Ivy, Susan D. / Niklaus, Thomas R. / Suter, Martin / Housley, Rupert A. / Bronk, Christopher R. / Van Klinken, Gert Haap / Hedges, Robert E. M.: *Altersbestimmung von Milligrammproben der Ötztaler Gletscherleiche mit der Beschleunigermassenspektrometrie-Methode (AMS)*. In: Höpfel, Frank / Platzer, Werner / Spindler, Konrad: *Der Mann im Eis*, Vol. 1, *Veröffentlichungen der Universität Innsbruck* 187, Innsbruck 1992, pp. 108–116.

Bortenschlager, Sigmar / Kofler, Werner / Oeggl, Klaus / Schoch, Werner: *Erste Ergebnisse der Auswertung der vegetabilischen Reste vom Hauslabjochfund.* In: Höpfel, Frank / Platzer, Werner / Spindler, Konrad: *Der Mann im Eis*, Vol 1, *Veröffentlichungen der Universität Innsbruck* 187, Innsbruck 1992, pp. 307–312.

Breitinger, Emil: *Zur Berechnung der Körperhöhe aus den langen Gliedmaßenknochen*. In: *Anthropologischer Anzeiger 14*, 1937, pp. 249–274.

Breunig, Peter: *^{14}C-Chronologie des vorderasiatischen, südost- und mitteleuropäischen Neolithikums. Fundamenta*, Series A, Vol. 13, Köln 1987.

Brothwell, Don: *The Bogman and the Archaeology of people. British Museum Publications*, London 1986.

Brunnacker, Karl / Heim, Roger / Huber, Bruno / Klötzli, Frank / Müller-Beck, Hansjürgen / Oertli, Henri J. / Oeschger, Hans / Schmid, Elisabeth / Schweingruber, Fritz / Villaret von Rochow, Margita / Volkart, Hans D. / Welten, Max / Wuthrich, Marguerite: *Seeberg Burgäschisee-Süd. Chronologie und Umwelt. Acta Bernensia* 2, Bern 1967, Part 4.

Burger, Ingrid: *Die chronologische Stellung der Fußschalen in den endneolithischen Kulturgruppen Mittel- und Südosteuropas*. In: *Vorzeit zwischen Main und Donau, Erlanger Forschungen*, Reihe A, Vol. 26, Erlangen 1980, pp. 11–45.

Burger, Ingrid: *Die Siedlung der Chamer Gruppe von Dobl, Gemeinde Prutting, Landkreis Rosenheim, und ihre Stellung im Endneolithikum Mitteleuropas, Materialhefte zur Bayrischen Vorgeschichte*, Series A, Vol. 56, Fürth 1988.

Capasso, Luigi / Capelli, Arnaldo / Frati, Luigi / Mariani-Costantini, Renato: *Notes on the Paleopathology of the Mummy from Hauslabjoch/Val Senales (Southern Tyrol, Italy)*. In: Höpfel, Frank / Platzer, Werner / Spindler, Konrad: *Der Mann im Eis*, Vol. 1, *Veröffentlichungen der Universität Innsbruck* 187, Innsbruck 1992, pp. 209–213.

Cornaggia Castiglioni, Ottavio: *La cultura di Remedello. Memorie della Società Italiana di Scienze Naturali*, Vol. 20, Fasc. 1, Milano 1971.

Cvetan-Žakula, Dragica: *Plašta Kisu u Hrvata Kajkavaca. (Summary: A traditional sheperd's garment in northern Croatia)*. In: *Etnološka tribina* 15, 1992, pp. 119–130.

Danieli, Ezio / Bolognese, Stefano: *L'uomo che venne dal ghiaccio*, Trento 1991.

Dehn, Wolfgang: *'Transhumance' in der westlichen Späthallstattkultur? Archäologisches Korrespondenzblatt* 2, 1972, pp. 125–127.

Diezel, P. B. / Hage, Walter / Jankuhn, Herbert / Klenk, E. / Schaefer, Ulrich /

Schlabow, Karl / Schütrumpf, Rudolf / Spatz, Hugo: *Zwei Moorleichen aus dem Domlandsmoor.* In: *Praehistorische Zeitschrift*, Vol. 36, 1958, pp. 118–219.
Dimitrijevič, Stojan: *Badenska Kultura.* In: *Praistorija Jugoslavenskih Zemalja* 3, Sarajevo 1979, pp. 183–235.
Drack, Walter: *Die frühen Kulturen mitteleuropäischer Herkunft.* In: *Archäologie der Schweiz*, Vol. 2, *Die jüngere Steinzeit.* Basel 1969, pp. 67–82.
Drenkhahn, Rosemarie / Germer, Renate: *Mumie und Computer. Katalog zur Sonderausstellung des Kestner-Museums Hannover vom 26.9. 1992 bis 19.1. 1993*, Hannover 1991.
Driehaus, Jürgen: *Die Altheimer Gruppe und das Jungneolithikum in Mitteleuropa.* Römisch-Germanisches Zentralmuseum zu Mainz, Mainz 1960.
Driesch, Angela von den / Peters, Joris / Stork, Marlies: *7000 Jahre Nutztierhaltung in Bayern.* In: *Bauern in Bayern – Von den Anfängen bis zur Römerzeit. Katalog des Gäubodenmuseums Straubing* Nr. 19, Straubing 1992, pp. 157–190.
Driesch, Angela von den / Peters, Joris: *Zur Ausrüstung des Mannes im Eis, Gegenstände und Knochenreste tierischer Herkunft.* In: *Der Mann im Eis*, Vol. 2, *Veröffentlichungen der Universität Innsbruck*, in course of publication.
Drößler, Rudolf: *Kunst der Eiszeit von Spanien bis Sibirien*, Wien 1980.
Eccher, Bernhard: *Privatrechtliche Verhältnisse bei komplexen Funden.* In: Höpfel Frank / Platzer, Werner / Spindler, Konrad: *Der Mann im Eis*, Vol. 1, *Veröffentlichungen der Universität Innsbruck* 187, Innsbruck 1992, pp. 36–42.
Egg, Markus: *Zur Ausrüstung des Toten vom Hauslabjoch, Gem. Schnals (Südtirol).* In: Höpfel, Frank / Platzer, Werner / Spindler, Konrad: *Der Mann im Eis*, Vol. 1, *Veröffentlichungen der Universität Innsbruck* 187, Innsbruck 1992, pp. 254–272.
Finsterwalder, Karl: *Tiroler Ortsnamenkunde*, Vols. 1 and 2, Innsbruck 1990.
Gaber, Othmar / Künzel, Karl-Heinz / Maurer, Herbert / Platzer, Werner: *Konservierung und Lagerung der Gletschermumie.* In: Höpfel, Frank / Platzer, Werner / Spindler, Konrad: *Der Mann im Eis*, Vol. 1, *Veröffentlichungen der Universität Innsbruck* 187, Innsbruck 1992, pp. 92–99.
Ganghofer, Ludwig: *Der mann im Salz*, Stuttgart 1906.
Geat, Cesare: *Aspekte des Kulturgüterschutzes im italienischen Recht.* In: Höpfel, Frank / Platzer, Werner / Spindler, Konrad: *Der Mann im Eis*, Vol. 1, *Veröffentlichungen der Universität Innsbruck* 187, Innsbruck 1992, pp. 43–49.
Giuliani, Stefano: *Das Verfahren bei menschlichen Funden von archäologischem Interesse nach italienischem Recht.* In: Höpfel, Frank / Platzer, Werner / Spindler, Konrad: *Der Mann im Eis*, Vol. 1, *Veröffentlichungen der Universität Innsbruck* 187, Innsbruck 1992, pp. 61–65.
Gleirscher, Paul: *Almwirtschaft in der Urgeschichte?* In: *Der Schlern 59*, 1985, pp. 116–124.
Gleirscher, Paul: *Zum frühen Siedlungsbild im oberen und mittleren Vinschgau mit Einschluß des Münstertales.* In: *Der Vinschgau und seine Nachbarräume*, Bozen 1993, pp. 35–50.
Gothe, Rainer / Schöl, Heidrun: *Hirschlausfliegen (Diptera, Hippoboscidae:*

Lipoptena cervi) in den Beifunden der Leiche vom Hauslabjoch. In: Höpfel, Frank / Platzer, Werner / Spindler, Konrad: *Der Mann im Eis*, Vol. 1, *Veröffentlichungen der Universität Innsbruck* 187, Innsbruck 1992, pp. 299–306.
Graupe, Friedrich / Scherer, Max: *Der Mann aus dem Eis*, Wien 1991.
Groenman-van Waateringe, Willy / Goedecker-Ciolek, Roswitha: *The Equipment made of hide and leather.* In: Höpfel, Frank / Platzer, Werner / Spindler, Konrad: *Der Mann im Eis*, Vol. 1, *Veröffentlichungen der Universität Innsbruck* 187, Innsbruck 1992, pp. 410–418.
Groenman-van Waateringe, Willy: *Die Entwicklung der Schuhmode in 2500 Jahren.* In: *Die Kunde* 25, 1974, pp. 111–121.
Groenman-van Waateringe, Willy: *Die Stellung der Lübecker Lederfunde im Rahmen der Entwicklung der mittelalterlichen Schuhmode.* In: *Lübecker Schriften zur Archäologie und Kulturgeschichte*, Vol. 4, Bonn 1980, pp. 169–174.
Groenman-van Waateringe, Willy: *Wederom prehistorisch schoisel uit Drenthe.* In: *Nieuwe Drentse Volksalmanak* 108, 1991, pp. 34–41.
Groenman-van Waateringe, Willy: *Die Lederfunde von Haithabu.* In: *Berichte über die Ausgrabungen in Haithabu*, No. 21, Neumünster 1984.
Guyan, Walter Ulrich: *Bogen und Pfeil als Jagdwaffe im 'Weier'.* In: *Die ersten Bauern*, Vol. 1, *Pfahlbaufunde Europas. Forschungsberichte zur Ausstellung im Schweizerischen Landesmuseum*, Zürich 1990, pp. 135–138.
Guyan, Walter Ulrich: *Zur Herstellung und Funktion einiger jungsteinzeitlicher Holzgeräte von Thayngen-Weier.* In: *Helvetia Antiqua, Festschrift Emil Vogt*, Zürich 1966, pp. 21–32.
Haid, Hans: *Aufbruch in die Einsamkeit – 5000 Jahre Überleben in den Alpen*, Rosenheim 1992.
Haid, Hans: *Mythos und Kult in den Alpen*, Bad Sauerbrunn 1992.
Hald, Margrethe: *Ancient danish textiles from bogs and burials.* In: *Publications of the National Museum of Denmark, Archaeological-Historical Series* Vol. 21, Copenhagen 1980.
Haller, Anton: *Das Similaun-Syndrom – Oecci Homo – Von der Entdeckung der Gletschermumie zum transdisziplinären Forschungsdesign*, Bottighofen am Bodensee 1992.
Hart Hansen, Jens Peder / Meldgaard, Jørgen / Nordqvist, Jørgen: *The Greenland Mummies.* The British Museum, London 1991.
Hauke, Herwig / Rottmar, Wilfried / Ottenschläger, Josef / Scholz, Walter / Kitzberger, Otto: *Der perfekte Tischler*, Part 1, Wien 1973.
Hayen, Hajo: *Die Moorleichen im Museum am Damm.* In: *Veröffentlichungen des Staatlichen Museums für Naturkunde und Vorgeschichte Oldenburg*, No. 6, Oldenburg 1987.
Heim, Michael / Nosko, Werner: *Die Ötztal-Fälschung–Anatomie einer archäologischen Groteske*, Reinbeck bei Hamburg 1993.
Henn, Rainer: *Auffindung und Bergung der Gletschlerleiche im Jahre 1991.* In: Höpfel, Frank / Platzer, Werner / Spindler, Konrad: *Der Mann im Eis*, Vol. 1, *Veröffentlichungen der Universität Innsbruck* 187, Innsbruck 1992, pp. 88–91.
Henn, Rainer: *Der Tote vom Hauslabjoch.* In: *Österreichische Hochschulzeitung* 43, 1991, No. 11, p. 13.

Hermann, Bernd / Grupe, Gisela / Hummel, Susanne / Piepenbrink, Hermann / Schutkowski, Holger: *Prähistorische Anthropologie – Leitfaden der Feld- und Labormethoden*, Berlin and Heidelberg 1990.

Hirschberg, Walter / Janata, Alfred: *Technologie und Ergologie in der Völkerkunde*, Berlin 1986.

Hoinkes-Wilfingseder, Barbara: *Zum strafrechtlichen Schutz archäologischer Funde*. In: Höpfel, Frank / Platzer, Werner / Spindler, Konrad: *Der Mann im Eis*, Vol. 1, *Veröffentlichungen der Universität Innsbruck* 187, Innsbruck 1992, pp. 56–60.

Höpfel, Frank / Platzer, Werner / Spindler, Konrad (Eds.): *Der Mann im Eis*, Vol. 1, *Bericht über das Internationale Symposium in Innsbruck 1992. Veröffentlichungen der Universität Innsbruck* 187, Innsbruck 1992.

Höpfel, Frank: *Der Mann vom Hauslabjoch – Vom Kriminalfall zum juristischen 'Krimi'*. In: Höpfel, Frank / Platzer, Werner / Spindler, Konrad: *Der Mann im Eis*, Vol. 1, *Veröffentlichungen der Universität Innsbruck* 187, Innsbruck 1992, pp. 29–35.

Itten, Marion: *Die Horgener Kultur*. In: *Archäologie der Schweiz*, Vol. 2, *Die jüngere Steinzeit*, Basel 1969, pp. 83–96.

Itten, Marion: *Die Horgener Kultur. Monographien zur Ur- und Frühgeschichte der Schweiz* Vol. 17, Basel 1970.

Jacomet, Stefanie / Brombacher, Christoph / Dick, Martin: *Ackerbau, Sammelwirtschaft und Umwelt*. In: *Die ersten Bauern*, Vol. 1, *Pfahlbaufunde Europas. Forschungsberichte zur Ausstellung im Schweizerischen Landesmuseum*, Zürich 1990, pp. 81–90.

Kalicz, Nándor: *Götter aus Ton – Das Neolithikum und die Kupferzeit in Ungarn*, Budapest 1970.

Kimmig, Wolfgang: *Friedingen an der Donau – Bemerkungen zu einer Höhensiedlung mit Funden der Horgen-Sipplinger Kultur*. In: *Fundberichte aus Baden-Württemberg* 1, 1974, pp. 82–102.

Klebelsberg, Raimund: *Das Ötztal – Natur and Bild*. In: *Schlern-Schriften* 229, Innsbruck 1963, pp. 1–22.

Kriesch, Elli G.: *Der Gletschermann und seine Welt*, Hamburg 1992.

Kubina, Peter: *Ötzi doch Österreicher?* In: *Eich- und Vermessungsmagazin Wien* 67, 1992, pp. 11–17.

Künzel, Karl-Heinz / Steinlechner, Martin / Gaber, Othmar / Platzer, Werner: *Morphologische Vergleichsstudie an Schädeln – Zur Schädel-CT-Rekonstruktion des Eismannes*. In: Höpfel, Frank / Platzer, Werner / Spindler, Konrad: *Der Mann im Eis*, Vol. 1, *Veröffentlichungen der Universität Innsbruck* 187, Innsbruck 1992, pp. 117–130.

Küster, Hansjörg: *Kulturpflanzenanbau in Südbayern seit der Jungsteinzeit*. In: *Bauern in Bayern – Von den Anfängen bis zu Römerzeit. Katalog des Gäubodenmuseums Straubing*, No. 19, Straubing 1992, pp. 137–156.

Ladurner-Parthanes, Matthias: *Die Algunder Menhire*. In: *Der Schlern* 26, 1952, pp. 310–325.

Lange, Joachim: *Vorläufige Befunde der Untersuchungen an Pelzlederproben*. In: Höpfel, Frank / Platzer, Werner / Spindler, Konrad: *Der Mann im Eis*, Vol. 1, *Veröffentlichungen der Universität Innsbruck* 187, Innsbruck 1992, pp. 419–434.

Lehner, Peter / Julen, Annemarie: *A man's bones with 16th-century weapons and coins in a glacier near Zermatt, Switzerland.* In: *Antiquity* 65, 1991, No. 247, pp. 269–273.
Leitner, Walter: *Die Urzeit.* In: *Geschichte des Landes Tirol*, Vol. 1, Bozen 1985, pp. 4–124.
Leroi-Gourhan, André: *L'Art des Cavernes – Atlas des Grottes Ornées Paleolithiques Françaises*, Paris 1984.
Lippert, Andreas / Spindler, Konrad: *Die Auffindung einer frühbronzezeitlichen Gletschermumie am Hauslabjoch in den Ötztaler Alpen (Gem. Schnals).* In: *Archäologie Österreichs*, 2, 1991, No. 2, pp. 11–17.
Lippert, Andreas: *Archäologische Nachuntersuchung am Tisenjoch.* In: *Archäologie Österreichs* 3, 1992, No. 2, pp. 36–37.
Lippert, Andreas: *Die erste archäologische Nachuntersuchung am Tisenjoch.* In: Höpfel, Frank / Platzer, Werner / Spindler, Konrad: *Der Mann im Eis*, Vol. 1, *Veröffentlichungen der Universität Innsbruck* 187, Innsbruck 1992, pp. 245–253.
Lorcke, Dietrich / Münzner, Hans / Walter, Edward: *Zur Rekonstruktion der Körpergröße eines Menschen aus den langen Gliedmaßenknochen.* In: *Deutsche Zeitschrift für gerichtliche Medizin* 42, 1953, pp 189–202.
Lüning, Jens: *Die Michelsberger Kultur – Ihre Funde in zeitlicher und räumlicher Gliederung.* In: *48. Bericht der Römisch-Germanischen Kommission*, 1968, pp. 1–350.
Lunz, Reimo: *Vor- und Frühgeschichte Südtirols*, Vol. 1, *Steinzeit*, Bruneck 1986.
Mahr, Adolf: *Das vorgeschichtliche Hallstatt*, Wien 1925.
Mairhofer, Johann: *Der Mann aus dem Eis oder 'Spiel mir das Lied vom Ötzi!'.* In: *Von der Wirklichkeit des Unwirklichen – Historische und gegenwärtige Aspekte der Volkserzählung in Tirol. Katalog zur Ausstellung des Instituts für Europäische Ethnologie der Universität Innsbruck vom 29. Juni 1992 bis 12. Juli 1992*, Innsbruck 1992, pp. 89–93.
Manouvrier, L.: *La détermination de la stature d'après les grands os des membres.* In: *Bulletin et Mémoires de la Societé d'Anthropologie du Paris, 2e série 4*, 1883, pp. 433–449.
Marinis, Raffaele C. de: *La più antica metallurgia nell'Italia settentrionale.* In: Höpfel, Frank / Platzer, Werner / Spindler, Konrad: *Der Mann im Eis*, Vol. 1, *Veröffentlichungen der Universität Innsbruck* 187, Innsbruck 1992, pp. 389–409.
Marinis, Raffaele C. de: *Età del Bronzo.* In: *Preistoria del Bresciano*, Brescia 1979, pp. 45–70.
Marzatico, Franco: *Scambi e commerci nel Trentino preromano.* In: *Economia Trentina*, No. 1 Trento 1986, pp. 75–104.
Matuschik, Irenäus: *Neolithische Siedlungen in Köfering und Alteglofsheim.* In: *Das archäologische Jahr in Bayern 1991*, 1992, pp. 26–29.
Matuschik, Irenäus: *Sengkofen-'Pfatterbreite', eine Fundstelle der Michelsberger Kultur im Bayrischen Donautal, und die Michelsberger Kultur im östlichen Alpenvorland.* In: *Bayerische Vorgeschichtsblätter 57*, 1992, pp. 1–31.
Mayer, Christian: *Bestattungen der Badener Kultur aus Österreich.* In: *Archaeologia Austriaca* 75, 1991, pp. 29–61.
Mayer, Eugen Friedrich: *Die Äxte und Beile in Österreich.* In: *Prähistorische Bronzefunde.* Section 9, Vol. 9, München 1977.

Messner, Reinhold: *Rund um Südtirol*, München 1992.

Meyer, Werner: *Der Söldner vom Theodulpaß und andere Gletscherfunde aus der Schweiz*. In: Höpfel, Frank / Platzer, Werner / Spindler, Konrad: *Der Mann im Eis*, Vol. 1, *Veröffentlichungen der Universität Innsbruck* 187, Innsbruck 1992, pp. 321–333.

Micozzi, Marc S.: *Postmortem Change in Human and Animal Remains – A Systematic Approach*, Springfield 1991.

Momias – Los secretos del pasado. Catalogo del Museo Arqueologico y Etnografico de Tenerife, Las Palmas 1992.

Müller-Karpe, Hermann: *Handbuch der Vorgeschichte*, Vol. 2, *Jungsteinzeit*, München 1968.

Müller-Karpe, Hermann: *Handbuch der Vorgeschichte*, Vol. 3, *Kupferzeit*, München 1974.

Müllner, Andreas / Platzer, Werner / Wilfing, Harald / Nedden, Dieter zur: *Zur Morphologie des Zahn- und Kieferapparates des Mannes vom Hauslabjoch – Vorläufige makroskopische Befunde und mögliche Interpretationen*. In: Höpfel, Frank / Platzer, Werner / Spindler, Konrad: *Der mann im Eis*, Vol. 1, *Veröffentlichungen der Universität Innsbruck* 187, Innsbruck 1992, pp. 227–233.

Nedden, Dieter zur / Wicke, Klaus: *Der Eismann aus der Sicht der radiologischen und computertomographischen Daten*. In: Höpfel, Frank / Platzer, Werner / Spindler, Konrad *Der Mann im Eis*, Vol. 1, *Veröffentlichungen der Universität Innsbruck* 187, Innsbruck 1992, pp. 131–148.

Nedden, Dieter zur / Wicke, Klaus: *The Similaun Mummy as Observed from the Viewpoint of Radiological and CT Data*. In: *The Similaun Mummy*, Vol. 1, *Report on the 1992 International Symposium in Innsbruck*, Separate Print, Innsbruck 1992, pp. 3–19.

Neubauer, Manfred: *Ötzi und die Staatsgrenze – Bericht über die Arbeiten zur Feststellung der Fundstelle in Bezug auf die Staatsgrenze Österreich-Italien am Hauslabjoch*. In: *Der Mann im Eis*, Vol. 2, *Veröffentlichungen der Universität Innsbruck*, in the course of publication.

Neubauer, Manfred: *Ötzi und die Staatsgrenze. Bericht über die Arbeiten zur Feststellung der Fundstelle in Bezug auf die Staatsgrenze Österreich – Italien am Hauslabjoch*. In: *Eich- und Vermessungsmagazin* Wien 67, 1992, pp. 5–11.

Niederer-Nelken, Loni: *Fundort Lötschental. Begleitheft zur gleichnamigen Ausstellung im Lötschentaler Museum*, Kippel 1991.

Nieszery, Norbert: *Bandkeramische Feuerzeuge*. In: *Archäologisches Korrespondenzblatt* 22, 1992, pp. 359–376.

Niklaus, Thomas R.: *CalibETH Version 1.5 – User's Manual*, Eidgenössische Technische Hochschule, Zürich 1991.

Oliver, Georges / Aaron, C. / Fully, Georges / Tissier, Henri: *New estimations of stature and cranial capacity in modern Man*. In: *Journal of Human Evolution* 7, 1978, pp. 513–534.

Ortner, Lorelies: *Von der Gletscherleiche zu unserem Urahn Ötzi – Benennungspraxis in der Presse*. In: *Deutsche Sprache* 21, 1993, in course of publication. In: *Der Mann im Eis*, Vol. 2, *Veröffentlichungen der Universität Innsbruck*, in the course of publication.

Ottaway, Barabara S.: *Neue Radiocarbondaten Altheimer und Chamer Siedlungsplätze in Niederbayern.* In: *Archäologisches Korrespondenzblatt* 16, 1986, pp. 141–147.
Patzelt, Gernot: *Gletscherbericht 1990/91 – Sammelbericht über die Gletschervermessungen des Österreichischen Alpenvereins im Jahre 1991.* In: *Mitteilungen des Österreichischen Alpenvereins* 47 (117), 1992, No. 2, pp. 17–22.
Patzelt, Gernot: *Neues vom Ötztaler Eismann. Mitteilungen des Österreichischen Alpenvereins* 47 (117), 1992, No. 2, pp. 23–24.
Pauli, Ludwig: *Die Alpen in Frühzeit und Mittelalter*, München 1980.
Paulsen, Harm: *Schußversuche mit einem Nachbau des Bogens von Koldingen, Ldkr. Hannover.* In: *Archäologische Mitteilungen aus Nordwestdeutschland*, Appendix vol. 4, 1990, pp. 298–305.
Pearson, Karl: *On the reconstruction of stature of prehistoric races.* In: *Transactions of the Royal Society, series A 192*, 1899, pp. 169–244.
Perini, Renato (Ed.): *Archeologia del Legno, Documenti dell'Età del Bronzo dall'Area Sudalpina. Quaderni della Sezione Archeologica Museo Provinciale d'Arte. Catalogo della Mostra a Trento settembre–ottobre 1988 aprile–ottobre 1989*, Trento 1988.
Perini, Renato: *Scavi Archeologici nella Zona Palafitticola di Fiavé-Carera*, Part 2. *Patrimonio storico e artistico del Trentino* 9, Trento 1987.
Pöder, Reinhold / Peintner, Ursula / Pümpel, Thomas: *Mykologische Untersuchungen an den Pilz-Beifunden der Gletschermumie vom Hauslabjoch.* In: Höpfel, Frank / Platzer, Werner / Spindler, Konrad: *Der Mann im Eis*, Vol. 1, *Veröffentlichungen der Universität Innsbruck* 187, Innsbruck 1992, pp. 313–320.
Probst, Ernst: *Deutschland in der Steinzeit*, München 1991.
Raetzel-Fabian, Dirk: *Phasenkartierung des mitteleuropäischen Neolithikums – Chronologie und Chorologie. British Archeological Reports*, International Series, Vol. 316, Oxford 1986.
Rageth, Jürgen: *Der Lago di Ledro im Trentino und seine Beziehungen zu den alpinen und mitteleuropäischen Kulturen.* In: *55. Bericht der Römisch-Germanischen Kommission* 1975, pp. 73–256.
Rast, Antoinette: *Die Verarbeitung von Bast.* In: *Die ersten Bauern*, Vol. 1, *Pfahlbaufunde Europas. Forschungsberichte zur Ausstellung im Schweizerischen Landesmuseum*, Zürich 1990, pp. 119–122.
Rast, Antoinette: *Jungsteinzeitliche Kleidung.* In: *Die ersten Bauern*, Vol. 1, *Pfahlbaufunde Europas. Forschungsberichte zur Ausstellung im Schweizerischen Landesmuseum*, Zürich 1990, pp. 123–126.
Rolle, Renate: *Die Skythenzeitlichen Mumienfunde von Pazyryk - Frostkonservierte Gräber aus dem Altaigebirge.* In: Höpfel, Frank / Platzer, Werner / Spindler, Konrad: *Der Mann im Eis*, Vol. 1, *Veröffentlichungen der Universität Innsbruck* 187, Innsbruck 1992, pp. 334–358.
Rollo, Franco: *Molecular investigations on mummified seeds and other plant remains.* In: Höpfel, Frank / Platzer, Werner / Spindler, Konrad: *Der Mann im Eis*, Vol. 1, *Veröffentlichungen der Universität Innsbruck* 187, Innsbruck 1992, pp. 198–208.
Rotter, Hans: *Ethische Aspekte zum Thema.* In: Höpfel, Frank / Platzer,

Werner / Spindler, Konrad: *Der Mann im Eis*, Vol. 1, *Veröffentlichungen der Universität Innsbruck* 187, Innsbruck 1992, pp. 24–28.

Rüttimann, Bettina: *Geräte aus Knochen*. In: *Archäologische Forschungen*, Schweizer Landesmuseum Zürich, Vol. 2, Zürich 1983, pp. 7–86.

Ruttkay, Elisabeth: *Typologie und Chronologie der Mondsee-Gruppe*. In: *Das Mondsee-Land – Geschichte und Kultur*, Linz 1981, pp. 269–294.

Salzani, Luciano: *Preistoria in Valpolicella*, Verona 1981.

Sauter, Friedrich / Jordis, Ulrich / Hayek, Erich: *Chemische Untersuchungen der Kittschäftungs-Materialien*. In: Höpfel, Frank / Platzer, Werner / Spindler, Konrad: *Der Mann im Eis*, Vol. 1, *Veröffentlichungen der Universität Innsbruck* 187, Innsbruck 1992, pp. 435–441.

Sauter, Friedrich / Stachelberger, Herbert: *Materialuntersuchungen an einem Begleitfund des 'Mannes vom Hauslabjoch' – Die 'schwarze Masse' aus dem 'Täschchen'*. In: Höpfel, Frank / Platzer, Werner / Spindler, Konrad: *Der Mann im Eis*, Vol. 1, *Veröffentlichungen der Universität Innsbruck* 187, Innsbruck 1992, pp. 442–453.

Sauter, Marc-R. / Gallay, Alain: *Les premières cultures d'origine méditerréenne*. In: *Archäologie der Schweiz*, Vol. 2, *Die jüngere Steinzeit*, Basel 1969, pp. 47–66.

Schibler, Jörg / Suter, Peter J.: *Jagd und Viehzucht im schweizerischen Neolithikum*. In: *Die ersten Bauern*, Vol. 1, *Pfahlbaufunde Europas, Forschungsberichte zur Ausstellung im Schweizerischen Landesmuseum*, Zürich 1990, pp. 91–103.

Schlabow, Karl: *Der Moorleichenfund von Peiting*. In: *Veröffentlichungen des Fördervereins Textilmuseum Neumünster e V.*, No. 2, Neumünster 1961.

Schmidt, Leopold: *Oberitalienische Strohmäntel*. In: *Wiener Zeitschrift für Volkskunde* 43, 1938, pp. 44–45.

Schweingruber, Fritz Hans: *Anatomie europäischer Hölzer*, Bern und Stuttgart 1990.

Seidler, Horst / Bernhard, Wolfram / Teschler-Nicola, Maria / Platzer, Werner / Nedden, Dieter zur / Henn, Rainer / Oberhauser, Andreas / Sjøvold, Torstein: *Some Anthropological Aspects of the Prehistoric Tyrolean Ice Man*. In: *Science 258*, 1992, pp. 369–616.

Seidler, Horst / Teschler-Nicola, Maria / Wilfing, Harald / Weber, Gerhard / Traindl-Prozhazka, Martina / Platzer, Werner / Nedden, Dieter zur / Henn, Rainer: *Zur Anthropologie des Mannes vom Hauslabjoch – Morphologische und metrische Aspekte*. In: Höpfel, Frank / Platzer, Werner / Spindler, Konrad: *Der Mann im Eis*, Vol. 1, *Veröffentlichungen der Universität Innsbruck* 187, Innsbruck 1992, pp. 149–162.

Sjøvold, Torstein: *Einige statistische Fragestellungen bei der Untersuchung des Mannes vom Hauslabjoch*. In: Höpfel, Frank / Platzer, Werner / Spindler, Konrad: *Der Mann im Eis*, Vol. 1, *Veröffentlichungen der Universität Innsbruck* 187, Innsbruck 1992, pp. 188–197.

Sjøvold, Torstein: *En förhistorisk man fran gränsen mellan Tyrolen och Sydtyrolen*. In: *Fornvännen* 87, 1992, pp. 109–114.

Sjøvold, Torstein: *Estimation of stature from long bones utilizing the line of organic correlation*. In: *Human Evolution* 5, 1990, pp. 431–447.

Sjøvold, Torstein: *Frost and found. The melting of an Alpine glacier brings scientists face to face with the Stone Age.* In: *Natural History* 4/93, 1993, pp. 60–64.

Sjøvold, Torstein: *Geschlechtsdiagnose am Skelett.* In: *Anthropologie – Handbuch der vergleichenden Biologie des Menschen,* Vol. 1: *Wesen und Methoden der Anthropologie,* Stuttgart und New York 1988, pp. 444–480.

Sjøvold, Torstein: *The Prehistoric Man from the Austrian–Italian Alps – Discovery, Description, and Current Research.* In: *Papers on Paleopathology, Paleopathology Association, Nineteenth Annual Meeting, 31 March and 1 April 1992, Las Vegas,* Las Vegas 1992, pp. 3–4.

Sjøvold, Torstein: *The Stone Age Iceman from the Alps: The find and the current status of investigation.* In: *Evolutionary Anthropology* 1, 1992, pp. 117–124.

Sjøvold, Torstein: *The Stone Age Man from the Austrian–Italian Alps: Discovery, Description and Current Research.* In: *Collegium Antropologicum,* Vol. 16, No. 1, Zagreb 1992, pp. 1–12.

Solzhenitsyn, Alexander: *Cancer Ward,* London 1970.

Sperl, Gerhard: *Das Beil vom Hauslabjoch.* In: Höpfel, Frank / Platzer, Werner / Spindler, Konrad: *Der Mann im Eis,* Vol. 1, *Veröffentlichungen der Universität Innsbruck* 187, Innsbruck 1992, pp. 454–459.

Spindler, Konrad: *Der Mann aus dem Eis.* In: *Archäologie in Deutschland,* No. 1, 1992, pp. 34–37.

Spindler, Konrad: *Der Mann im Eis.* In: *Sandoz-Bulletin* 99, 1992, pp. 21–29.

Spindler, Konrad: *L'Homme du Glacier.* In: *Sandoz-Bulletin* 99, 1992, pp. 21–30.

Stead, I. M. / Bourke, J. B. / Brothwell, Don: *Lindow Man – The Body in the Bog,* London 1986.

Street, Martin: *Jäger und Schamanen – Bedburg-Königshoven – Ein Wohnplatz am Niederrhein vor 10 000 Jahren.* Römisch-Germanisches Zentralmuseum, Mainz 1989.

Suter, Peter J. / Schifferdecker, François: *Das Neolithikum im schweizerischen Mittelland.* In: *Antiqua* 15, Basel 1986, pp. 34–44.

Tecchiati, Umberto: *'Prähistorische Bronzefunde' conservati al Museo Civico di Rovereto (Trento) – Le Asce.* In: *Annali dei Musei Civici Rovereto* 7, 1991, pp. 3–36.

Tiefenbrunner, Friedrich: *Bakterien und Pilze, ein Problem für unseren ältesten Tiroler?* In: Höpfel, Frank / Platzer, Werner / Spindler, Konrad: *Der Mann im Eis* Vol. 1, *Veröffentlichungen der Universität Innsbruck* 187, Innsbruck 1992, pp. 100–107.

Trotter, Mildred / Gleser, Goldine C.: *Estimation of stature from long bones of American Whites and Negroes.* In: *American Journal of Physical Anthropology, New Series 10,* 1952, pp. 463–514.

Trotter, Mildred / Gleser, Goldine C.: *A re-evaluation of stature based on measurements taken during life and of long bones after death.* In: *American Journal of Physical Anthropology, New Series 16,* 1958, pp. 79–124.

Urbon, Benno: *Spanschäftung für Lanzen und Pfeile.* In: *Fundberichte aus Baden-Württemberg* 16, 1991, pp. 127–131.

Voss, Hella: *Die große Jagd – Von der Vorzeit bis zur Gegenwart – 30 000 Jahre Jagd in der Kunst.* München 1961.

Wahl, Joachim / König, Günter: *Anthropologisch-traumatologische Unter-*

suchung der menschlichen Skelettreste aus dem bandkeramischen Massengrab bei Talheim, Kreis Heilbronn. In: *Fundberichte aus Baden-Württemberg* 12, 1987, pp. 67–193.

Währen, Max: *Brot und Getreide in der Urgeschichte.* In: *Die ersten Bauern,* Vol. 1, *Pfahlbaufunde Europas. Forschungsberichte zur Ausstellung im Schweizerischen Landesmuseum,* Zürich 1990, pp. 117–118.

Weber, Karl: *Eine Mumie als Denkmal.* In: Höpfel, Frank / Platzer, Werner / Spindler, Konrad: *Der Mann im Eis,* Vol. 1, *Veröffentlichungen der Universität Innsbruck* 187, Innsbruck 1992, pp. 50–55.

Weber, Karl: *Ist der 'Otzi' ein Denkmal? – Kulturgüterschutzrechtliche Aspekte von archäologischen Funden.* In: *Österreichische Juristen-Zeitung* 47, 1992, No. 20, pp. 673–677.

Weiner, Jürgen: *Der Lousberg in Aachen – Feuersteinbergbau in der jungsteinzeit. – Ein Führer zur prähistorischen Abteilung des Stadtgeschichtlichen Museums Burg Frankenberg Aachen,* Aachen-Eilendorf 1984.

Weiner, Jürgen: *Mit Stahl, Stein und Zunder – Die in Vergessenheit geratene Technik des Feuerschlages.* In: *Pulheimer Beiträge zur Geschichte und Heimatkunde 5,* 1981, pp. 13–18.

Weisgerber, Gerd: *Der Kupfermann – Indianischer Bergbau vor Kolumbus.* In: *Amerika 1492–1992. Neue Welten – Neue Wirklichkeiten,* Braunschweig 1992, pp. 159–167.

Weiss, Gabriele: *Zur Archäologie des Todes.* In: *Mitteilungen der Anthropologischen Gesellschaft in Wien* 113, 1983, pp. 27–32.

Wesselkamp, Gerhard: *Die organischen Reste der Cortaillod-Schichten. Die neolithischen Ufersiedlungen von Twann,* Vol. 5, Bern 1980.

Wesselkamp, Gerhard: *Neolithische Holzartefakte aus Schweizer Seeufersiedlungen.* Freiburg i. Br. 1992.

Wilrich, Cordula / Wortmann, Gabriele / Wortmann, Franz-Josef: *Beitrag zur taxonomischen Einstufung verschiednener Federkeratine durch vergleichende Auswertung ihrer Elektropherogramme.* In: *Der Mann im Eis,* Vol. 2, *Veröffentlichungen der Universität Innsbruck,* in the course of publication.

Winiger, Josef: *Beinerne Doppelspitzen aus dem Bielersee.* In: *Jahrbuch der Schweizerischen Gesellschaft für Ur- und Frühgeschichte 75,* 1992, pp. 65–99.

Winiger, Josef: *Feldmeilen-Vorderfeld – Der Übergang von der Pfyner zur Horgener Kultur. Antiqua* 8, Basel 1981.

Winkler, Ivo: *Autonome Südtiroler Zuständigkeiten bei archäologischen Funden.* In: Höpfel, Frank / Platzer, Werner / Spindler, Konrad: *Der Mann im Eis,* Vol. 1, *Veröffentlichungen der Universität Innsbruck* 187, Innsbruck 1992, pp. 81–87.

Wittig, Manfred / Wortmann, Gabriele: *Untersuchungen an Haaren aus den Begleitfunden des Eismannes vom Hauslabjoch – Vorläufige Ergebnisse.* In: Höpfel, Frank / Platzer, Werner / Spindler, Konrad: *Der Mann im Eis,* Vol. 1, *Veröffentlichungen der Universität Innsbruck* 187, Innsbruck 1992, pp. 273–298.

Wölfli, Willy: *Möglichkeiten und Grenzen der Beschleunigermassenspektrometrie in der Archäologie.* In: *10 Jahre Beschleunigermassenspektrometrie in der Schweiz. PSI-Proceedings 92–04,* Villingen 1992, pp. 31–44.

Wyss, René: *Das jungsteinzeitliche Jäger-Bauerndorf von Egozwil 5 im Wauwilermoos*. In: *Archäologische Forschungen*, Schweizer Landesmuseum Zürich, Zürich 1976, pp. 5–116.

Wyss, René: *Ein Netzbeutel – Zur Thematik des Fernhandels*. In: *Die ersten Bauern*, Vol. 1, *Pfahlbaufunde Europas. Forschungsberichte zur Ausstellung im Schweizerischen Landesmuseum*, Zürich 1990, pp. 131–133.

Wyss, René: *Geräte aus Holz*. In: *Archäologische Forschungen*. Schweizerisches Landesmuseum Zürich, Vol. 2, Zürich 1983, pp. 87–160.

Zissernig, Elisabeth: *Der Mann vom Hauslabjoch – Von der Entdeckung bis zur Bergung*. In: Höpfel, Frank / Platzer, Werner / Spindler, Konrad: *Der Mann im Eis*, Vol. 1, *Veröffentlichungen der Universität Innsbruck* 187, Innsbruck 1992, pp. 234–244.

Zürcher, Andreas C.: *Urgeschichtliche Fundstellen Graubündens*. In: *Schriftenreihe des Rätischen Museums Chur*, No. 27, Chur 1982.

Index

Page references in *italic* type indicate illustrations in the text.

Index

Index

Index

Index

Index